Random Light Beams

Theory and Applications

Random Light Beams
Theory and Applications

Olga Korotkova

CRC Press
Taylor & Francis Group
Boca Raton London New York

CRC Press is an imprint of the
Taylor & Francis Group, an **informa** business

CRC Press
Taylor & Francis Group
6000 Broken Sound Parkway NW, Suite 300
Boca Raton, FL 33487-2742

First issued in paperback 2019

ISBN-13: 978-1-4398-1950-0 (hbk)
ISBN-13: 978-0-367-37939-1 (pbk)

Library of Congress Cataloging-in-Publication Data

Korotkova, Olga.
   Random light beams : theory and applications / Olga Korotkova.
      pages cm
   "A CRC title."
   Includes bibliographical references and index.
   ISBN 978-1-4398-1950-0 (hardcover : alk. paper)
   1. Light. 2. Light--Wave-length. 3. Light--Transmission. I. Title.

QC455.K67 2014
535.5--dc23
                                    2013026449

Visit the Taylor & Francis Web site at
http://www.taylorandfrancis.com

and the CRC Press Web site at
http://www.crcpress.com

Contents

List of Figures

Foreword

"..... I'd gladly be locked up in a dungeon ten fathoms below ground, if in return I could find out one thing: What is light?"

Galileo Galilei

I wrote this book during my first several years of professorship at the Department of Physics of the University of Miami, FL. Being heavily involved in the research topics discussed in this text since the time of my dissertation, I recently started to realize that certain subtleties that at some point seemed to have transparent explanations could readily escape from memory. My own need for this monograph became apparent. It also was so for my group of graduate and undergraduate students as well as visiting scholars. Even if technically it is not my first book (my Ph.D. thesis has recently been published in Germany), I perceive it as being such. It is also dearer to me since while working on this text I have discovered tons of facts in areas of optics that I had thought before I knew well. Moreover, it was a remarkable revelation that namely the author of the book seems to learn more than everybody else from it. By no means is this text designed as a self-sufficient account of classical statistical optics. It is only meant to be an upgrade for a particular direction, the one relating to various phenomena associated with beam-like fields which are random in nature.

In Chapter 1 the essential mathematical and physical concepts needed for deeper understanding of the main text are reviewed. Various classes of deterministic paraxial beams are introduced in Chapter 2. Some of these beams are used in subsequent chapters as building blocks for random beams. Chapter 3 concerns random scalar beams, which were explored in depth several decades ago and discussed in detail in other books, for instance, the fundamental text *Optical Coherence and Quantum Optics* by L. Mandel and E. Wolf and *Coherent Mode Representation in Optics* by A. Ostrovsky. Hence, we will only briefly present the well-established issues relating to scalar beams and focus on the findings not yet reflected in the literature. In Chapter 4 we introduce electromagnetic random beams and point out the matters relating to their generation, propagation in free space and various media, transmission through optical systems, etc. Some of this analysis can be found in a fairly new book titled *Introduction to the Theories of Coherence and Polarization of Light* by E. Wolf. This book, however, treats a greater variety of aspects and goes deeper into details. Chapters 5 to 8 discuss some of the applications that benefit from the use of random beams. In particular, in Chapter 5 the interaction of beams with deterministic optical systems is explored and the examples are given of light propagation through the human eye, laser resonators, negative phase materials, etc. Chapters 6 and 7 are concerned with propagation of random beams in random media, such as the atmosphere and ocean, as well as with optical systems operating in their presence. Finally, Chapter 8 is devoted to scattering of random beams from collections of scatterers and thin random layers, such as bio-tissue slices.

While being the sole author of this monograph I am indebted to my advisors, colleagues and students who through both short in-office discussions and profound e-mail correspondence stimulated my interest in this diverse and fast-growing field. At times entire articles of my graduate students N. Farwell, S. Sahin, and Z. Tong were used as book sections and uncountable number of times their comments and ideas were included. I am particularly grateful for the financial support of our research group provided in recent years by the US Air Force Research Office (A. Nachman, K. Miller), the US Office of Naval Research (R. Malek-Madani) and also by the Physics Department and College of Arts and Sciences at the University of Miami. The last but not least acknowledgment goes to my colleagues and co-authors of more than a hundred peer-reviewed papers including Y. Cai, G. Gbur, F. Gori, D. Zhao, J. Pu, E. Wolf, L. Andrews, R. Phillips, Y. Baykal, S. Avramov-Zamurovic, C. Nelson, E. Shchepakina, Z. Mei and many others for their perpetual guidance and shared insight into statistical optics, mathematical modeling and optical engineering.

Olga Korotkova
January 2013

Preface

"In the beginning God created the heavens and the earth. The earth was formless and void, and darkness was over the surface of the deep, and the Spirit of God was moving over the surface of the waters. Then God said, "Let there be light"; and there was light."

The Old Testament, Genesis.

What could be a common ground between a purely scientific book and the Holy Bible? When I came across this very first page of Genesis several years ago I was amazed by the fact that even in the Bible everything starts from the problem of light-matter interactions. In the very first page light is introduced as the essential tool for dealing with natural environments: information transfer and sensing.

Another thought I had on reading this passage that there is no description of light in the Bible, it is referred to as just "light". It might seem that light cannot be modified or optimized; it must be taken for granted. However, in possession of all the variety of modern light sources, modulators and detectors one may wonder whether it is possible to create "better" light fields than the natural ones for interacting with the same old "the heavens and the earth"? In other words, one may wonder whether the natural light given to us is really the optimal possible kind of light, or can we do better?

This book explores some of the natural and man-made light fields, stochastic in nature and limited in space to narrow channels, which will be referred to as stochastic beam-like fields or, simply, stochastic beams. We restrict our attention to this subset of all electromagnetic fields because they exhibit unique phenomena on propagation and interaction with various media. The majority of such phenomena are only possible due to this combination of random nature and spatial confinement. For instance, it has been found that stochastic beams can exhibit spectral and polarization changes on propagation in free space, even though until recently it was believed that only interaction with media, such as scatterers or turbulence, can modify these properties. Moreover, a beam-like structure alone or the fluctuating behavior of a wave alone cannot result in such effects. In addition, the state of coherence of a wave, which may, roughly speaking, be associated with the degree of its randomness, can only change on propagation in vacuum if the wave is having a beam-like structure.

How common are the light fields of our interest? Due to energy limitations, all electromagnetic fields found in nature or artificially made are confined to certain regions in space and are random. Even laser radiation is not an exception: this seemingly perfect type of wave has internal temporal and spatial fluctuations, even if not so well pronounced as those in other types of waves.

The last but not the least argument for trying to know well and to be able to control random light is the fact that about eighty percent of overall human perception of the environment is attributed to vision, the other twenty percent being shared by the rest of the senses. And what we perceive by our eyes is the light radiated and scattered by objects, not the objects themselves. In order to know the world around us it is crucial to learn the ways in which the light may behave.

Symbol Description

$\star$	convolution	a	amplitude
$.^{\dagger}$	complex conjugation	A	element of ray-transfer matrix
$.^{*}$	Hermitian adjoint		
$\tilde{\ }$	Fourier transform	$A(\cdot)$	angular correlation function
$.^{(B)}$	Bessel beam		
$.^{(BI)}$	modified Bessel beam	$\mathbf{A}(\cdot)$	angular correlation matrix
$.^{(COS)}$	Cosine beam	A_0	initial amplitude
$.^{(G)}$	Gaussian/Gaussian Schell-model beam	a_A, a_B	amplitudes of annular beam
$.^{(norm)}$	normalized version	$a^{(eHG)}$	angular spectrum of elegant Hermite-Gaussian beam
$.^{(PL)}$	plane wave		
$.^{T}$	matrix transpose		
$2h_f$	interfocal separation	$a^{(eLG)}$	angular spectrum of elegant Hermite-Gaussian beam
arg	argument of a complex number		
curl	curl operator	$a^{(G)}$	angular spectrum of Gaussian beam
Det	determinant of matrix		
div	divergence operator	$A^{(G)}$	angular correlation function of Gaussian Schell-model beam
grad	gradient operator		
Im	imaginary part		
Re	real part	$a^{(HG)}$	angular spectrum of Hermite-Gaussian beam
∂	partial derivative		
∇^2	Laplacian	$A^{(J)}$	angular correlation function of J-Bessel beam
$\cdot_{\perp}$	transverse part of vector		
$\cdot_{(AB)}$	absorber	$a^{(LG)}$	angular spectrum of Laguerre-Gaussian beam
$\cdot_{(L)}$	linear polarization		
$\cdot_{(LC)}$	left circular polarization	$A^{(LG)}$	angular correlation function of Laguerre-Gaussian correlated beam
$\cdot_{(LE)}$	left elliptical polarization		
$\cdot_{(P)}$	polarizer		
$\cdot_{(RC)}$	right circular polarization	A_x, A_y	parameters of electromagnetic Gaussian Schell-model beam
$\cdot_{(RE)}$	right elliptical polarization		
$\cdot_{(RO)}$	rotator		
$\cdot_{(RT)}$	retarder	A_c	element of ray-transfer matrix in crystalline lens
$\langle\rangle_d$	ensemble average over realizations of device	A_ξ	aperture averaging factor
$\langle\rangle_{EC}$	continuous ensemble average	$\mathbf{B}$	magnetic induction
		B	element of ray-transfer matrix
$\langle\rangle_{ED}$	discrete ensemble average		
$\langle\rangle_M$	ensemble average over realizations of random medium	B_c	element of ray-transfer matrix in crystalline lens
$\langle\rangle_S$	ensemble average over realizations of scatterer	b_G	auxiliary parameter of Gaussian Schell-model beam
$\langle\rangle_T$	temporal average		

b_I intensity correlation coefficient

B_I covariance function of intensity

B_n covariance function of refractive index

$B_{\alpha\beta}$ parameter of electromagnetic Gaussian Schell-model beam

BER Bit-error rate

c velocity of light in vacuum

C element of ray-transfer matrix

C_0 normalization factor of multi-Gaussian Schell-model beam

C_c element of ray-transfer matrix in crystalline lens

c_I scintillation index

C_F correlation function of scattering potential

C_n correlation function of refractive index

$\mathbf{C}_n$ covariance matrix

c_p scintillation flux

C_Z auto-covariance function

C_ϵ correlation function of dielectric permittivity

C_ϵ^2 relative strength of fluctuating dielectric permittivity

$C_F^{(QH)}$ correlation function of scattering potential of quasi-homogeneous medium

$C_F^{(SM)}$ correlation function of scattering potential of Gaussian Schell-model medium

C_n^2 refractive index structure parameter

$\tilde{C}_n^2$ generalized refractive-index structure parameter

C_p^m even Ince polynomials of order p and degree m

Ce_0 even radial Mathieu function of first kind

Co_0 odd radial Mathieu function of first kind

$\mathbf{D}$ electric displacement

D element of ray-transfer matrix

d_A area of a pinhole

D_c element of ray-transfer matrix in crystalline lens

D_f particle size distribution

D_Z structure function

$\mathbf{E}$ electric field

$\mathcal{E}$ complex amplitude of electric vector

$\mathbf{E}^{(f)}$ free electric field

$\mathbf{E}^{(t)}$ transmitted electric field

$\mathbf{E}^{(i)}$ incident electric field

$E^{(1)}, E^{(2)}$ integrals of Rytov theory

f frequency, auxiliary function

$F(\mathbf{r}; \omega)$ scattering potential

$\mathcal{F}[\cdot]$ Fourier transform

$\mathcal{F}^{-1}[\cdot]$ inverse Fourier transform

$F^{(E)}$ scattering potential of ellipsoid

$F^{(G)}$ scattering potential of Gaussian sphere

$F^{(HMG)}$ scattering potential of hollow sphere

$F^{(MG)}$ scattering potential of semi-soft sphere

$F^{(S)}$ scattering potential of hard sphere

$f_\varpi(\boldsymbol{\rho})$ window function

Fey_0 even radial Mathieu function of second kind

Foy_0 odd radial Mathieu function of second kind

G Green's function

g_f 2D Gaussian correlated array

$g_c(z)$ transverse gradient parameter of crystalline lens

$G_X(\omega)$ energy spectral density

$\mathbf{H}$ magnetic vector

$H(\cdot,\cdot)$	auxiliary function in condition for genuine cross-spectral density		Gaussian Schell-model beam
$h_a(\cdot)$	amplitude spread function of optical system	n	index of refraction
		n_0	the mean value of index of refraction
H_p	Hermite polynomial of order p	N_m	Neumann function
$H_m^{(1)}$	Hankel function of first kind	O	order of magnitude
		$o(\cdot)$	amplitude transparency of object
$H_m^{(2)}$	Hankel function of second kind	p	power
$H_s(\cdot)$	Heaviside step function	$p(\cdot)$	auxiliary function in condition for genuine cross-spectral density
i	complex unit		
I	intensity	$\mathbf{P}$	polarization of the medium
I_0	initial intensity	$\mathcal{P}$	degree of cross-polarization
I_n	Modified Bessel function	$p_A(I)$	probability density function of intensity
$\mathbf{j}$	electric current density		
$\mathbf{J}$	transpose antisymmetric matrix	p_E	encircled energy
		P_e	equivalent power of crystalline lens
$J^{(\infty)}$	radiant intensity of far zone		
j_m	spherical Bessel function	p_F	fractional power
J_m	Bessel function of the first kind	P_m	Legendre polynomial
		p_n	n-th order joint probability density function
k	wave number in medium		
$\mathbf{k}$	wave vector	$p_A^{(G)}$	Gamma model for intensity probability density function
K	propagator		
$\mathbf{K}$	momentum transfer vector		
k_0	wave number in vacuum	$p_A^{(GG)}$	Gamma-Gamma model for intensity probability density function
$\mathcal{L}$	transformation matrix for scattered far field		
l_0	inner scale	$p_A^{(GL)}$	Gamma-Laguerre model for intensity probability density function
L_0	outer scale		
l_c	typical correlation width of bio-tissue		
		$P_{\mathbf{u}}^M$	random plane wave
L_m^n	Laguerre polynomial of order m, and azimuthal index n	pdf	probability density function
		Pr	principle value of contour integral
$\mathbb{M}$	Mueller matrix of optical system		
		r	magnitude of $\mathbf{r}$
$\mathbf{M}$	Mueller matrix of optical element or medium	$\mathbf{r}$	position vector in 3D space
		R	radius of curvature of Gaussian beam
$\mathfrak{M}$	pair-scattering matrix		
$\widetilde{\mathbf{M}}_{\alpha\beta}^{-1}$	4×4 matrix characterizing electromagnetic multi-	$\mathbf{R}^{-1}$	transverse wave front-curvature matrix

$R^{(G)}$	generalized radius of curvature of Gaussian Schell-model beam	s_3	classic Stokes parameter of monochromatic field
R_g	radius of core in GRIN fiber	S_0	generalized Stokes parameter of monochromatic field
$\widehat{R}_{\alpha\beta}$	radii of curvature of electromagnetic Gaussian Schell-model beam in $ABCD$ system	S_1	generalized Stokes parameter of monochromatic field
		S_2	generalized Stokes parameter of monochromatic field
$R_\varpi(\boldsymbol{\rho})$	2D array of random variables	S_3	generalized Stokes parameter of monochromatic field
$R_m^{(MG)}$	generalized radius of curvature of multi-Gaussian Schell-model beam	$S_0^{(f)}$	generalized Stokes parameter of free field
$R_{\alpha\beta}^{(G)}$	phase front radius of curvature of electromagnetic Gaussian Schell-model beam	$S_1^{(f)}$	generalized Stokes parameter of free field
		$S_2^{(f)}$	generalized Stokes parameter of free field
$R_{m\alpha\beta}^{(MG)}$	radius of curvature of electromagnetic multi-Gaussian Schell-model beam	$S_3^{(f)}$	generalized Stokes parameter of free field
		$S_0^{(t)}$	transmitted Stokes parameter
S	spectral density	$S_1^{(t)}$	transmitted Stokes parameter
$\mathfrak{S}$	scattering matrix		
$S^{(\infty)}$	spectral density of random beam in far zone	$S_2^{(t)}$	transmitted Stokes parameter
$S^{(BG)}$	spectral density of Bessel-Gaussian Schell-model beam	$S_3^{(t)}$	transmitted Stokes parameter
		$S_0^{(i)}$	incident Stokes parameter
$S^{(f)}$	spectral density of free field	$S_1^{(i)}$	incident Stokes parameter
$S^{(G)}$	spectral density of Gaussian Schell-model beam	$S_2^{(i)}$	incident Stokes parameter
$S^{(IB)}$	spectral density of I-Bessel-correlated beam	$S_3^{(i)}$	incident Stokes parameter
		S_N	normalized spectral density
$S^{(MG)}$	spectral density of multi-Gaussian Schell-model beam	S_p	Poynting vector
		$S_X(\omega)$	power spectral density
		$\bar{S}$	anti-spectral density
$S^{(p)}$	spectral density of polarized part	SNR_A	averaged signal-to-noise ratio
s_0	classic Stokes parameter of monochromatic field	SNR_c	free-space signal-to-noise ratio
s_1	classic Stokes parameter of monochromatic field	t	time
		T	averaging time interval, period
s_2	classic Stokes parameter of monochromatic field	$\mathbb{T}$	Jones matrix of optical system

$\mathbf{T}$	Jones matrix of optical element
$\tilde{\mathbf{T}}$	4×4 matrix of rough target's surface
$\mathfrak{T}_{kl}^{pq}$	transmission coefficient
Tr	trace of matrix
$\mathbf{u}$	unit vector
$U(t)$	analytic signal
$U^{(AN)}$	annular field
$U^{(CE)}$	cusp-elliptical field
$U^{(CG)}$	circular cusp-Gaussian field
$U^{(CGE)}$	elliptical cusp-Gaussian field
$U^{(CGR)}$	rectangular cusp-Gaussian field
$U^{(DHE)}$	elliptical dark-hollow field
$U^{(DHR)}$	rectangular dark-hollow field
$U^{(EG)}$	elliptical Gaussian field
$U^{(FG)}$	flat-Gaussian field
$U^{(G)}$	Gaussian field
$U^{(i)}$	incident scalar field
$U^{(s)}$	scattered scalar field
$U^{(t)}$	total scattered scalar field
$U_{mn}^{(eHG)}$	elegant Hermite-Gaussian field
$U_{mn}^{(eLG)}$	elegant Laguerre-Gaussian field
$U_{mn}^{(HG)}$	Hermite-Gaussian field
$U^{(IG)}$	Ince-Gaussian field
$U_n^{(JBG)}$	J-Bessel-Gaussian field
$U^{(LG)}$	Laguerre-Gaussian field
$U^{(MG)}$	multi-Gaussian field
$U^{(MGE)}$	elliptical multi-Gaussian field
$U^{(MGR)}$	rectangular multi-Gaussian field
$U^{(p)}$	paraxial field
$U^{(SG)}$	super-Gaussian field
$\mathcal{V}$	fringe visibility
v_m	velocity of light in medium
w	spatial radius of Gaussian beam
W	cross-spectral density function

$\mathbf{W}$	cross-spectral density matrix
$W^{(\infty)}$	cross-spectral density function in far zone
$W^{(BG)}$	cross-spectral density function of Bessel-Gaussian Schell-model beam
$\mathbf{W}^{(f)}$	cross-spectral density matrix of the free field
$W^{(G)}$	cross-spectral density function of Gaussian Schell-model beam
$W^{(IB)}$	cross-spectral density function of I-Bessel-correlated beam
$W^{(J)}$	cross-spectral density function of J_0-Bessel-correlated beam
$W^{(LG)}$	cross-spectral density function of Laguerre-Gaussian Schell-model beam
$W^{(MG)}$	cross-spectral density function of multi-Gaussian Schell-model beam
$\mathbf{W}^{(MG)}$	cross-spectral density matrix of multi-Gaussian Schell-model beam
$W^{(NUC)}$	cross-spectral density function of non-uniformly correlated beam
$\mathbf{W}^{(NUC)}$	cross-spectral density matrix of non-uniformly correlated beam
$\mathbf{W}^{(p)}$	cross-spectral density matrix of polarized part
$\mathbf{W}^{(t)}$	transmitted cross-spectral density matrix
$\mathbf{W}^{(u)}$	cross-spectral density matrix of unpolarized part
$\mathbf{W}^{(\iota)}$	incident cross-spectral density matrix
w_0	waste radius of Gaussian beam
w_{0A}	outer waste radius of dark-hollow beams

w_{0B}	inner waste radius of dark-hollow beams	α_P	parameter of a parabolic beam
w_{0x}, w_{0y}	beam waste radii components	α_{Ax}, α_{Ay}	absorption coefficients of absorber
w_{0xy}	mixed beam waste radius	$\beta(\omega)$	parameter of J_0-Bessel-correlated beam
w_b, w_c	parameters of non-uniformly correlated beam	β_b	parameter of Bessel-Gaussian correlated beam
w_I	parameter of I_0-Bessel correlated beam	Γ	mutual coherence function
w_K	relative strength of temperature-salinity fluctuations	$\boldsymbol{\Gamma}$	beam coherence-polarization matrix
W_n	cross-spectral density function of coherent mode	$\Gamma^{(4)}$	fourth-order field correlation function
$\mathbf{W}_{\alpha\beta}^{(MG)}$	cross-spectral degree density matrix of multi-Gaussian Schell-model beam	$\boldsymbol{\Gamma}^{(4)}$	fourth-order correlation matrix
		γ_0	parameter of non-uniformly correlated beam
x	first Cartesian coordinate	Γ_f	Gamma-function
$x(t)$	deterministic real function	Γ_X	auto-correlation function
$X(t)$	real random process	$\Gamma_{Z1,Z2}$	cross-correlation function
$\hat{x}$	unit vector	$\gamma_\alpha, \gamma_\beta$	parameters of electromagnetic non-uniformly correlated beam
$x_m(t)$	m-th realization of a real random process		
$x_{TR}(t)$	truncated version of a function	γ_ϖ	root-mean-square width of phase correlation
y	second Cartesian coordinate	δ	root-mean-square width of spectral degree of coherence
$Y(t)$	real random process		
$\hat{y}$	unit vector	$\delta(\cdot)$	delta-function
$y_m(t)$	mth realization of a real random process	$\boldsymbol{\delta}^2$	transverse correlation matrix of root-mean-square width
z	third Cartesian or Cylindrical coordinate	$\Delta^{(G)}$	generalized spreading coefficient of Gaussian Schell-model beam
$Z(t)$	complex random process		
$z_m(t)$	m-th realization of a complex random process	δ_b	parameter of Bessel-Gaussian correlated beam
Z_f	impedance of free-space	δ_I	parameter of I-Bessel correlated beam
z_R	Rayleigh range	δ_l	parameter of Laguerre-Gaussian correlated beam
z_ζ	normalized distance		
α	slope of the power spectrum	δ_n	relative refractive index
$\alpha_g(\omega)$	radial gradient of GRIN fiber	δ_T	root-mean-square width of correlation function of target surface roughness

$\delta_{\alpha\beta}$ parameter of electromagnetic Gaussian Schell-model beam

δ_{ϖ}^2 correlation width of random array

$\Delta_m^{(MG)}$ generalized spreading coefficient of multi-Gaussian Schell-model beam

$\Delta_{\alpha\beta}^{(G)}$ spreading coefficient of electromagnetic Gaussian Schell-model beam

$\Delta_{m\alpha\beta}^{(MG)}$ spreading coefficient of electromagnetic multi-Gaussian Schell-model beam

$\widehat{\Delta}_{\alpha\beta}$ expansion coefficients of electromagnetic Gaussian Schell-model beam in $ABCD$ system

Δw transverse spread

$\Delta\theta$ divergence angle

$\Delta\omega$ bandwidth

ϵ degree of ellipticity of polarization ellipse

ϵ_I ellipticity parameter of Ince-Gaussian beam

ϵ_m dielectric permittivity

ε_K dissipation rate of turbulent kinetic energy per unit mass

η spectral degree of coherence for electromagnetic beam

η_K Kolmogorov inner scale in oceanic power spectrum

$\eta_{\alpha\beta}$ degree of correlation between α and β electric field components

θ azimuthal angle

Θ scattering angle

θ_D divergence angle

$\Theta_{\alpha\beta}$ parameter of electromagnetic quasi-homogeneous beam

$\Theta(\omega)$ ratio of y and x field components

θ_{max} angle for maximum intensity

$\theta_{max}^{(BG)}$ angle for maximum intensity of Bessel-Gaussian correlated beam

$\theta_{max}^{(LG)}$ angle for maximum intensity of Laguerre-Gaussian correlated beam

κ 3D vector of spatial frequencies

λ wavelength

Λ root-mean-square wavelength width

$\lambda^{(x)}, \lambda^{(y)}$ eigenvalues of diagonal components of cross-spectral density matrix

λ_0 wavelength in vacuum

λ_1 shifted wavelength

λ_{0I} central wavelength of intensity

$\lambda_{0\delta}$ central wavelength of correlation width

$\lambda_{0\sigma}$ central wavelength of beam width

Λ_I root-mean-square spectral width

Λ_δ root-mean-square correlation width

Λ_σ root-mean-square beam width

λ_n eigenvalues

Λ_{nm} eigenvalues of off-diagonal components of cross-spectral density matrix

$\lambda_n^{(G)}$ 1D eigenvalues of Gaussian Schell-model beam

$\lambda_n^{(J)}$ eigenvalues of J_0-Bessel-correlated beam

$\lambda_{nm}^{(G)}$ 2D eigenvalues of Gaussian Schell-model beam

μ spectral degree of coherence

$\mu^{(\infty)}$ spectral degree of coherence in far zone

$\mu^{(BG)}$ spectral degree of coherence of Bessel-Gaussian Schell-model beam

$\mu^{(G)}$ spectral degree of coherence of Gaussian Schell-model beam

$\mu^{(IB)}$ spectral degree of coherence of I-Bessel-correlated beam

$\mu^{(J)}$ spectral degree of coherence of J_0-Bessel-correlated beam

$\mu^{(LG)}$ spectral degree of coherence of Laguerre-Gaussian correlated beam

$\mu^{(MG)}$ spectral degree of coherence of multi-Gaussian Schell-model beam

$\mu^{(NUC)}$ spectral degree of coherence of non-uniformly correlated beam

μ_F degree of spatial correlation of random medium

μ_m magnetic permeability

μ_n spectral degree of coherence of mode n

$\mu_{\alpha\beta}^{(MG)}$ spectral degree of correlation of multi-Gaussian Schell-model beam

ξ parameter of I-Bessel-correlated beam

ξ_c amplitude transmission function of a circular aperture

Π_e electric Hertz potential

ϖ 2D phase distribution on SLM

ϖ_0^2 maximum correlation of random array

ρ magnitude of $\boldsymbol{\rho}$

$\boldsymbol{\rho}$ position vector in 2D space

$\tilde{\boldsymbol{\rho}}$ 4D vector

ρ_0 coherence radius of spherical wave in random medium

ρ_c electric charge density

$\bar{\rho}$ hard radius of circular detector

$\boldsymbol{\rho}_d$ difference vector

$\boldsymbol{\rho}_s$ half-sum vector

$\bar{\rho}_g$ soft Gaussian radius of detector

ϱ_λ wavelength shift

ϱ_ω frequency shift

σ root-mean-square width of spectral density

$\boldsymbol{\sigma}^2$ transverse spectral density matrix of root-mean-square widths

Σ_d area of detector

σ_m specific conductivity

σ_x, σ_y parameters of electromagnetic Gaussian Schell-model beam

ς first cylindrical coordinates

ς_1, ς_2 major and minor semi-axes of polarization ellipse

τ time delay

$\tau(\cdot)$ amplitude profile function

τ_δ twist phase

Υ parameter of Hermite-Gaussian beams

ϕ polar angle

Φ_ϵ power spectrum of dielectric permittivity

φ phase

Φ_n power spectrum of refractive index

φ_p orientation of polarizer

φ_r angle of rotator

$\varphi_{Rx}, \varphi_{Ry}$ phases of retarder

χ shape angle of polarization ellipse

χ_T dissipation rate of mean-square temperature

ψ orientation angle of polarization ellipse

Ψ	complex phase perturbation	ω	angular frequency
ψ_G	Gouy phase of Gaussian beam	Ω	root-mean-square angular frequency width
ψ_{HG}	Gouy phase of Hermite-Gaussian beam	$\boldsymbol{\Omega}$	ray-transfer matrix
ψ_{LG}	Gouy phase of Laguerre-Gaussian beam	ω_1	shifted angular frequency
ψ_n	eigenfunctions	ω_T	root-mean-square width of target extent
$\psi_n^{(G)}$	1D eigenfunctions of Gaussian Schell-model beam	$\omega_{c\alpha\beta}$	parameter of electromagnetic non-uniformly correlated beam
$\psi_n^{(J)}$	eigenfunctions of J_0-Bessel-correlated beam	$\bar{\omega}$	central frequency
$\psi_{nm}^{(G)}$	2D eigenfunctions of Gaussian Schell-model beam	$\wp$	spectral degree of polarization

1

Introduction

CONTENTS

1.1 Brief history

Stochastic (random) light fields have been studied by scientists at least from the time of Sir Isaac Newton. It was observed first by an unaided eye then with the help of optical devices that natural light scattered from rough surfaces, such as the sea surface, can form random intensity patterns, sometimes of various colors. Such patterns were named "speckles". Speckles formed by natural light are still of interest for science and art.

It later became known that speckle introduces deleterious effects for imaging systems operating in the presence of random media, such as atmospheric turbulence, for instance. In particular, the atmosphere introduces limitation to the quality of telescopic images of celestial objects. It was first realized by D. Fried how postprocessing techniques may be used for reduction of the speckle noise [1]. Later several other speckle-related techniques were developed that made it possible to obtain the high-quality information about the object, for instance speckle interferometry [2] and aperture masking [3], [4]. Thus,

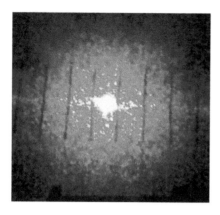 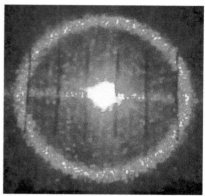

FIGURE 1.1
Intensity distributions in transverse cross-sections of typical stochastic beams produced with the help of the spatial light modulator. (left) random beam with flat-top intensity profile; (right) ring-shaped beam.

early encounters with speckle were often associated with negative, harmful phenomena.

With the invention of lasers in the early 1960s monochromatic beam-like fields have been produced which we will briefly discuss in Chapter 2. Soon after that spatially randomized laser light with predictable statistical properties has found uses in active imaging systems [5], [6].

At about the same time, in the 1960s and 1970s the classical coherence theory was developed [7]–[10] and various laws for propagation, transmission and scattering of speckled fields have been formulated. It was then discovered that the statistical properties of speckles can be controlled not only in a certain cross-section of the field but also at an arbitrary distance from it, provided the properties of the medium in which the field travels are known.

The idea of producing speckles with controllable directionality belongs to Collett and Wolf [11]. Namely, the authors came to the conclusion that on propagation in vacuum an optical field preserving its narrow spatial distribution, should not necessarily be deterministic, according to a general belief, but instead could be random. Collett-Wolf beams are the most basic examples of fluctuating optical beams that this book will be concerned with. Figure 1.1 shows transverse (with respect to direction of propagation) cross-sections of actual random beams.

Computer simulations provide a very useful tool for analyzing the phenomena occurring with random beams. Figure 1.2 includes simulations of a typical random beam as it travels at two distances from the source.

From the early 1980s up to now a large body of research articles have been published on various aspects of generation of scalar stochastic fields and their interaction with various media and optical systems. These studies will be

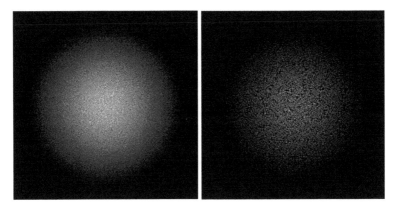

FIGURE 1.2
Simulated transverse cross-sections of a typical stochastic beam (left) at 10 m
and (right) at 50 m from the source.

discussed in detail in Chapter 3. At that time the treatment of random beams
was scalar since it was not yet understood how to incorporate the polarization
properties of light into the propagation laws.

Various measurements of speckle fields have revealed that their polarization states may vary with propagation. In his seminal paper, James showed
how the evolution of polarization properties in a model stochastic field propagating in vacuum can be predicted [12]. Not much later the first class of
stochastic beam-like fields with arbitrary polarization states was described
by Gori et al. [13]. It was followed by predictions and measurements of electromagnetic random beams on interaction with various optical devices and
media. The results of these studies are summarized in Chapter 4. Some fundamental concepts and examples relating to electromagnetic random beams
can also be found in Refs. [14] and [15].

Among the practical applications of random beams the easiest ones to
deal with are based on their interaction with image-forming linear systems.
In Chapter 5 we will discuss the outcomes of such interactions relating to the
human eye, a telescopic system, optical resonators and "perfect lens" systems
formed with layers of positive and negative phase materials [16].

Several applications of scalar and electromagnetic stochastic beams developed during the last two decades involve propagation in linear random media,
such as atmospheric and oceanic turbulence. We devote Chapter 6 to the description of the main phenomena associated with random beams on passage
in such natural media. Perhaps the most far-reaching use of random beams
is in the Free-Space Optical (FSO) communication systems which operate in
the atmosphere. The feasibility of this application has been thoroughly analytically and experimentally investigated. In Chapter 7 we discuss the basic
principles of operation of a laser communication link and some calculations

used in its performance assessment. We also outline the advantages of using
stochastic beams instead of laser beams for communications through turbulent channels which were first pointed by Banakh et al. [17] (see also [18]).
On the other hand, as we will also demonstrate, inverse problems of random
media, including sensing and identification of targets embedded in random
media, can be also effectively approached with the help of stochastic fields.
We will outline the principles of operation for active Light Detection And
Ranging (LIDAR) systems and show how stochastic beams can be employed
there for efficient target recognition.

A relatively recent, but important practical interest lies in the area of
scattering of random fields, and beams in particular, from discrete scattering
collections and continuous, random thin layers. The results of the weak scattering theory presented in Chapter 8 can potentially improve remote sensing
and imaging techniques of environmental and medical optics.

1.2   Preliminary mathematics

1.2.1   Random processes

The area of mathematics that analyzes the behavior of fields varying with
time in a random manner is known as the *theory of random processes* [19].
By a field we mean the distribution in space of a certain physical quantity
(scalar field) or the finite set of such quantities (vector field). In both cases
the field can be either real or complex, in general, but for our purposes it is
almost always complex, since optical disturbances are characterized by their
amplitudes and phases.

Among several possible ways of characterizing a random process is list-
ing all of its possible realizations. Such an approach is doomed for all but
a few trivial processes. A more realistic method is based on specification
of the statistical moments of all orders. Statistical moments provide the
wide range of information on how the process behaves "on average". To illustrate this idea we assume, without loss of generality, that a real scalar
random process $X(t)$ can assume any functional form from the ensemble
$\{x(t)\} = \{x_1(t), x_2(t), ..., x_m(t), ..., x_M(t)\}$ of its realizations (see Fig. 1.3).
The first statistical moment is called the *mean value* of the process and can
be evaluated with the help of the *time-averaging* procedure by the expression

$$\langle x_m \rangle_T = \lim_{T \to \infty} \frac{1}{T} \int_{-T/2}^{T/2} x_m(t)dt, \qquad (1.1)$$

for any fixed realization $m = 1, ..., M$ and sufficiently long time interval T. An
alternative type of averaging that uses all the realizations of the process at

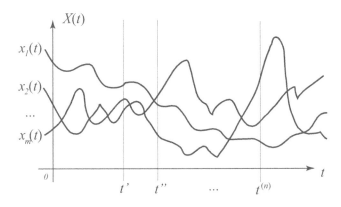

FIGURE 1.3
Illustration of realizations of a random process.

a fixed time instant, say, t', called the *ensemble averaging*, is defined by the integral

$$\langle x(t') \rangle_{EC} = \int \xi p_1(\xi, t') d\xi, \qquad (1.2)$$

ξ being the variable of integration. Here $p_1(\xi, t')$ is called the *probability density function*, returning the probability that random process $X(t')$ has value in the interval $(\xi, \xi + d\xi)$ at time instant t'. An alternative, discrete form of ensemble averaging is also used at times:

$$\langle x(t') \rangle_{ED} = \lim_{M \to \infty} \frac{1}{M} \sum_{k=1}^{M} x_m(t'). \qquad (1.3)$$

The higher-order statistical moments of the field can be defined with any type of averaging listed above. For example, the generalization of the continuous ensemble average to the n-th moment results in expression

$$\langle x(t')x(t'')...x(t^{(n)}) \rangle_{EC}$$
$$= \int \int ... \int \xi_1 \xi_2 ... \xi_n p_n(\xi_1, \xi_2, ..., \xi_n; t', t'', ..., t^{(n)}) d\xi_1 d\xi_2 ... d\xi_n, \qquad (1.4)$$

where $p_n(\xi_1, \xi_2, ..., \xi_n; t', t'', ..., t^{(n)})$ is the *joint probability density function* returning the probability with which $X(t^{(m)})$ has value in the interval $(\xi_m, \xi_m + d\xi_m)$ at time instant $t^{(m)}$. If the joint probability density functions p_n and, hence, the moments defined in Eq. (1.4), are known for any n the random process is fully characterized.

The following hierarchy of random processes can be established with respect to the properties of their statistical moments [6].

- *Ergodic processes* constitute the most restrictive class of random processes. To introduce an ergodic process we first define a more general average compared to that in Eq. (1.4). For arbitrary function f we set:

$$\langle f[x(t'), x(t''), ..., x(t^{(n)})]\rangle_{EC} = \int \int ... \int f(\xi_1, \xi_2, ..., \xi_n)$$
$$\times p_n[\xi_1, \xi_2, ..., \xi_n; t', t'', ..., t^{(n)}]d\xi_1 d\xi_2 ... d\xi_n. \tag{1.5}$$

  The same can be done using the time average. Then the process is called ergodic if for any realization m and any order n the time and the ensemble averages are equal, i.e., if

$$\langle f[x_m(t'), x_m(t''), ..., x_m(t^{(n)})]\rangle_T = \langle f[x(t'), x(t''), ..., x(t^{(n)})]\rangle_{EC}. \tag{1.6}$$

- *Stationary processes* make a wider than ergodic class for which probability density functions $p_n(\xi_1, \xi_2, ..., \xi_n; t', t'', ..., t^{(n)})$ are independent of the time origin for all orders n. In particular, for the first two probability densities we then must have:

$$p_1(\xi, t') = p_1(\xi);$$
$$p_2(\xi_1, \xi_2, t', t'') = p_2(\xi_1, \xi_2, t' - t''). \tag{1.7}$$

- *Wide-sense stationary processes* is yet a less restrictive type of random processes compared with the listed above. Here we require:

$$\langle x(t')\rangle_{EC} = const.;$$
$$\langle x(t')x(t'')\rangle_{EC} = g(t'' - t'), \tag{1.8}$$

  for some function g.

- *Processes with stationary increments*, the broadest class so far, requires the difference $x(t') - x(t'')$ to be stationary, for all time instants t' and t''. In particular, the processes for which the first-order moment changes with time as a linear function fall under this category.

In dealing with complex random processes, defined as $Z(t) = X(t) + iY(t)$, where i is the complex unit, the concepts introduced for real processes are readily generalized: the probability density functions must now depend on complex variables $z_1, z_2, ..., z_m, ..., z_M$. In this case the statistical moments are defined with the help of complex conjugates, denoted by $*$. For instance,

$$\langle z^*(t')z^*(t'')...z(t^{(n-1)})z(t^{(n)})\rangle_{EC}$$
$$= \int \int ... \int \chi_1^* \chi_2^* ... \chi_{n-1} \chi_n p_n(\chi_1, \chi_2, ..., \chi_n; t', t''...t^{(n)})d^2\chi_1 d\chi_2^2 ... d^2\chi_n. \tag{1.9}$$

Here χ is the complex variable of integration and $d^2\chi_n$ is the differential unit of the complex plane.

1.2.2 Spectral representation of random processes

Let us first consider a deterministic real function $x(t)$ satisfying any of two inequalities:

$$\int_{-\infty}^{\infty} |x(t)|dt < \infty, \tag{1.10}$$

or

$$\lim_{T\to\infty} \frac{1}{T} \int_{-T/2}^{T/2} x^2(t)dt < \infty. \tag{1.11}$$

In the former case function $x(t)$ has the direct Fourier transform, i.e., the integral

$$\mathcal{F}[x(t)] = \tilde{x}(\omega)$$
$$= \int_{-\infty}^{\infty} x(t)\exp(i\omega t)dt \tag{1.12}$$

exists, where ω is the variable in Fourier space, which also plays the part of the angular frequency in optics. The quantity G_X defined as

$$G_X(\omega) = \left| \int_{-\infty}^{\infty} x(t)\exp(i\omega t)dt \right|^2 \tag{1.13}$$

is called the *energy spectral density* and determines average energy per angular frequency ω. In the latter case function $x(t)$ has finite power but might not have the Fourier transform. Then its truncated version

$$x_{TR}(t) = \left\{ \begin{array}{ll} x(t), & -\frac{T}{2} \le t \le \frac{T}{2}, \\ 0, & t < -\frac{T}{2}, t > \frac{T}{2} \end{array} \right. \tag{1.14}$$

does have the Fourier transform. On taking the limit as $T \to \infty$ we arrive at the definition of the *power spectral density* of the signal:

$$S_X(\omega) = \lim_{T\to\infty} \frac{1}{T} \left| \int_{-\infty}^{\infty} x_{TR}(t)\exp(i\omega t)dt \right|^2. \tag{1.15}$$

On passing from the ordinary real functions to random processes, which are generally complex, we must note that often the limit in Eq. (1.15) does not exist, as, for instance, for sample functions of a stationary process. Hence, it is convenient to define the energy density and the power density of a random process by the formulas:

$$G_X(\omega) = \langle \left| \int_{-\infty}^{\infty} x(t) \exp(i\omega t) dt \right|^2 \rangle_T,$$

$$S_X(\omega) = \lim_{T \to \infty} \frac{1}{T} \langle \left| \int_{-\infty}^{\infty} x(t) \exp(i\omega t) dt \right|^2 \rangle_T. \tag{1.16}$$

Further, the *autocorrelation function* of a random process defined by the double integral

$$\Gamma_X(t', t'') = \langle x(t')x(t'') \rangle_T$$

$$= \iint_{-\infty}^{\infty} \xi_1 \xi_2 p(\xi_1, \xi_2, t', t'') d\xi_1 d\xi_2, \tag{1.17}$$

determines the time-averaged degree of similarity of random process $X(t)$ at time instances t' and t''. For complex-valued processes formula (1.17) generalizes to form

$$\Gamma_Z(t', t'') = \langle z^*(t')z(t'') \rangle_T$$

$$= \iint_{-\infty}^{\infty} \chi_1^* \chi_2 p(\chi_1, \chi_2, t', t'') d\chi_1 d\chi_2. \tag{1.18}$$

The following relation exists between the autocorrelation function and the power spectral density: for a process that is at least wide-sense statistically stationary, they constitute the Fourier transform pair, i.e.,

$$S_Z(\omega) = \mathcal{F}[\Gamma_Z(\tau)],$$
$$\Gamma_Z(\tau) = \mathcal{F}^{-1}[S_Z(\omega)], \tag{1.19}$$

where $\tau = t'' - t'$. These relations are known as the *Wiener-Khintchine theorem* and have a tremendous practical outcome: the distribution of power over frequencies can be determined via the autocorrelation function, which is directly measured or easily calculated analytically.

The *autocovariance function* $C_Z(t', t'')$ and the *structure function* $D_Z(t', t'')$ of the process can be defined by the expressions

$$C_Z(t', t'') = \langle [z(t') - \langle z(t') \rangle_T][z(t'') - \langle z(t'') \rangle_T] \rangle_T$$
$$= \Gamma_Z(t', t'') - \langle z(t') \rangle_T \langle z(t'') \rangle_T \tag{1.20}$$

and

$$D_Z(t', t'') = \langle [z(t') - z(t'')]^2 \rangle_T$$
$$= \langle z^2(t') \rangle_T + \langle z(t'') \rangle_T - 2\Gamma_Z(t', t''), \tag{1.21}$$

respectively. If the structure function $D_Z(t', t'')$ depends only on the time delay $\tau = t'' - t'$ it might be of great importance, even for certain random processes that are not wide-sense stationary. This is the case for processes with stationary increments. For instance, atmospheric fluctuations in temperature, and, hence, in the refractive index, can be well modeled for short time intervals as a process with stationary increments because the mean temperature slowly changes during the day-night cycle. Hence, for atmospheric applications the knowledge of the structure functions is a must [18].

It can be readily verified that the structure function relates to the power spectral density by the formula:

$$D_Z(\tau) = 2 \int_{-\infty}^{\infty} S_Z(\omega)[1 - \cos(\omega\tau)]d\omega. \tag{1.22}$$

Another group of functions greatly important for our subsequent discussion are the *cross-correlation functions*, defined for two processes, Z_1 and Z_2:

$$\Gamma_{Z1Z2}(t', t'') = \langle z_1^*(t')z_2(t'')\rangle_T, \tag{1.23}$$

and, in particular,

$$\Gamma_{Z1Z2}(\tau) = \langle z_1^*(t)z_2(t + \tau)\rangle_T. \tag{1.24}$$

Random processes are called *jointly wide-sense stationary* if the cross-correlation function takes the form

$$\Gamma_{Z_1Z_2}(t', t'') = \Gamma_{Z_1Z_2}(\tau), \tag{1.25}$$

i.e., it depends only on time difference $\tau = t' - t''$.

The counterparts of the cross-correlation functions in frequency domain defined as

$$W_{Z_1Z_2}(\omega) = \lim_{T\to\infty} \frac{1}{T}\langle[\tilde{z}_{1TR}^*(\omega)\tilde{z}_{2TR}(\omega)]\rangle_{EC}, \tag{1.26}$$

are known as the *cross-spectral density functions*, where $\tilde{z}_{TR}$ is the Fourier transform of a truncated process [see Eq. (1.14)]. The cross-spectral density function is a measure of similarity between the two processes at a fixed frequency. It obeys the following property:

$$W_{Z_1Z_2}(\omega) = W_{Z_2Z_1}^*(\omega), \tag{1.27}$$

and, in particular, for any real-valued processes X and Y:

$$W_{XY}(-\omega) = W_{XY}(\omega). \tag{1.28}$$

The generalized Wiener-Khintchine theorem relates the cross-correlation function and the cross-spectral density function of two processes as a Fourier transform pair:

$$\begin{aligned} W_{Z_1Z_2}(\omega) &= \mathcal{F}[\Gamma_{Z_1Z_2}(\tau)], \\ \Gamma_{Z_1Z_2}(\tau) &= \mathcal{F}^{-1}[W_{Z_1Z_2}(\omega)]. \end{aligned} \tag{1.29}$$

We will see in the following sections that these relations together with Eq. (1.19) play the crucial part in coherence phenomena of light fields.

1.2.3 Analytic representation of complex signals

In this section we will introduce a convenient mathematical tool for dealing with quantities quadratic in the field, such as intensity, known as the *analytic signal*. Let us consider a real, scalar field $X(t)$ at a time instant t and at a fixed position $\mathbf{r} = (x, y, z)$ in space, which we will omit. If $\tilde{X}(\omega)$ is its Fourier transform and $H_s(\omega)$ is a Heaviside step function then the piecewise continuous function

$$
\tilde{U}(\omega) = \begin{cases} 2\tilde{X}(\omega), & \omega > 0, \\ \tilde{X}(\omega), & \omega = 0, \\ 0, & \omega < 0, \end{cases} \tag{1.30}
$$
$$
= \tilde{X}(\omega) \cdot 2H_s(\omega)
$$

contains only the non-negative frequency components of $X(t)$. Since $X(t)$ is Hermitian we can express $\tilde{X}(\omega)$ as

$$
\tilde{X}(\omega) = \begin{cases} \dfrac{1}{2}\tilde{U}(\omega), & \omega > 0, \\ \tilde{U}(\omega), & \omega = 0, \\ \dfrac{1}{2}\tilde{U}^*(-\omega), & \omega < 0. \end{cases} \tag{1.31}
$$

Then we introduce the *analytic signal representation* $U(t)$ of a real field $X(t)$ as the inverse Fourier transform of $\tilde{X}(\omega)$, i.e.,

$$
U(t) = X(t) + i\left[X(t) \star \frac{1}{\pi t}\right], \tag{1.32}
$$

where $\star$ stands for convolution. To arrive at Eq. (1.32) we have used the facts that the inverse Fourier transforms of $\tilde{X}(\omega)$ and $2H_s(\omega)$ are $X(t)$ and $\delta(t) + i/(\pi t)$, respectively, where $\delta(t)$ is the Dirac delta-function. It can be shown that the real field $X(t)$ and its imaginary counterpart, say, $Y(t)$ of the analytic signal $U(t)$, are related as the Hilbert transform pair, i.e., as

$$
Y(t) = \frac{1}{\pi}Pr\int_{-\infty}^{\infty} \frac{X(t')}{t' - t}dt',
$$
$$
X(t) = -\frac{1}{\pi}Pr\int_{-\infty}^{\infty} \frac{Y(t')}{t' - t}dt', \tag{1.33}
$$

where Pr stands for the *principle value* of a contour integral [20]. Equations (1.33) are also known as Kramers-Kronig relations. Neither $X(t)$ nor $U(t)$ can be measured at optical frequencies.

Unlike the idealistic monochromatic fields whose time variation is of the form

$$X(t) = a\cos(\omega t - \varphi), \qquad (1.34)$$

where a is a constant amplitude, φ is a constant phase, ω is the angular frequency of oscillation, the realistic, quasi-monochromatic fields have variation of the form

$$X(t) = a(t)\cos[\bar{\omega}t - \varphi(t)], \qquad (1.35)$$

where $a(t) = |U(t)|$ and $\varphi(t) = \arg\{U(t)\}$ are called the *amplitude* of the envelope and the *instantaneous phase* of the envelope, $\bar{\omega}$ is the mean frequency that is much greater than the signal bandwidth $\Delta\omega$. Since the left and right sides of Eq. (1.35) have a different number of functions the choice of $a(t)$ and $\varphi(t)$ is not unique. With the help of the analytic signal representation

$$U(t) = X(t) + iY(t) \qquad (1.36)$$

such ambiguity is removed. The complex analytic signal is sometimes expressed in the polar coordinate system, i.e., as

$$U(t) = a(t)\exp[i\varphi(t)]\exp[i\bar{\omega}t]. \qquad (1.37)$$

The envelope $a(t)\exp[i\varphi(t)]$ varies with time much slower than $\exp[i\bar{\omega}t]$.

1.2.4 Gaussian random processes

As we have previously stated, for the complete description of a random process the joint probability density functions of all orders must be specified, which is generally impossible in practice. The only process for which such a requirement is substantially weakened is Gaussian. For a real-valued Gaussian process the joint probability density function of the n-th order has the form

$$p_n(\mathbf{x}) = \frac{1}{(\sqrt{2\pi})^n\sqrt{\mathbf{C}_n}}\exp\left[-\frac{1}{2}(\mathbf{x}_n - \langle\mathbf{x}\rangle_T)^T\mathbf{C}_n^{-1}(\mathbf{x}_n - \langle\mathbf{x}\rangle_T)\right], \qquad (1.38)$$

where superscript T stands for matrix transposition, to be distinguished with time average denoted by $\langle\cdot\rangle_T$, and

$$\begin{aligned}
\mathbf{x}_n(t) &= \begin{bmatrix} x(t') & x(t'') & \cdots & x(t^{(n)}) \end{bmatrix}^T, \\
\langle\mathbf{x}\rangle_T &= \begin{bmatrix} \langle x\rangle_T & \langle x\rangle_T & \cdots & \langle x\rangle_T \end{bmatrix}^T,
\end{aligned} \qquad (1.39)$$

while $\mathbf{C}_n$ is the $n \times n$ covariance matrix with elements

$$\begin{aligned}
c_{lp}^2 &= \langle[x(t^{(l)}) - \langle x(t^{(l)})\rangle_T][x(t^{(p)}) - \langle x(t^{(p)})\rangle_T]\rangle_T, \\
&\qquad (l, p = 1, ..., n).
\end{aligned} \qquad (1.40)$$

In the case of the Gaussian process it can be shown that a statistical moment of any order can be written as a combination of the second-order moments. For instance, the fourth-order moment reduces to the sum

$$\langle x(t')x(t'')x(t''')x(t^{IV})\rangle_T = \Gamma_X(t'',t')\Gamma_X(t^{IV},t''')$$
$$+ \Gamma_X(t''',t')\Gamma_X(t^{IV},t'') + \Gamma_X(t''',t'')\Gamma_X(t^{IV},t').$$

$$(1.41)$$

We note that if a Gaussian process is wide-sense stationary it is stationary automatically, due to the fact that the n-th order probability density function and, hence, all moments are such that the mean $\langle \mathbf{x}_n \rangle = const$ and the covariance matrix $\mathbf{C}_n$ depends on time differences.

A complex random process is called the *complex Gaussian process* if its real and imaginary parts are joint Gaussian processes. Suppose $U(t)$ is an analytic signal representation of the real-valued Gaussian process $X(t)$, i.e., $U(t) = X(t) + iY(t)$. Then $Y(t)$ is also the Gaussian random process, because of relations (1.33) and, hence, $U(t)$ itself is the complex Gaussian process. Nevertheless, the converse is not true, i.e., not every complex Gaussian process is the analytic signal for some real Gaussian process.

The complex Gaussian process is called *circular* if for any four-time instances the following relations hold

$$X(t^{(m)}) = Y(t^{(m)}) = 0, \qquad (m = 1, 2, 3, 4);$$
$$X(t^{(m)})X(t^{(n)}) = X(t^{(m)})X(t^{(n)}), \quad (m, n = 1, 2, 3, 4);$$
$$X(t^{(m)})Y(t^{(n)}) = -X(t^{(m)})Y(t^{(n)}), \quad (m, n = 1, 2, 3, 4).$$

$$(1.42)$$

Analytic signal $U(t)$ of a real Gaussian process $X(t)$ with zero mean can be shown to be a circular Gaussian process [6]. The fourth-order moment of $U(t)$ then reduces to the sum

$$\langle U^*(t')U^*(t'')U(t''')U(t^{IV})\rangle_T = \Gamma_U(t''',t')\Gamma_U(t^{IV},t'')$$
$$+ \Gamma_U(t''',t'')\Gamma_U(t^{IV},t').$$

$$(1.43)$$

This formula is of importance in statistical optics since, if evaluated at $t' = t'' = t''' = t^{IV}$ it relates the average intensity of a fluctuating field with the contrast (variance) of fluctuation. Also, for $t' = t'''$ and $t'' = t^{IV}$ it relates intensity-intensity fluctuations with the autocorrelation function.

1.3 Preliminary optics

1.3.1 Maxwell's, wave and Helmholtz equations

An electromagnetic field is a disturbance arising in some region of space containing electric charges. The following set of four Maxwell's equations describes the behavior of the electromagnetic field at a position specified by vector $\mathbf{r} = (x, y, z)$ (which we omit for brevity until later) on its interaction with the medium where it is located [9]:

$$\text{curl}\mathbf{H} - \frac{1}{c}\dot{\mathbf{D}} = \frac{4\pi}{c}\mathbf{j};$$

$$\text{curl}\mathbf{E} + \frac{1}{c}\dot{\mathbf{B}} = 0; \tag{1.44}$$

$$\text{div}\mathbf{D} = 4\pi\rho_c;$$

$$\text{div}\mathbf{B} = 0.$$

Here dot denotes the time derivative, $\mathbf{E}$ is the *electric field*, $\mathbf{B}$ is the *magnetic induction*, $\mathbf{D}$ is the *electric displacement*, $\mathbf{H}$ is the *magnetic vector*, $\mathbf{j}$ is the *electric current density*, ρ_c is the *electric charge density*, $c = 299,792,458$ m/s is the *velocity of light in vacuum*, and curl and div are the operators of differential calculus. Maxwell's equations lead to a unique solution for electric and magnetic field vectors, $\mathbf{E}$ and $\mathbf{H}$, for given $\mathbf{j}$ and ρ_c if the following so-called *material equations* hold:

$$\mathbf{j} = \sigma_m\mathbf{E}, \quad \mathbf{D} = \epsilon_m\mathbf{E}, \quad \mathbf{B} = \mu_m\mathbf{H}, \tag{1.45}$$

where σ_m, ϵ_m and μ_m are called the *specific conductivity*, the *dielectric permittivity* and the *magnetic permeability* of the medium, respectively. In terms of conducting properties, one distinguishes between dielectrics if $\sigma \ll 1$ and conductors, otherwise. Magnetic properties are present if $\mu_m \neq 1$; in particular, the material is paramagnetic for $\mu_m > 1$ and diamagnetic for $\mu_m < 1$. Equations (1.45) accurately describe interaction between fields and media if the fields are sufficiently weak, otherwise inclusion of the nonlinear terms is required.

In the absence of charges and currents, i.e., if $\mathbf{j} = 0$ and $\rho_c = 0$ it is possible to uncouple the differential equations (1.44) for electric and magnetic fields, $\mathbf{E}$ and $\mathbf{H}$, with the help of several identities of multivariate calculus [9], obtaining equations

$$\nabla^2\mathbf{E} + \frac{\epsilon_m\mu_m}{c^2}\ddot{\mathbf{E}} + (\text{grad}\ln\mu_m) \times \text{curl}\mathbf{E} + \text{grad}(\mathbf{E} \cdot \text{grad}\ln\epsilon_m) = 0, \tag{1.46}$$

and

$$\nabla^2\mathbf{H} - \frac{\epsilon_m\mu_m}{c^2}\ddot{\mathbf{H}} + (\text{grad}\ln\epsilon_m) \times \text{curl}\mathbf{H} + \text{grad}(\mathbf{H} \cdot \text{grad}\ln\mu_m) = 0, \tag{1.47}$$

where ∇^2 denotes the Laplacian, and double dot above the vector field denotes its second time-derivative.

For a homogeneous medium, i.e., if $\text{grad}\ln\epsilon_m = 0$ and $\text{grad}\ln\mu_m = 0$, Eqs. (1.46) and (1.47) reduce to the *wave equations*

$$\nabla^2\mathbf{E} - \frac{\epsilon_m\mu_m}{c^2}\ddot{\mathbf{E}} = 0, \tag{1.48}$$

and

$$\nabla^2 \mathbf{H} - \frac{\epsilon_m \mu_m}{c^2} \ddot{\mathbf{H}} = 0, \tag{1.49}$$

where factor $\epsilon_m \mu_m / c^2$ can be related to the wave velocity v_m in the given medium by the formula

$$\frac{\epsilon_m \mu_m}{c^2} = \frac{1}{v_m^2}. \tag{1.50}$$

In vacuum, where $\epsilon_1 = \mu_1 = 1$, we find that $v_m \equiv c$. The ratio n of wave velocities in vacuum and in the medium

$$n = \frac{c}{v_m} \tag{1.51}$$

is called the *refractive index*. It follows from Eqs. (1.50) and (1.51) that

$$n = \sqrt{\epsilon_m \mu_m}. \tag{1.52}$$

In particular, for non-magnetic materials ($\mu_m = 1$) the electric permittivity is directly related with the refractive index:

$$n = \sqrt{\epsilon_m}, \tag{1.53}$$

where both quantities are complex numbers, in general. The index of refraction can be expressed as:

$$n = n_{(r)} + i n_{(i)}, \tag{1.54}$$

where its real part $n_{(r)}$ is typically greater than one but can sometimes attain values smaller than one and also can be negative for left-hand materials [16]. The imaginary part $n_{(i)}$ is negative for absorbing materials and positive for the gain media. The real and the imaginary parts of n are related via Kramers-Kronig relations [see (1.33)], which can be used for the determination of real part $n_{(r)}$ from an absorption spectrum of the material measured at different wavelengths.

Let us now consider a scalar wave equation:

$$\nabla^2 X(\mathbf{r}, t) - \frac{1}{v_m^2} \ddot{X}(\mathbf{r}, t) = 0, \tag{1.55}$$

which is obeyed by each Cartesian component $X(\mathbf{r}, t)$ of the electric and magnetic fields at a point specified by a position vector $\mathbf{r} = (x, y, z)$. Consider, in particular the case when the temporal behavior of the wave is harmonic:

$$X(\mathbf{r}, t) = a(\mathbf{r}) \cos[\omega t - \varphi(\mathbf{r})], \tag{1.56}$$

where $a(\mathbf{r})$ and $\omega t - \varphi(\mathbf{r})$ are the position-dependent amplitude and phase, respectively, both a and φ being real scalar functions; ω is the angular frequency, as before, $\omega = 2\pi f$, f is ordinary *frequency*, $f = 1/T$, with cycle *period* T. The distance traveled by the wave in the medium during one cycle is called the *wavelength*:

$$\lambda = vT. \tag{1.57}$$

In vacuum the wavelength is determined by the relation

$$\lambda_0 = cT = n\lambda. \tag{1.58}$$

Expression (1.56) can also be written as

$$X(\mathbf{r}, t) = Re[U(\mathbf{r}) \exp(-i\omega t)], \tag{1.59}$$

where Re is the real part and

$$U(\mathbf{r}) = a(\mathbf{r}) \exp[i\varphi(\mathbf{r})], \tag{1.60}$$

is the so-called *complex amplitude* of the wave. On substituting from the previous two equations into Eq. (1.55) we find that the complex amplitude $U(\mathbf{r})$ obeys the Helmholtz equation:

$$\nabla^2 U(\mathbf{r}) + \frac{\omega^2}{v_m^2} U(\mathbf{r}) = 0. \tag{1.61}$$

Here

$$\frac{\omega}{v_m} = k = nk_0 \tag{1.62}$$

with $k_0 = \omega/c$ and $k = nk_0$ are called the *wave numbers* in vacuum and in the medium with refractive index n, respectively.

1.3.2 Angular spectrum representation and beam conditions

The invention of the laser has made it possible to efficiently control the directionality of radiated light fields. The concept of an optical beam then acquired a very definite meaning. By a beam-like field we intuitively imply a radiated field that occupies a relatively narrow region about its axis, independently of how far it propagates. For propagation in free space it is sufficient to consider the behavior of a diffracted field in the far zone of the source, i.e., in the region where it finally "stabilizes". However, the concept of the beam meets difficulties in the situations when radiation interacts with media, in which the far zone cannot be strictly defined. For instance, when a field propagates in optical turbulence, the turbulence-induced diffraction acts on it in a continuous way, resulting in a tremendous spreading starting from relatively short propagation distances. Hence, in a random medium, even a beam-like field, according to its free-space related definition, ceases to be such starting from some distance. Therefore, the rigorous definition of the beam-like optical fields should be limited to free-space analysis.

Even though there exist several ways of introducing the beam-like fields throughout the book we will mostly use the so-called *angular spectrum representation*. According to such representation a monochromatic scalar field $U(\mathbf{r}; \omega)$ generated in the plane $z = 0$ and radiated into the half-space $z \geq 0$

(see Fig. 1.4), can be decomposed, at a point with position vector $\mathbf{r} = (x, y, z)$, as ([7], Section 3.2.2)

$$U(x, y, z; \omega) = \int_{-\infty}^{\infty} \int_{-\infty}^{\infty} a(k_x, k_y; \omega) \exp[i(k_x x + k_y y + k_z z)] dk_x dk_y. \quad (1.63)$$

Here the wave vector $\mathbf{k} = (k_x, k_y, k_z)$ is such that

$$k_z = \begin{cases} \sqrt{k^2 - k_x^2 - k_y^2}, & k_x^2 + k_y^2 \le k^2; \\ i\sqrt{k_x^2 + k_y^2 - k^2}, & k_x^2 + k_y^2 > k^2, \end{cases} \quad (1.64)$$

where $k = |\mathbf{k}| = \sqrt{k_x^2 + k_y^2 + k_z^2}$. The waves corresponding to k_z as defined by the first and second lines of Eq. (1.64) are called *homogeneous* and *evanescent*, respectively. The homogeneous waves propagate into the half-space $z > 0$ while the evanescent waves decay exponentially with increasing distance z from the source plane vanishing at distances on the order of a wavelength. Therefore, the evanescent waves are significant only very close to the source and do not contribute to the far field. The quantities $a(k_x, k_y; \omega)$ in Eq. (1.63) are the amplitudes of the plane waves, which can be related to the field $U(x', y', 0; \omega)$ in the source plane $z = 0$ by the formula

$$a(k_x, k_y; \omega) = \frac{1}{(2\pi)^2} \int_{-\infty}^{\infty} \int_{-\infty}^{\infty} U(x', y', 0; \omega) \exp[-i(k_x x' + k_y y')] dx' dy', \quad (1.65)$$

being its two-dimensional Fourier transform, while k_x and k_y can then be regarded as spatial frequencies. At this point we stress that while Eq. (1.65) is the Fourier transform relationship, Eq. (1.63) is not.

Using the method of stationary phase for double integrals it can be proven that the field radiated into the far zone to a point specified by position vector $\mathbf{r} = r\mathbf{u}$, where $\mathbf{u} = (u_x, u_y, u_z)$ is the unit vector in the direction $\mathbf{r}$, i.e., $\mathbf{u} = \mathbf{r}/r$, has the form ([7], Section 3.3)

$$U^{(\infty)}(r\mathbf{u}; \omega) = -2\pi i k u_z a(k_x, k_y; \omega) \frac{\exp(ikr)}{r}, \quad (1.66)$$

as $kz \to \infty$, where $k_x = k u_x$ and $k_y = k u_y$.

In order for a field to have the beam-like structure one simply needs to require that its spectral amplitudes $a(k_x, k_y; \omega)$ are negligible for all directions except the ones that are very close to the optical axis. Hence, in Eq. (1.66) one must set

$$|a(k_x, k_y; \omega)| \ll 1, \quad \text{unless} \quad k_x^2 + k_y^2 \ll k^2, \quad (1.67)$$

retaining only the plane waves that propagate close to the z-axis. In other words, the angular spectra of beam-like fields only contain low portions of the spatial frequency components k_x and k_y. First of all, this implies that all evanescent waves and some of the homogeneous waves must be excluded from the angular spectrum. On dividing inequality $k_x^2 + k_y^2 \ll k^2$ by k^2 we find that, alternatively,

$$u_x^2 + u_y^2 \ll 1. \tag{1.68}$$

Using the inequality (1.67) in the first line of Eq. (1.64) and after employing in the last step the binomial approximation, we arrive at the expression

$$k_z = \sqrt{k^2 - (k_x^2 + k_y^2)} \approx k \left[1 - \frac{1}{2} \left(\frac{k_x^2 + k_y^2}{k^2} \right) \right]. \tag{1.69}$$

On substituting this approximation for k_z into Eq. (1.63) we finally obtain the formula

$$U(x, y, z; \omega) = \exp(ikz) \int\limits_{-\infty}^{\infty} \int\limits_{-\infty}^{\infty} a(k_x, k_y; \omega) \exp[i(k_x x + k_y y)]$$

$$\times \exp\left[-i\frac{z}{2k}(k_x^2 + k_y^2) \right] dk_x dk_y. \tag{1.70}$$

Relation (1.70), together with Eq. (1.65) provide the free-space propagation law for any optical beam-like field. In fact, these two formulas can be combined by direct substitution of $a(k_x, k_y; \omega)$ and simplified. The result is the four-folded integral:

$$U(x, y, z; \omega) = \frac{\exp(ikz)}{(2\pi)^2} \int\limits_{-\infty}^{\infty} \int\limits_{-\infty}^{\infty} \int\limits_{-\infty}^{\infty} \int\limits_{-\infty}^{\infty} U(x', y', 0; \omega) \exp\left[-i\frac{z}{2k}(k_x^2 + k_y^2) \right]$$

$$\times \exp\{i[k_x(x - x') + k_y(y - y')]\} dk_x dk_y dx' dy'. \tag{1.71}$$

After evaluation of the integrals with respect to k_x and k_y, with the help of the formula

$$\int\limits_{-\infty}^{\infty} \exp[-q_1^2 t^2] \exp[-iq_2 t] dt = \frac{\sqrt{\pi}}{q_1} \exp[-q_2^2/4q_1^2], \quad Re(q_1) > 0, \tag{1.72}$$

where q_1 and q_2 are constants, we finally arrive at the propagation law

$$U(x, y, z; \omega) = -\frac{ik}{2\pi z} \exp(ikz) \int\limits_{-\infty}^{\infty} \int\limits_{-\infty}^{\infty} U(x', y', 0; \omega)$$

$$\times \exp\left[-\frac{ik}{2z}[(x - x')^2 + (y - y')^2] \right] dx' dy'. \tag{1.73}$$

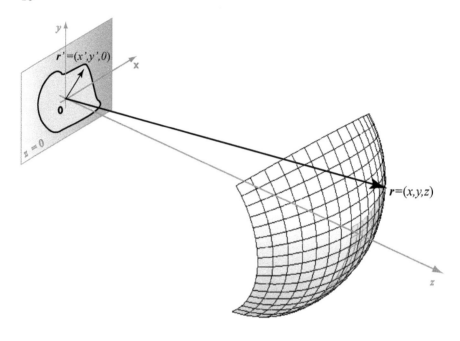

FIGURE 1.4
Notation relating to beam propagation.

Thus, provided that an optical field generated in the plane $z = 0$ satisfies conditions expressed by Eq. (1.67), its free-space propagation formula can be found from approximation (1.73). This formula also applies to the case of any constant, generally complex values of k being capable of predicting the beam evolution in absorpting/gain media and media with a negative refractive index.

There is an alternative way of introducing beam-like solutions of the Helmholtz equation. Let us first represent the monochromatic field $U(\mathbf{r};\omega)$ by a product

$$U(\mathbf{r};\omega) = U^{(p)}(\mathbf{r};\omega)\exp(ikz). \qquad (1.74)$$

For physically realizable beam-like fields $U^{(p)}(\mathbf{r};\omega)$ generally depends on x, y, and z but should vary slowly with z, which can be expressed by inequality

$$\left|\frac{\partial^2 U^{(p)}(\mathbf{r};\omega)}{\partial z^2}\right| \ll 2k\left|\frac{\partial U^{(p)}(\mathbf{r};\omega)}{\partial z}\right|. \qquad (1.75)$$

Using this condition, known as the *paraxial approximation*, in Eq. (1.61) we find that $U^{(p)}(\mathbf{r};\omega)$ is the solution of equation

$$\frac{\partial^2 U^{(p)}(\mathbf{r};\omega)}{\partial x^2} + \frac{\partial^2 U^{(p)}(\mathbf{r};\omega)}{\partial y^2} + 2i\frac{\partial U^{(p)}(\mathbf{r};\omega)}{\partial z} = 0. \qquad (1.76)$$

This equation is known as the *parabolic equation* of the Schrödinger type. It admits solutions in 17 coordinate systems that are not spatially invariant but preserve their structural form. In the Cartesian coordinate system the solutions of Eq. (1.76) coincide with the solution given by Eq. (1.73). However, the reader should be careful in following this approach without checking condition (1.67), since not all the solutions of the paraxial equation represent optical beam-like fields (see also [7], Section 5.6.1). We leave the discussion of known classes of paraxial beams until Chapter 2.

1.3.3 Exact beams

We have previously shown that monochromatic light waves oscillating at frequency ω and propagating in free space to a point with position vector $\mathbf{r}$ obey the Helmholtz equation [see Eq. (1.61)]

$$\nabla^2 U(\mathbf{r};\omega) + k^2 U(\mathbf{r};\omega) = 0, \qquad (1.77)$$

and derived its approximate solution in the Cartesian coordinates (1.73). However, the Helmholtz equation possesses "exact" solutions, i.e., the wave fields that satisfy it without approximation (1.67), at the same time being able to resemble beam-like structures.

It is known that the Helmholtz equation has the so-called *separable solutions*, i.e., solutions in n dimensions in the form of the product

$$f(x_1, x_2, ..., x_n) = f_1(x_1) \cdot f_2(x_2) \cdot ... \cdot f_n(x_n), \qquad (1.78)$$

in 11 orthogonal coordinate systems [21], [22]. Only four of them: Cartesian, circular cylindrical, elliptic cylindrical and parabolic cylindrical coordinate systems possess translation symmetries, which allows separation of the Helmholtz equation into transverse and longitudinal parts. This type of separability implies that the solutions for the transverse part are independent from the longitudinal coordinate, and hence such fields exhibit spatial invariance on propagation, hence the terms "propagation-invariant" or "non-diffracting" optical fields can be often met. The exact, separable solutions of the Helmholtz equation have been found for all four mentioned coordinate systems. Perhaps the discovery of the fact that Maxwell's equations may have solutions in the form of beams that are free of diffraction effects belongs to Sheppard and Wilson [23]. It was then found that the spatial part of the electric field of a typical non-diffracting beam can have the form of the J_0-Bessel function. Later Durnin et al. had established [24] (see also [25], [26]) that practical realizability of such beams is somewhat limited, the limitation stemming from the fact that the carried energy of such a field is infinite. After the seminal J_0-Bessel

20

beam and its extension to Bessel beams of higher orders several other classes of diffraction-free beams were introduced, such as modified Bessel-correlated beams [27], Mathieu beams [28, 29] and parabolic beams [30]. All such exact fields will be discussed in detail here since some of them resemble beam-like structures and can be used for construction of a wide variety of random beams.

Although the exact and the paraxial beams can be visually very similar in terms of the spatial content, they carry a fundamental structural difference. While the exact beams are purely non-diffractive and self-invariant, they cannot be produced in the laboratory because they should carry infinite energy. On the other hand, the paraxial beams diffract, changing their spatial size and shape on propagation, however they are practically readily realizable.

In order for a monochromatic field $U(\mathbf{r}; \omega)$ propagating along the positive z-direction to be non-diffracting it must satisfy the Helmholtz equation (1.77) and its spatial part must have the form

$$U(\mathbf{r}; \omega) = U_\perp(x, y; \omega) \exp(-ik_z z), \qquad (1.79)$$

where $U_\perp(x, y; \omega)$ is the transverse amplitude profile, independent of z. Equation (1.79) implies that the changes in transverse, (x, y), and longitudinal, z, wave profiles are not coupled.

We will now briefly discuss the geometrical interpretation of the non-diffracting fields in terms of their angular spectra (see Fig. 1.5). The non-diffracting field $U(\mathbf{r}; \omega)$ can be expressed by a reduced Whittaker integral [31]

$$U(\mathbf{r}; \omega) = \exp[-ik_z z] \int_{-\pi}^{\pi} a(\phi) \exp[-ik_\perp (x \cos \phi + y \sin \phi)]d\phi, \qquad (1.80)$$

in the cylindrical coordinate system, where $a(\phi)$ represents the rotationally symmetric angular spectrum, $k_\perp = \sqrt{k_x^2 + k_y^2}$ is the magnitude of the transverse part of the wave number, obeying relation $k^2 = k_\perp^2 + k_z^2$ [32]. The magnitude of the transverse and the longitudinal components of the wave vector $\mathbf{k}$ can be expressed as $k_\perp = k_0 \sin \theta_0$, $k_z = k_0 \cos \theta_0$, and hence the integral in Eq. (1.80) can be considered as a superposition of the plane waves on a sphere $|\mathbf{k}| = k_0$, known as a McCutchen sphere [33]. The wave vectors of theses plane waves must lie on a cone with central angle θ_0, such that $\tan \theta_0 = k_\perp/k_z$. The angular spectrum amplitude $a(\phi)$ selects certain plane waves from this set and determines the shape of the non-diffracting field.

1.3.3.1 Plane waves and cosine beams

The simplest non-diffracting field is the well-known homogeneous plane wave, being the solution of Eq. (1.77) in the Cartesian coordinate system. Indeed, using the method of separation of variables and assuming that the solution is in the form of the product

$$\begin{aligned} U(\mathbf{r}; \omega) &= U(x, y, z; \omega) \\ &= X(x; \omega)Y(y; \omega)Z(z; \omega), \end{aligned} \qquad (1.81)$$

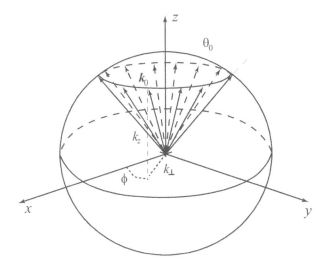

FIGURE 1.5
McCutchen sphere.

we find, on substituting from Eq. (1.81) into Eq. (1.77), that

$$U^{(PL)}(\mathbf{r};\omega) = \exp[-i(k_x x + k_y y + k_z z)], \tag{1.82}$$

where we assumed that the generally complex constant amplitude is unity and $k_z = \sqrt{k^2 - (k_x^2 + k_y^2)}$. Evanescent waves are omitted from discussion about beams since typically the interest for the beam-like fields lies in the regime of intermediate or large propagation distances, whereas evanescent waves vanish in the region of several wavelengths from the source [7]. The angular spectrum of a homogeneous plane wave has the form

$$a(\phi) = \delta(\phi - \phi_0), \tag{1.83}$$

where ϕ_0 is a constant. More generally, for the superposition of M plane waves the angular spectrum becomes

$$a(\phi) = \sum_{m=1}^{M} \delta(\phi - \phi_m). \tag{1.84}$$

In particular, if chosen to lie in the vertices of a polygonal with an even number of facets the plane waves result in the diffraction-free field exhibiting the rotationally symmetric kaleidoscopic patterns. In the simplest case, the sum of two symmetrically oriented plane waves results in the cosine beam:

$$a(\phi) = \delta(\phi - \pi/2) + \delta(\phi + \pi/2). \tag{1.85}$$

In the real space the cosine beam, for instance having constant values in x direction, is represented as

$$U^{(COS)}(\mathbf{r};\omega) = \frac{1}{2}[\exp(ik_\perp y) + \exp(-ik_\perp y)]. \tag{1.86}$$

1.3.3.2 Bessel beams

The family of Bessel beams appears as a solution of the homogeneous Helmholtz equation (1.77) in the circular cylindrical coordinate system:

$$x = r\cos\phi; \quad y = r\sin\phi; \quad z = z, \tag{1.87}$$

where radius r and polar angle ϕ are such that $r \in [0,\infty)$, $\phi \in [0,2\pi]$. On assuming that the field is a product

$$U(r,\phi,z;\omega) = R(z)\Phi(\phi)\exp(-ik_z z), \tag{1.88}$$

and noting that Φ must be a periodic function:

$$\Phi(\phi) = \exp(im\phi), \quad m = 0,1,2,..., \tag{1.89}$$

we find, on substituting from Eqs. (1.88) and (1.89) into Eq. (1.77), that the radial part $R(r)$ of the field satisfies the Bessel equation

$$\frac{d^2 R(r)}{dr^2} + \frac{1}{2}\frac{1}{r}\frac{dR(r)}{dr} + k_\perp^2 R(r)\left(1 - \frac{m^2}{k_\perp^2 r^2}\right) = 0. \tag{1.90}$$

The general solution of this equation has the form

$$R_m(r) = w_1 J_m(k_\perp r) + w_2 N_m(k_\perp r), \tag{1.91}$$

where w_1 and w_2 are the weighting coefficients, J_m is the m-th order Bessel function, and N_m is the m-th order Neumann function. In particular, the J_m-Bessel family of functions gives the solutions of the form

$$U_m^{(B)}(r,\phi,z;\omega) = J_m(k_\perp r)\exp[i(m\phi - k_z z)], \tag{1.92}$$

where $0 < k_\perp \le k$. The 0-th order member of this family is the well-known J_0-Bessel beam [24]:

$$U^{(B)}(r,\phi,z;\omega) = J_0(k_\perp r)\exp[-ik_z z]. \tag{1.93}$$

A quasi-Bessel beam, i.e., an ideal Bessel beam truncated by an aperture was realized in the laboratory [25] and was shown to have better divergence properties compared to typical laser beams [34], [35].

The Neumann functions alone may not be considered as a physical solution because of their singularity at the optical axis. However, combinations of Bessel and Neumann functions can form physically meaningful solutions, yielding Hankel functions of the m-th order [32]:

$$H_m^{(1)}(r, \phi; \omega) = J_m(k_\perp r) + iN_m(k_\perp r),$$
$$H_m^{(2)}(r, \phi; \omega) = J_m(k_\perp r) - iN_m(k_\perp r). \tag{1.94}$$

The real amplitude of the rotationally symmetrical beam is defined by expression

$$\bar{U}^{(H)}(r, z; \omega) = \frac{1}{2}[U(r, z; \omega) + U^*(r, z; \omega)], \tag{1.95}$$

where

$$U(r, z; \omega) = \frac{1}{2}[H_0^{(1)}(k_\perp r) + H_0^{(2)}(k_\perp r)] \exp(-ik_z z). \tag{1.96}$$

Hence, applying the last two equations we get

$$\bar{U}^{(H)}(r, z; \omega) = \frac{1}{2}[\bar{U}_0^{(1)}(r, z; \omega) + \bar{U}_0^{(2)}(r, z; \omega)], \tag{1.97}$$

where

$$U_0^{(1)}(r, z; \omega) = J_0(k_\perp r) \cos(k_z z) + N_0(k_\perp r) \sin(k_z z),$$
$$U_0^{(2)}(r, z; \omega) = J_0(k_\perp r) \cos(k_z z) - N_0(k_\perp r) \sin(k_z z). \tag{1.98}$$

In the angular spectrum domain the J_m-Bessel beams can be represented by the expression

$$a(\phi) = \exp(im\phi), \tag{1.99}$$

and, in particular, for the J_0 Bessel beam it is just a unity.

In Fig. 1.6 the intensity $|U_0^{(B)}|^2$ of a J_0-Bessel beam (left) and $|U_1^{(B)}|^2$ of J_1-Bessel beam (right) as functions of x and y (in millimeters). Unlike the J_0-Bessel beam the higher-order Bessel beams possess the zero value of intensity at the optical axis. In addition, the higher-order Bessel beams carry the phase singularity at the origin, which is called the *optical vortex*.

The transverse part of the wave number, $k_\perp$ is not limited to real numbers [27]. On setting

$$k_\perp = ik'_\perp, \tag{1.100}$$

we find that the field takes the form

$$U^{(BI)}(r, \phi, z; \omega) = \exp[i(m\phi - k_z z)]I_m(k'_\perp r), \tag{1.101}$$

where I_m is the modified Bessel function of order m. It is important to observe that $k'_\perp$ remains real just like $I_m(k'_\perp r)$, and $k'_\perp$ relates to k_z as

$$k_z^2 = k^2 + k'^2_\perp. \tag{1.102}$$

The modified Bessel beams have a minimum intensity center, and unlike ordinary Bessel beams, represent propagating solutions along the z-axis for any real value of $k'_\perp$.

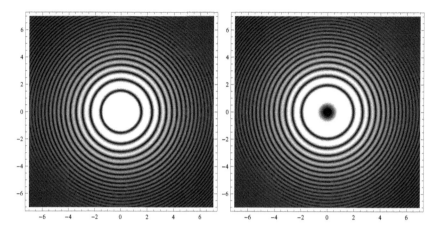

FIGURE 1.6
The density plots of the transverse intensity distributions of J_0-Bessel beam (left) and J_1-Bessel beam (right) with $k_\perp=1$.

1.3.3.3 Mathieu beams

Another coordinate system in which the Helmholtz equation has separable solutions is the elliptic-cylindrical system [28], defined as

$$x = h_f \cosh \xi \cos \eta, \quad y = h_f \sinh \xi \sin \eta, \quad z = z, \qquad (1.103)$$

where $\xi \in [0, \infty)$ and $\eta \in [0; 2\pi)$ are the radial and angular variables, $2h_f$ being the interfocal separation, i.e., the distance between the foci of an ellipse placed at the plane (x, y) of the Cartesian coordinate system.

In this case the Helmholz equation separates into a longitudinal part, which has a solution with the $\exp(ik_z z)$ dependence, and a transverse part, whose solution $U_\perp(\xi, \eta; \omega)$ obeys equation

$$\frac{\partial^2 U_\perp(\xi, \eta; \omega)}{\partial \xi^2} + \frac{\partial^2 U_\perp(\xi, \eta; \omega)}{\partial \eta^2} + \frac{h^2 k_\perp^2}{2}(\cosh 2\xi - \cos 2\eta)U_\perp(\xi, \eta; \omega) = 0,$$

$$(1.104)$$

which can be split into the radial and angular Mathieu differential operators. The 0-th order fundamental traveling wave solutions are

$$U^{(1)}(\xi, \eta, z, q; \omega) = [Ce_0(\xi, q) + iFey_0(\xi, q)]ce_0(\eta, q) \exp[ik_z z],$$
$$U^{(2)}(\xi, \eta, z, q; \omega) = [Ce_0(\xi, q) - iFey_0(\xi, q)]ce_0(\eta, q) \exp[ik_z z],$$
$$(1.105)$$

where Ce_0 and Fey_0 are the even radial Mathieu functions of the first and second kinds, respectively, ce_0 is the angular Mathieu function, and $q = h^2 k_\perp^2/4$ is the ellipticity parameter of the system. The expressions in the brackets are also known as the first and second Mathieu-Hankel functions of the 0-th order. The higher-order mode solutions of Eq. (1.104), just like the higher-order

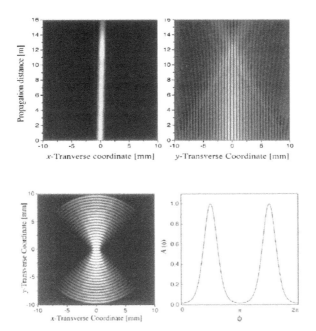

FIGURE 1.7
Longitudinal (top) and transverse (bottom, left) intensity distributions of a
typical Mathieu beam; angular spectrum (bottom, right), from Ref. [28].

Bessel beams, possess optical vortices. Equations (1.105) represent traveling
conical waves, modulated azimuthally, slanted outward for $U^{(1)}$ and inward
for $U^{(2)}$. In the infinite space, where both solutions coexist and overlap, the
field takes the form

$$U(\xi, \eta, z, q; \omega) = Ce_0(\xi, q)ce_0(\eta, q)\exp(ik_z z), \qquad (1.106)$$

and is named as the 0-th order Mathieu beam [28].

The angular spectrum of the m-th order Mathieu beam has the form

$$a(\phi, q) = ce_m(\phi, q) + ise_m(\phi, q), \qquad (1.107)$$

where ce and se are the angular Mathieu functions and q is the ellipticity
parameter [28, 29]. Figure 1.7 shows cross-sections of a typical apertured 0-th
order Mathieu beam and its angular spectrum.

1.3.3.4 Parabolic beams

The fourth coordinate system in which the Helmholz equation is separable
into transverse and longitudinal parts is called *parabolic-cylindrical*. It relates

to the Cartesian system by the transformation

$$x + iy = \frac{1}{2}(\eta + i\xi)^2, \quad z = z, \tag{1.108}$$

where $\eta \in (-\infty, \infty)$, $\xi \in [0, \infty)$, and $z \in (-\infty, \infty)$. While the longitudinal part of the Helmholtz equation has the $\exp(-ik_z z)$ dependence, the transverse part $U_\perp(\eta, \xi; \omega)$ splits into equations

$$
\begin{aligned}
\frac{d^2\Phi(\eta)}{d\eta^2} + (k_\perp^2 \eta^2 + 2k_\perp \alpha_P)\Phi(\eta) &= 0, \\
\frac{d^2 R(\xi)}{d\xi^2} + (k_\perp^2 \xi^2 - 2k_\perp \alpha_P)R(\xi) &= 0,
\end{aligned}
\tag{1.109}
$$

with the separation constant $2k_\perp \alpha_P \in (-\infty, \infty)$. Via the change of variables $2k_\perp \xi^2 = q^2$, Eqs. (1.109) can be transformed into the canonical form of the parabolic cylinder ordinary differential equation

$$\frac{d^2 P}{dq^2} + (q^2/4 - \alpha_P)P = 0. \tag{1.110}$$

Using the Frobenius method it is possible to establish that the two solutions of this equation near the origin (named even and odd) have the forms

$$
\begin{aligned}
U_e(\eta, \xi, \alpha_P) &= \frac{1}{\pi\sqrt{2}}|\Gamma_1|^2 P_e(\sqrt{2k_\perp}\xi, \alpha_P)P_e(\sqrt{2k_\perp}\eta, -\alpha_P) \\
U_o(\eta, \xi, \alpha_P) &= \frac{\sqrt{2}}{\pi}|\Gamma_3|^2 P_o(\sqrt{2k_\perp}\xi, \alpha_P)P_o(\sqrt{2k_\perp}\eta, -\alpha_P),
\end{aligned}
\tag{1.111}
$$

where $\Gamma_1 = \Gamma_f(1/4 + i\alpha_P/2)$, $\Gamma_3 = \Gamma_f(3/4 + i\alpha_P/2)$, Γ_f is the Gamma function, and

$$P(v, \alpha_P) = \sum_{n=0}^{\infty} c_n \frac{v^n}{n!}, \quad c_{n+2} = \alpha_P c_n - \frac{n(n-1)c_{n-2}}{4}. \tag{1.112}$$

The series P_e and P_o in Eq. (1.111) are obtained on setting $c_0 = 1$, $c_1 = 0$ and $c_0 = 0$, $c_1 = 1$, respectively.

The angular spectrum of the parabolic beam can be expressed through the elementary functions as

$$
\begin{aligned}
A_e(\phi, \alpha_P) &= \frac{1}{2\sqrt{\pi|\sin\phi|}}\exp(i\alpha_P \ln|\tan(\phi/2)|), \\
A_o(\phi, \alpha_P) &= \begin{cases} -\frac{1}{i}A_e(\phi, \alpha_P), & \phi \in (-\pi, 0) \\ \frac{1}{i}A_e(\phi, \alpha_P), & \phi \in (0, \pi). \end{cases}
\end{aligned}
\tag{1.113}
$$

A typical parabolic beam in its transverse and longitudinal cross-sections, as well as its phase distribution are shown in Fig. 1.8.

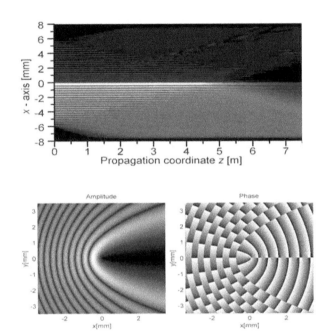

FIGURE 1.8
Longitudinal (top) and transverse (bottom, left) intensity distributions of a typical parabolic beam; phase distribution (bottom, right), from Ref. [30].

1.3.4 Vectorial nature of optical fields: polarization

1.3.4.1 Polarization ellipse

Let us consider the transverse, two-dimensional, electric field-vector of a monochromatic plane wave at a position $\mathbf{r}$ and time instant t propagating in free space along the positive z-direction [9]:

$$\mathbf{E}(\mathbf{r}, t) = \mathcal{E}(x, y) \exp[i(kz - \omega t)], \tag{1.114}$$

where the amplitude $\mathcal{E}$ is generally a complex vector. Such a vector determines the state of the wave's polarization and in the Cartesian coordinate system (x, y) can be expressed as

$$\begin{aligned} \mathcal{E}(x, y) &= \varepsilon_x \hat{x} + \varepsilon_y \hat{y} \\ &= a_x \exp[i\varphi_x]\hat{x} + a_y \exp[i\varphi_y]\hat{y}, \end{aligned} \tag{1.115}$$

where a_x and a_y are the amplitudes and φ_x and φ_y are the phases of the two Cartesian components.

We will now show that in general on free-space propagation, as time goes on, the tip of the electric field traces an ellipse in the plane (x, y). Indeed, on taking the real parts of vector components in Eq. (1.114) we find that:

$$\begin{aligned} E_x &= \varepsilon_x \cos(kz - \omega t), \\ E_y &= \varepsilon_y \cos(kz - \omega t). \end{aligned} \tag{1.116}$$

On eliminating argument $(kz - \omega t)$ from these equations we arrive at the quadratic form

$$\left(\frac{E_y}{a_y}\right)^2 + \left(\frac{E_y}{a_y}\right)^2 - 2\left(\frac{E_x}{a_x}\right)\left(\frac{E_y}{a_y}\right)\cos\varphi = \sin^2\varphi, \tag{1.117}$$

where $\varphi = \varphi_y - \varphi_x$. Equation (1.117) is the general equation of an ellipse placed at the origin of the coordinate system (x, y) (see Fig. 1.9).

Two important special cases follow from Eq. (1.117): linear polarization state, if $\varphi = n\pi$, $n = 0, \pm 1, \pm 2, \ldots$ and circular polarization state, if $a_x = a_y$ and $\varphi = \pm(n + 1/2)\pi$. In the former special case the electric field components oscillate in phase or completely out of phase (with the lag of π). In the latter case the electric field components have the phase lag of a multiple of $\pi/2$, in addition to having the equal amplitudes.

Another polarization characteristic of the monochromatic field is its helicity. In all cases different from linear polarization states the phase lag φ may take a positive or negative value. In this respect, *left* polarization (counterclockwise rotation) corresponds to $\varphi > 0$ and *right* polarization (clockwise rotation) corresponds to $\varphi < 0$.

The orientation of the polarization ellipse is conventionally characterized

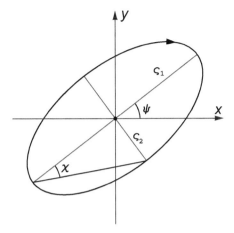

FIGURE 1.9
Polarization ellipse of a monochromatic optical field.

by an angle, say ψ, between the major semi-axis of the ellipse and the positive x-direction (see Fig. 1.9), which can be readily derived to be [9]:

$$\psi = \frac{1}{2} \arctan\left(\frac{2a_x a_y \cos\varphi}{a_x^2 - a_y^2}\right). \tag{1.118}$$

The magnitudes of the major and the minor semi-axes of the ellipse, ς_1 and ς_2, can be expressed as:

$$\begin{aligned}
\varsigma_1 &= \sqrt{a_x^2 \cos^2\psi + a_y^2 \sin^2\psi + a_x a_y \sin(2\psi)\cos\varphi}, \\
\varsigma_2 &= \sqrt{a_x^2 \cos^2\psi + a_y^2 \sin^2\psi - a_x a_y \sin(2\psi)\cos\varphi}.
\end{aligned} \tag{1.119}$$

The shape of the ellipse can be characterized by a ratio

$$\epsilon = \frac{\varsigma_2}{\varsigma_1}, \tag{1.120}$$

returning value 1 for circularly polarized light and value 0 for linearly polarized light. Alternatively it can be given by angle χ (see Fig. 1.9)

$$\chi = \arctan\left(\frac{\varsigma_2}{\varsigma_1}\right). \tag{1.121}$$

1.3.4.2   Jones calculus

A useful approach for describing the interaction of light with linear polarization-modulating devices was developed by Jones [36]. According to this treatment, the time-independent transverse electric field vector can be written as a two-dimensional vector

$$
\begin{bmatrix} \varepsilon_x \\ \varepsilon_y \end{bmatrix} = \begin{bmatrix} a_x \exp(i\varphi_x) \\ a_y \exp(i\varphi_y) \end{bmatrix}.
\tag{1.122}
$$

Since only the phase difference φ, but not the actual values of φ_x and φ_y of the field components themselves determine the state of polarization, the vector in (1.122) can be written as

$$
\begin{bmatrix} \varepsilon_x \\ \varepsilon_y \end{bmatrix} = \exp(i\varphi_x) \begin{bmatrix} a_x \\ a_y \exp(i\varphi) \end{bmatrix},
\tag{1.123}
$$

where, as above, $\varphi = \varphi_y - \varphi_x$.

Further, the common phase factor $\exp(i\varphi_x)$ can be neglected and the Jones vector can be written as

$$
\begin{bmatrix} \varepsilon_x \\ \varepsilon_y \end{bmatrix} = \begin{bmatrix} \cos\phi \\ \sin\phi \exp(i\varphi) \end{bmatrix},
\tag{1.124}
$$

where

$$
\cos\phi = \frac{a_x}{\sqrt{a_x^2 + a_y^2}}, \quad \sin\phi = \frac{a_y}{\sqrt{a_x^2 + a_y^2}}.
\tag{1.125}
$$

In particular, for linearly polarized light vector (1.124) reduces to

$$
\begin{bmatrix} \varepsilon_x \\ \varepsilon_y \end{bmatrix}_{(L)} = \begin{bmatrix} \cos\phi \\ \sin\phi \end{bmatrix},
\tag{1.126}
$$

while for the right and left circularly polarized light we have

$$
\begin{bmatrix} \varepsilon_x \\ \varepsilon_y \end{bmatrix}_{(RC)} = \frac{1}{\sqrt{2}} \begin{bmatrix} 1 \\ -i \end{bmatrix}, \quad \begin{bmatrix} \varepsilon_x \\ \varepsilon_y \end{bmatrix}_{(LC)} = \frac{1}{\sqrt{2}} \begin{bmatrix} 1 \\ +i \end{bmatrix}.
\tag{1.127}
$$

In the general case of elliptical polarization the right and left elliptical polarization states acquire the forms:

$$
\begin{bmatrix} \varepsilon_x \\ \varepsilon_y \end{bmatrix}_{(RE)} = \frac{1}{\sqrt{q_1^2 + q_2^2 + q_3^2}} \begin{bmatrix} q_1 \\ q_2 - iq_3 \end{bmatrix},
\tag{1.128}
$$

$$
\begin{bmatrix} \varepsilon_x \\ \varepsilon_y \end{bmatrix}_{(LE)} = \frac{1}{\sqrt{q_1^2 + q_2^2 + q_3^2}} \begin{bmatrix} q_1 \\ q_2 + iq_3 \end{bmatrix},
\tag{1.129}
$$

where

$$
q_1 = a_x, \quad \sqrt{q_2^2 + q_3^2} = a_y, \quad \mp\tan^{-1}\left(\frac{q_3}{q_2}\right) = \phi.
\tag{1.130}
$$

Jones vectors can be effectively used for describing the modification of the polarization state of light by linear non-image-forming devices. In order to relate the polarization states of the incident and the transmitted light waves, 2×2 transmission matrices can be employed. Then the transformation becomes:

$$\begin{bmatrix} \varepsilon_x \\ \varepsilon_y \end{bmatrix} = \begin{bmatrix} T_{11} & T_{12} \\ T_{21} & T_{22} \end{bmatrix} \begin{bmatrix} \varepsilon'_x \\ \varepsilon'_y \end{bmatrix}, \tag{1.131}$$

where prime denotes the components of the electric vector before transmission.

For instance, for a polarizer oriented at angle φ_p with respect to the positive x-direction the Jones matrix has the form:

$$\mathbf{T}_{(PL)} = \begin{bmatrix} \cos^2 \varphi_p & \cos \varphi_p \sin \varphi_p \\ \cos \varphi_p \sin \varphi_p & \sin^2 \varphi_p \end{bmatrix}. \tag{1.132}$$

Polarizers change the direction of oscillations in the incident electric field to φ_p and reduce its intensity by a factor of $\cos^2(\varphi - \varphi_p)$.

A phase retarder introduces phase shifts in one or both components of the electric field vector being generally characterized by the Jones matrix

$$\mathbf{T}_{(RT)} = \begin{bmatrix} \exp(i\varphi_{Rx}) & 0 \\ 0 & \exp(i\varphi_{Ry}) \end{bmatrix}. \tag{1.133}$$

Among the most useful retarders are the half-wave plates $(\varphi_{Rx} - \varphi_{Ry} = \pm\pi)$ and the quarter-wave plates $(\varphi_{Rx} - \varphi_{Ry} = \pm\pi/2)$.

Optical absorbers modify the amplitudes of the electric field components and are accounted for by matrix

$$\mathbf{T}_{(AB)} = \begin{bmatrix} \exp(-\alpha_{Ax}) & 0 \\ 0 & \exp(-\alpha_{Ay}) \end{bmatrix}. \tag{1.134}$$

Finally, rotators are capable of changing the axis of the optical field oscillation without its intensity change. A rotation of the field through angle φ_r is represented by transformation

$$\mathbf{T}_{(RO)} = \begin{bmatrix} \cos \varphi_r & -\sin \varphi_r \\ \sin \varphi_r & \cos \varphi_r \end{bmatrix}. \tag{1.135}$$

1.3.4.3 Stokes vectors

An alternative description of polarization states of light fields, either deterministic or random, was proposed by Stokes and later further developed Mueller and Chandrasekhar (see Ref. [37]). Since Mueller-Stokes calculus has widespread applications for random fields we postpone this discussion to Chapter 4, where we present recently developed generalized and unified Jones-Stokes-Mueller calculus. In this section we will only briefly review the Stokes representation for monochromatic plane waves. In this case the set of

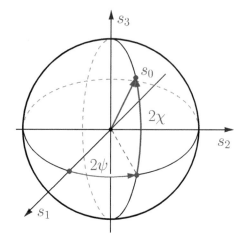

FIGURE 1.10
Poincaré sphere.

four Stokes parameters is defined by the formulas

$$
\begin{aligned}
s_0 &= a_x^2 + a_y^2, \\
s_1 &= a_x^2 - a_y^2, \\
s_2 &= 2a_x a_y \cos\varphi, \\
s_3 &= 2a_x a_y \sin\varphi,
\end{aligned}
\tag{1.136}
$$

constrained by the relation

$$
s_0^2 = s_1^2 + s_2^2 + s_3^2.
\tag{1.137}
$$

In terms of angles χ and ψ of the polarization ellipse the last three Stokes parameters can be written as

$$
\begin{aligned}
s_1 &= s_0 \cos 2\chi \cos 2\psi, \\
s_2 &= s_0 \cos 2\chi \sin 2\psi, \\
s_3 &= 2s_0 \sin 2\psi,
\end{aligned}
\tag{1.138}
$$

i.e., they can be viewed as Cartesian coordinates of a point on a sphere with radius s_0 and with polar and azimuthal angles 2ψ and 2χ in the spherical coordinate system. The sphere with points on its surface corresponding to the states of polarization of the plane wave is known as the Poincaré sphere (see Fig. 1.10).

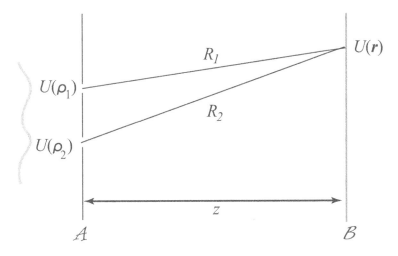

FIGURE 1.11
Young's interference experiment.

1.3.5 Spatial interference in light fields

The classic direct method for measuring the strength of spatial interference in optical fields is the double-pinhole experiment first performed by Young [38]. Electromagnetic waves and light, in particular, due to their wave-like nature, are capable of forming the intensity interference fringes on superposition. Figure 1.11 shows a typical interferometric setup. Suppose a monochromatic light field $U(\boldsymbol{\rho})$ impinges onto an opaque screen $\mathcal{A}$ with two pinholes at positions $\boldsymbol{\rho}_1$ and $\boldsymbol{\rho}_2$, passes through the pinholes and propagates to an observation screen $\mathcal{B}$, where the intensity pattern of the superposition is recorded. In case when $z \gg \lambda$ the distribution of the intensity $|U(\mathbf{r})|^2$ of the superposed waves on the screen $\mathcal{B}$ obeys the following law

$$
\begin{aligned}
|U(\mathbf{r})|^2 &= |d_1 U(\boldsymbol{\rho}_1)|^2 + d_1^* U^*(\boldsymbol{\rho}_1) d_2 U(\boldsymbol{\rho}_2) \\
&\quad + d_1 U(\boldsymbol{\rho}_1) d_2^* U^*(\boldsymbol{\rho}_2) + |d_2 U(\boldsymbol{\rho}_2)|^2 \\
&= |d_1 U(\boldsymbol{\rho}_1)|^2 + 2Re[d_1^* U^*(\boldsymbol{\rho}_1) d_2 U(\boldsymbol{\rho}_2)] + |d_2 U(\boldsymbol{\rho}_2)|^2,
\end{aligned}
\tag{1.139}
$$

where

$$
d_{1,2} = \frac{-iz}{\lambda R_{1,2}} dA_{1,2}
\tag{1.140}
$$

are geometrical factors, $R_{1,2}$ are the distances from the pinholes to point $\mathbf{r}$, and $dA_{1,2}$ are the areas of two pinholes and z is the distance between the screens. On setting $U_1 = a \exp[i(\omega t + \phi_1)]$, $U_2 = a \exp[i(\omega t + \phi_2)]$ and assuming that $d_1 \approx d_2 \approx d$ we find that Eq. (1.139) reduces to the form

$$
|U(\mathbf{r})|^2 \approx 2d^2 a^2 [1 + \cos(\phi_1 - \phi_2)].
\tag{1.141}
$$

Here the term in the square bracket determines the value of the intensity at position $\mathbf{r}$ depending on the discrepancy between phases ϕ_1 and ϕ_2 of the illumination at two pinholes. In cases when $\phi_1 = \phi_2$ and $\phi_1 - \phi_2 = \pi$ the maximum and the minimum intensity values are attained, $4d^2a^2$ and 0, respectively.

A historical review [39] of Young's interference experiment explores how the extension of relation (1.141) to random, scalar and electromagnetic beam-like fields was developed. We will outline these generalizations in Chapters 3 and 4 while introducing the coherence states of light beams.

Bibliography

[1] D. Fried, "Optical resolution through a randomly inhomogeneous medium for very long and very short exposures," *J. Opt. Soc. Am.* **56**, 1372–1379 (1966).

[2] A. Labeyrie, "Attainment of diffraction limited resolution in large telescopes by Fourier analysing speckle patterns in star images," *Astronomy and Astrophysics* **6**, 85L (1970).

[3] J. Baldwin, "Closure phase in high-resolution optical imaging," *Nature* **320**, 595–597 (1986).

[4] J. Baldwin, "The first images from optical aperture synthesis," *Nature* **328**, 694–696 (1987).

[5] C. Dainty Ed., *Laser Speckle and Related Phenomena*, Springer Verlag, 1984.

[6] G. W. Goodman, *Statistical Optics*, Wiley, 1985.

[7] L. Mandel and E. Wolf, *Optical Coherence and Quantum Optics*, Cambridge University Press, 1995.

[8] C. Brosseau, *Fundamentals of Polarized Light: A Statistical Optics Approach*, Wiley, 1998.

[9] M. Born and E. Wolf, *Principles of Optics*, Cambridge University Press, 7th Edition, 1999.

[10] G. W. Goodman, *Speckle Phenomena in Optics: Theory and Applications*, Roberts & Company, 2007.

[11] E. Collett and E. Wolf, "Is complete spatial coherence necessary for the generation of highly directional light beams?" *Opt. Lett.* **2**, 27–29 (1978).

[12] D. F. V. James, "Change of polarization of light beams on propagation in free space," *J. Opt. Soc. Am. A* **11**, 1641–1643 (1994).

[13] F. Gori, M. Santarsiero, G. Piquero, R. Borghi, A. Mondello, and R. Simon, "Partially polarized Gaussian Schell-model beams," *J. Opt. A: Pure and Appl. Optics* **3**, 1–9 (2001).

[14] E. Wolf, *Introduction to Theories of Coherence and Polarization of Light*, Cambridge University Press, 2007.

[15] G. Gbur and T. Visser, "The structure of partially coherent fields," in *Progress in Optics*, E. Wolf Ed. **55**, 285–341 (2010).

[16] V. G. Veselago, "Electrodynamics of substances with simultaneously negative electrical and magnetic permeabilities," *Sov. Phys. Usp.* **10**, 509–514 (1968).

[17] V. A. Banakh, V. M. Buldakov, and V.L. Mironov, "Intensity fluctuations of a partially coherent light beam in a turbulent atmosphere," *Opt. Spectrosk.* **54**, 1054-1059 (1983).

[18] L. C. Andrews and R. L. Phillips, *Laser Beam Propagation in the Turbulent Atmosphere*, 2nd edition, SPIE Press, 2005.

[19] A. Papoulis and S. U. Pillai, *Probability, Random Variables and Stochastic Processes*, 4th edition, McGraw-Hill, 2002.

[20] E. C. Titchmarsh, *The Theory of Functions*, Oxford University Press, 1939.

[21] P. M. Morse and H. Feshbach, *Methods of Theoretical Physics*, Vol. 1, McGraw-Hill 1953.

[22] W. Miller, Jr., *Symmetry and Sepoaration of Variables*, Addison-Wesley, 1977.

[23] C. J. Sheppard and T. Wilson, "Gaussian-beam theory of lenses with annular aperture," *IEEE J. Microwaves Optics and Acoustics* **2**, 105–112 (1978).

[24] J. Durnin, "Exact solutions for nondiffracting beams. I. The scalar theory," *J. Opt. Soc. Am. A* **4**, 651–654 (1987).

[25] J. Durnin, J. J. Miceli, Jr., and J. H. Eberly, "Diffraction-free beams," *Phys. Rev. Lett.* **58**, 1499–1501 (1987).

[26] G. Indebetouw, "Nondiffracting optical fields: some remarks on their analysis and synthesis," *J. Opt. Soc. Am. A* **6**, 150–152 (1989).

[27] S. Ruschin, "Modified Bessel non-diffracting beams," *J. Opt. Soc. Am. A* **11**, 3224–3228 (1994).

[28] J. C. Gutierrez-Vega, M. D. Iturbe-Castillo, and S. Chavez-Cerda, "Alternative formulation for invariant optical fields: Mathieu beams," *Opt. Lett.* **25**, 1493–1495 (2000).

[29] J. C. Gutierrez-Vega, M. D. Iturbe-Castillo, G. A. Ramirez, E. Tepichin, R. M. Rodriguez-Dagnino, S. Chavez-Cerda, and G. H. C. New, "Experimental demonstration of optical Mathieu beams," *Opt. Commun.* **195**, 35–40 (2001).

[30] M. A. Bandres, J. C. Gutierrez-Vega, and S. Chavez-Cerda, "Parabolic nondiffracting optical wave fields," *Opt. Lett.* **29**, 44–46 (2004).

[31] E. T. Whittaker and G. N. Watson, *A Course of Modern Analysis*, Cambridge University Press, 1927.

[32] Z. Bouchal, "Nondiffracting optical beams: physical properties, experiments, and applications," *Chechoslovak J. of Phys.* **53**, 537–624 (2003).

[33] C. W. McCutchen, "Generalized aperture and the three-dimensional diffraction image," *J. Opt. Soc. Am.* **54**, 240 (1964).

[34] P. Sprangle and B. Hafizi, "Comment on nondiffracting beams," *Phys. Rev Lett.* **66**, 837 (1987).

[35] J. Durnin, J. J. Jr. Miceli, and J. H. Eberly, "Reply to comment on nondiffracting beams," *Phys. Rev Lett.* **66**, 838 (1987).

[36] R.C. Jones, "A new calculus for the treatment of optical systems. I. Description and discussion of the new calculus," *J. Opt. Soc. Am.* **31**, 488–493 (1941).

[37] S. Chandrasekhar, *Radiative Transfer*, Dover, 1960.

[38] T. Young, *A Course of Lectures on Natural Philosophy and the Mechanical Arts*, Vols. I and II, J. Johnson, London, England, 1807.

[39] E. Wolf, "The influence of Young's interference experiment on the development of statistical optics", *Progress in Optics*, Vol. **50**, E. Wolf, ed., North-Holland, 2007, pp. 251–274.

2

Deterministic paraxial beams

CONTENTS

Before we begin consideration of random beams in the subsequent chapters we will first make a brief overview of deterministic beam-like fields. Unlike the non-diffracting fields, discussed in Chapter 1, which are exact solutions of the Helmholz equation, the beams of this chapter satisfy the paraxial wave equation with certain restrictions on the wave vector. They exhibit spatial spreading. Our interest in such fields is primarily dictated by the need to model the random beams. For instance, as we will see in Chapters 3 and 4, a paraxial field may be used for modeling the intensity of a random beam, its correlation function or as an orthogonal coherent mode of the correlation function.

2.1 Basic family of Gaussian beams

A conventional starting point for introducing the basic types of deterministic optical beam-like fields is solving the Helmholtz equation within the parax-ial approximation [1]. Virtually all monographs on laser optics include such treatment [2]–[6]. Often the solutions may be found by their direct substi-tution into the governing differential equation. However, because of the fact that not all the solutions of the paraxial equation represent beams, it may be preferential to employ the alternative technique known as the *angular spec-trum representation*, that we have already introduced in Chapter 1. As we have seen, the fields in their angular spectrum representation, if obey sim-ple conditions (1.67), are guaranteed to represent the beams. Therefore, the angular spectrum can be used directly in deciding whether or not a given field is a highly-directional beam. In this section we will employ the angular spectrum, together with other transformations and optimization procedures, for obtaining the expressions for basic deterministic beams-like fields: Gaussian, Hermite-Gaussian and Laguerre-Gaussian [7]. All these beams possess the structural stability, i.e. their shapes remain invariant on propagation, with a possible modification of parameters. On introducing these beam classes we will illustrate how simple modulation of the angular spectrum can lead to novel classes of physical beams in spatial domain. Before finishing this chapter we will also mention several other families of deterministic beams, such as Bessel-Gaussian, for example.

2.1.1   Fundamental Gaussian beam

We begin by recalling that any scalar optical field propagating close to the z axis may be represented via its angular spectrum [see Eq. (1.63)]. In addition, under conditions (1.67), such a field obeys the following propagation law [see (1.70)]

$$U(x, y, z; \omega) = \exp(ikz) \int\limits_{-\infty}^{\infty} \int\limits_{-\infty}^{\infty} a(k_x, k_y; \omega) \exp[i(k_x + k_y)]$$

$$\times \exp\left[-i\frac{z}{2k}(k_x^2 + k_y^2)\right] dk_x dk_y, \tag{2.1}$$

where $a(k_x, k_y; \omega)$ is the amplitude of the plane wave in the angular spectrum expansion defined in Eq. (1.65). Each plane wave carries energy proportional to $|a(k_x, k_y; \omega)|^2$. The two principle quantities that may characterize the total energy flux are the *divergence angle* and the *transverse spread*, are defined by the expressions [7]:

$$\Delta\theta(z; \omega) = \frac{1}{4\pi^2} \int\limits_{-\infty}^{\infty} \int\limits_{-\infty}^{\infty} (k_x^2 + k_y^2)|a(k_x, k_y; \omega)|^2 dk_x dk_y, \tag{2.2}$$

$$\Delta w(z, \omega) = \int\limits_{-\infty}^{\infty} \int\limits_{-\infty}^{\infty} (x^2 + y^2)|U(x, y, z; \omega)|^2 dx dy. \tag{2.3}$$

We note that the former quantity characterizes the energy flow in spatial frequency domain and the latter does so in the direct space domain. On substituting from Eq. (1.71) into Eq. (2.3) for $|U(x, y, z; \omega)|^2$ we find, after integrations with respect to x and y, that

$$\Delta w(z; \omega) = \frac{1}{4\pi^2} \int\limits_{-\infty}^{\infty} \int\limits_{-\infty}^{\infty} \left(\left|\frac{\partial a(k_x, k_y; \omega)}{\partial k_x}\right|^2 + \left|\frac{\partial a(k_x, k_y; \omega)}{\partial k_y}\right|^2 \right) dk_x dk_y, \tag{2.4}$$

where ∂ stands for the partial derivative. The expression for the lowest-order Gaussian beam field can be determined by minimizing the product of the divergence and the transverse spread, i.e. by setting

$$\Delta\theta(z;\omega)\Delta w(z;\omega) \to \min. \qquad (2.5)$$

In other words, the Gaussian beam is the mode with minimum uncertainty in the sense of the Heisenberg's principle of quantum mechanics, since in the product (2.5) the two terms describe the spreads in the phase space and in the real space, respectively. The fact that such a product is minimal implies that the Gaussian beam has minimal diffraction and dispersion among all the deterministic optical beams.

Let us now derive the expressions for the Gaussian mode in the angular and direct spaces from condition (2.5). In fact, due to the $x-y$ symmetry it is sufficient to minimize the products along the two directions separately. For instance, in x direction we have

$$\frac{1}{4\pi^2} \left(\int_{-\infty}^{\infty} k_x^2 |a(k_x,k_y;\omega)|^2 dk_x \right) \left(\int_{-\infty}^{\infty} \left| \frac{\partial a(k_x,k_y;\omega)}{\partial k_x} \right|^2 dk_x \right) \to \min. \qquad (2.6)$$

Recall the Cauchy-Schwartz inequality, which holds for two arbitrary square-integrable complex functions, say f and g [8]:

$$\int_{-\infty}^{\infty} |f(\xi)|^2 d\xi \int_{-\infty}^{\infty} |g(\xi)|^2 d\xi \geq \left| \int_{-\infty}^{\infty} f^*(\xi)g(\xi)d\xi \right|^2, \qquad (2.7)$$

where ξ is a variable of integration. It is important to note that the inequality (2.7) reduces to equality if and only if functions f and g are proportional to each other. Applied for the product (2.6) the Cauchy-Schwartz inequality (2.7) leads to the result

$$\frac{1}{4\pi^2} \left(\int_{-\infty}^{\infty} k_x^2 |a(k_x,k_y;\omega)|^2 dk_x \right) \left(\int_{-\infty}^{\infty} \left| \frac{\partial a(k_x,k_y;\omega)}{\partial k_x} \right|^2 dk_x \right)$$

$$\geq \left| \frac{1}{4\pi^2}\frac{1}{2} \left(k_x a^*(k_x,k_y;\omega)\frac{\partial a(k_x,k_y;\omega)}{\partial k_x} + k_x a(k_x,k_y;\omega)\frac{\partial a^*(k_x,k_y;\omega)}{\partial k_x} \right) dk_x \right|^2$$

$$= \left| \frac{1}{8\pi^2} \int_{-\infty}^{\infty}\int_{-\infty}^{\infty} \frac{\partial |a(k_x,k_y;\omega)|^2}{\partial k_x} k_x dk_x \right|^2 = \frac{1}{64\pi^4} \left| \int_{-\infty}^{\infty} |a(k_x,k_y;\omega)|^2 dk_x \right|^2. \qquad (2.8)$$

The rules of partial integration were used here together with the assumption that $a(k_x,k_y;\omega)$ vanishes as $k_x \to \pm\infty$. In Eq. (2.8) the last expression is a multiple of the total energy carried by the field at a fixed value of y, which must be conserved when the beam propagates in free space, due to its rotational symmetry. In this limiting case the product of interest attains its minimum value, being equal to the first line of (2.8), which can only occur in the case when $k_x a(k_x, k_y; \omega)$ and $\partial a(k_x, k_y; \omega)/\partial k_x$ are proportional to each other, as a consequence of the Schwartz inequality. The only function that satisfies this condition is a Gaussian function. On carrying out a similar argument for the product in the y-direction one arrives at the following form of the angular spectrum representation

$$a^{(G)}(k_x, k_y; \omega) = \exp\left[-\frac{w_0^2}{4}(k_x^2 + k_y^2)\right], \tag{2.9}$$

where $w_0^2/4$ is a proportionality factor.

Further, on substituting from Eq. (2.9) into Eq. (2.1) the beam-like field, corresponding to the angular spectrum (2.9) becomes:

$$U(x, y, z; \omega) = \frac{\exp(ikz)}{4\pi^2} \int\limits_{-\infty}^{\infty} \int\limits_{-\infty}^{\infty} \exp\left[-\frac{w_0^2}{4}(k_x^2 + k_y^2)\right] \exp[i(k_x + k_y)]$$
$$\times \exp\left[-i\frac{z}{2k}(k_x^2 + k_y^2)\right] dk_x dk_y. \tag{2.10}$$

The integrals with respect to k_x and k_y can be performed with the help of the formula

$$\frac{1}{2\pi} \int\limits_{-\infty}^{\infty} \exp\left[i\xi q_1 - \frac{q_2^2}{2}\xi^2\right] d\xi = \frac{1}{q_2\sqrt{2\pi}} \exp\left[-\frac{q_1}{2q_2^2}\right], \tag{2.11}$$

where q_1, q_2 are some constants. Thus, after omitting the constant factor π, we arrive at the following expression for the lowest-order Gaussian beam field:

$$U^{(G)}(x, y, z; \omega) = \frac{1}{w_0^2 + 2iz/k} \exp\left[ikz - \frac{x^2 + y^2}{w_0^2 + 2iz/k}\right]. \tag{2.12}$$

On introducing the non-dimensional coordinate z_ζ by the formula

$$z_\zeta = \frac{\lambda z}{\pi w_0^2} \tag{2.13}$$

and using the relation

$$\frac{1}{1 + iz_\zeta} = \frac{1 - iz_\zeta}{1 + z_\zeta^2} = \frac{\exp[-i\arctan(z_\zeta)]}{\sqrt{1 + z_\zeta^2}} \tag{2.14}$$

it is possible to rewrite the formula (2.12) in the form

$$U^{(G)}(x, y, z; \omega) = \frac{1}{w_0\sqrt{1 + z_\zeta^2}} \exp\left[ikz - \frac{(1 - iz_\zeta)\rho^2}{w_0^2(1 + z_\zeta^2)} - i\arctan(z_\zeta)\right], \tag{2.15}$$

where $\rho = (x, y)$, $\rho^2 = x^2 + y^2$, and w_0 is omitted in the numerator. The field in Eq. (2.15) represents the fundamental Gaussian mode. Indeed, the intensity of this mode has the rotationally symmetric, Gaussian profile

$$|U^{(G)}(\rho, z; \omega)|^2 = \frac{1}{w_0^2(1 + z_\zeta^2)} \exp\left[-\frac{2\rho^2}{w_0^2(1 + z_\zeta^2)}\right] \tag{2.16}$$

in any cross-section perpendicular to the propagation axis z. All three parameters entering expression (2.15) are of importance. The phase parameter

$$\psi_G(z_\zeta) = \arctan(z_\zeta) \tag{2.17}$$

is known as the *Gouy's phase* and describes a sharp phase transition from $-\pi/2$ to $\pi/2$ as the beam goes through the plane $z_\zeta = 0$, which is called the *waist plane* of the beam. This shift is also the consequence of the uncertainty principle ([9], p. 267). Except in the vicinity of the waist of the beam this factor varies slowly with z and can be approximately treated as a constant.

The second parameter called the *spatial radius* of the Gaussian beam is defined in any plane perpendicular to its direction of propagation. It is given by the formula

$$w(z_\zeta) = w_0\sqrt{1 + z_\zeta^2}. \tag{2.18}$$

The spatial radius corresponds to the transverse distance at which the intensity of the electric field $|U|^2$ drops to the value of $e^2/2$ of its maximum, attained on the optical axis. The third important parameter of the Gaussian beam is called the *phase front radius of curvature* and is found from determining the lines in (ρ, z) space with constant phase, i.e. from condition

$$kz + \frac{z_\zeta\rho^2}{w^2(z_\zeta) - \psi_G(z_\zeta)} kz + \frac{z_\zeta\rho^2}{w^2(z_\zeta)} \approx const, \tag{2.19}$$

leading to the set of parabolic curves with the radius of curvature

$$R(z_\zeta) = \frac{kw_0^2}{2} \frac{\left(1 + z_\zeta^2\right)}{z_\zeta}. \tag{2.20}$$

We stress that this definition is valid only at distances sufficiently far from the waist of the beam. On expressing the Gaussian beam in terms of the Gouy's phase, the spatial radius and the radius of curvature we obtain its alternative form

$$U^{(G)}(x, y, z; \omega) = \frac{1}{w(z_\zeta)} \exp\left[-\frac{\rho^2}{w^2(z_\zeta)} + ik\left(z + \frac{\rho^2}{2R(z_\zeta)}\right) - i\psi_G(z_\zeta)\right]. \tag{2.21}$$

On finishing the discussion about the fundamental Gaussian beam we will introduce the important parameter z_R known as the *Rayleigh range*

$$z_R = \frac{\pi w_0^2}{\lambda}. \tag{2.22}$$

Hence, the normalized propagation distance becomes

$$z_\zeta = \frac{z}{z_R}. \tag{2.23}$$

Further, the Gouy's phase, the spatial radius and the radius of curvature can be also expressed via the Rayleigh range as

$$\psi_G(z) = \arctan\left(\frac{z}{z_R}\right), \tag{2.24}$$

$$w(z) = w_0 \sqrt{1 + \left(\frac{z}{z_R}\right)^2}, \tag{2.25}$$

and

$$R(z) = z\left[1 + \left(\frac{z_R}{z}\right)^2\right]. \tag{2.26}$$

It follows from Eq. (2.25) that $w(z) = \sqrt{2}w_0$ for $z = z_R$. Therefore, the Rayleigh range can be considered as the propagation distance at which the spatial radius is increased by a factor of $\sqrt{2}$ compared to that at the waist. Further, for $z \gg z_R$

$$w \approx w_0 \frac{z}{z_R} = \frac{\lambda z}{\pi w_0}, \tag{2.27}$$

i.e it increases with propagation distance as a linear function. Angle $\Delta\theta(\infty)$ at which the spatial radius grows for $z \gg z_R$ is given by the formula

$$\theta_D = \Delta\theta(\infty) = \frac{\lambda}{\pi w_0}. \tag{2.28}$$

In Fig. 2.1 the major parameters of the Gaussian beam are presented. The dashed line with the slope corresponding to the divergence angle $\theta_D = \Delta\theta(\infty)$ defines the asymptotic behavior of the beam for $z \gg z_R$.

The straightforward extension of the Gaussian circular to Gaussian elliptical field profile, is described by the distribution (see also [10])

$$U^{(EG)}(x,y,0) = \exp\left[-\left(\frac{x^2}{w_{0x}^2} + \frac{2xy}{w_{0xy}^2} + \frac{y^2}{w_{0y}^2}\right)\right], \tag{2.29}$$

where $(x, y, 0)$ are the Cartesian coordinates of the point in the waist plane and w_{0x}, w_{0y}, w_{0xy} are the beam waists along x, y and mixed directions. In Fig. 2.2 we show the intensity profiles of a circularly Gaussian and an elliptical Gaussian beams.

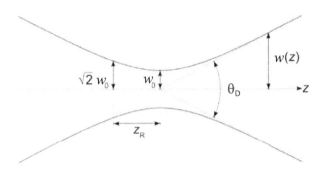

FIGURE 2.1
Parameters of a Gaussian beam.

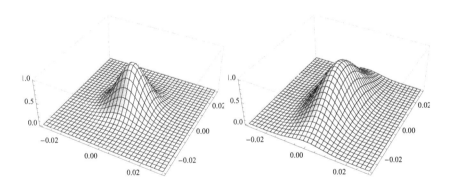

FIGURE 2.2
Intensity distributions of typical circular (left) and elliptical (right) Gaussian beams, as functions of x and y [m], at $z = 0$, with $w_0 = 1$ cm.

2.1.2 Hermite-Gaussian beams

The Hermite-Gaussian modes are generalizations of the fundamental Gaussian mode possessing rectangular symmetry. Their free-space spatial evolution under paraxial conditions belongs to Refs. [11] and [12]. However, just like for the Gaussian beams, we will derive the expressions for such beams in spatial domain from their angular spectrum expansions [7]. In order to do so it suffices to modulate the angular spectrum representation of the Gaussian beam by monomials k_x^m and k_y^n:

$$a^{(eHG)}(k_x, k_y; \omega) = k_x^m k_y^n \exp\left[-\frac{w_0^2}{4}(k_x^2 + k_y^2)\right], \qquad (2.30)$$

where $a^{(eHG)}$ denotes the plane wave representation of the *elegant Hermite-Gaussian beams*. Therefore their spatial representation can be obtained from the expression

$$U_{mn}^{(eHG)}(x, y, z; \omega) = \frac{\exp(ikz)}{4\pi^2} \int\limits_{-\infty}^{\infty} \int\limits_{-\infty}^{\infty} (ik_x)^m (ik_y)^n \exp[\Upsilon(k_x, k_y)] dk_x dk_y,$$

$$(2.31)$$

where

$$\Upsilon(k_x, k_y) = \exp\left[-\frac{w_0^2}{4}(1 + iz_\zeta)(k_x^2 + k_y^2)\right] \exp[i(k_x x + k_y y)], \qquad (2.32)$$

and the additional factor i^{m+n} is used for convenience. Because of the relations

$$ik_x \exp[\Upsilon(k_x, k_y)] = \frac{\partial}{\partial x} \exp[\Upsilon(k_x, k_y)],$$
$$ik_y \exp[\Upsilon(k_x, k_y)] = \frac{\partial}{\partial y} \exp[\Upsilon(k_x, k_y)], \qquad (2.33)$$

the elegant Hermite-Gaussian modes are just the partial derivatives of the fundamental Gaussian mode. Namely, the following relation holds:

$$U_{mn}^{(eHG)}(x, y, z; \omega) = \frac{\partial^{m+n}}{\partial x^m \partial y^n} U^{(G)}(x, y, z; \omega). \qquad (2.34)$$

After performing the differentiations and using the definition of the Hermite polynomials,

$$H_p(x) = (-1)^p \exp(x^2) \frac{d^p}{dx^p} \exp(-x^2), \qquad (2.35)$$

we arrive at the expression

$$U_{mn}^{(eHG)}(x, y, z; \omega) = \frac{1}{w(z_\zeta)^{(m+n)/2+1}} H_m\left(\frac{x}{w_0\sqrt{1 + iz_\zeta}}\right) H_n\left(\frac{y}{w_0\sqrt{1 + iz_\zeta}}\right)$$

$$\times \exp\left[ikz - \frac{\rho^2}{w_0^2(1 + iz_\zeta)} - i\psi_{HG}(z_\zeta)\right], $$

$$(2.36)$$

where

$$\psi_{HG}(z_\zeta) = \left(1 + \frac{m+n}{2}\right)\arctan z_\zeta \qquad (2.37)$$

is the generalized Gouy's phase. Representation (2.36) involves complex arguments both in the amplitude and in the phase. The beams in this form are called *elegant Hermite-Gaussian* to distinguish them from similar family known as the *Hermite-Gaussian beams*.

In order to obtain the representation of the Hermite-Gaussian modes in which the amplitude and the phase terms are separated one may use the same idea of the Gaussian angular spectrum's modulation, using, instead of polynomials $(ik_x)^m$ and $(ik_y)^n$, the operators $\left(ik_x + q^{-1}\partial/\partial k_x\right)^m$ and $\left(ik_y + q^{-1}\partial/\partial k_y\right)^n$, where q is an arbitrary constant, i.e.

$$a^{(HG)}(k_x, k_y; \omega) = \left(ik_x + q^{-1}\partial/\partial k_x\right)^m \left(ik_y + q^{-1}\partial/\partial k_y\right)^n \exp\left[-\frac{w_0^2}{4}(k_x^2 + k_y^2)\right].$$
$$(2.38)$$

Substituting such an angular spectrum to the field representation leads to the formula

$$U_{mn}^{(HG)}(x, y, z; \omega) = \frac{\exp(ikz)}{4\pi^2} \int\limits_{-\infty}^{\infty}\int\limits_{-\infty}^{\infty} \left(ik_x + \frac{1}{q}\frac{\partial}{\partial k_x}\right)^m \left(ik_y + \frac{1}{q}\frac{\partial}{\partial k_y}\right)^n$$
$$\times \exp[\Upsilon(k_x, k_y)]dk_x dk_y.$$
$$(2.39)$$

The expression on the right side of (2.39) is the solution of the paraxial wave equation because the operators in the brackets are independent from x and y and commute with $\partial/\partial x$ and $\partial/\partial y$. After performing lengthy integrations and using some identities of differential calculus one can show that Eq. (2.39) reduces to [7]

$$U_{mn}^{(HG)}(x, y, z; \omega) = \frac{1}{w(z_\zeta)} H_m\left(\frac{\sqrt{2}x}{w(z_\zeta)}\right) H_n\left(\frac{\sqrt{2}y}{w(z_\zeta)}\right)$$
$$\times \exp\left[ikz - \frac{\rho^2}{w_0^2(1 + iz_\zeta)} - i\psi_{HG}(z_\zeta)\right].$$
$$(2.40)$$

Figure 2.3 shows the intensity profiles $|U^{(HG)}(x, y, 0; \omega)|^2$ of several first Hermite-Gaussian modes at the waist plane.

2.1.3 Laguerre-Gaussian beams

Laguerre-Gaussian modes can also be viewed as a generalization of the fundamental Gaussian beam. Unlike the Hermite-Gaussian modes they have rotational symmetry along the propagation axis and the phase that carries a

48

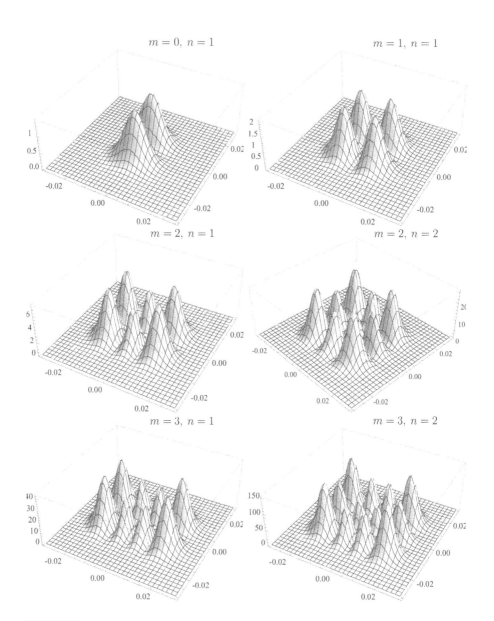

FIGURE 2.3

Intensity distributions of several first beams in the Hermite-Gaussian family as functions of x and y, [m], at $z = 0$, with $w_0 = 1$ cm.

vortex. We will first derive the expression for the *elegant Laguerre-Gaussian modes* for which we will modulate the angular spectrum of the fundamental Gaussian beam as [7], [13]

$$a^{(eLG)}(k_x, k_y; \omega) = (k_x + ik_y)^m (k_x - ik_y)^{n+m} \exp\left[-\frac{w_0^2}{4}(k_x^2 + k_y^2)\right], \quad (2.41)$$

where n and m are integers. Hence the corresponding electric field can be expressed as

$$U^{(eLG)}(x, y, z; \omega) = \frac{\exp(ikz)}{4\pi^2} \int_{-\infty}^{\infty} \int_{-\infty}^{\infty} (k_x + ik_y)^m (k_x - ik_y)^{n+m}$$
$$\times \exp[\Upsilon(k_x, k_y)] dk_x dk_y. \quad (2.42)$$

Similarly to the steps taken in the derivation of the elegant Hermite-Gaussian modes we find that

$$U_{mn}^{(eLG)}(x, y, z; \omega) = (\partial_x + i\partial_y)^m (\partial_x - i\partial_y)^{n+m} U^{(G)}(x, y, z; \omega). \quad (2.43)$$

For the new set of variables $\chi_s = x + iy$, $\chi_d = x - iy$ the partial derivatives become $\partial_x + i\partial_y = 2\partial_{\chi_d}$ and $\partial_x - i\partial_y = 2\partial_{\chi_s}$ leading to relation

$$U_{mn}^{(eLG)}(x, y, z; \omega) = \partial_{\chi_d}^m \partial_{\chi_s}^{n+m} U^{(G)}(x, y, z; \omega). \quad (2.44)$$

Further, with the help of the Laguerre polynomials

$$L_m^n(\xi) = \frac{\exp(\xi)\xi^{-n}}{m!} \frac{d^m}{d\xi^m} \left[\exp(-\xi)\xi^{n+m}\right] \quad (2.45)$$

and relation $\rho^2 = x^2 + y^2 = \chi_s \chi_d$ we find, by differentiating the fundamental Gaussian mode, that

$$\partial_{\chi_d}^m \partial_{\chi_s}^{n+m} U^{(G)}(x, y, z; \omega) = (-\chi_s)^{-n-m} m! \left[\frac{\chi_s \chi_d}{w_0^2(1 + iz_\zeta)}\right]^n \left[\frac{\chi_s}{w_0^2(1 + iz_\zeta)}\right]^m$$
$$\times L_m^n \left[\frac{\chi_s \chi_d}{w_0^2(1 + iz_\zeta)}\right] U^{(G)}(x, y, z; \omega).$$
$$(2.46)$$

Finally, neglecting the constant factor we obtain the elegant Laguerre-Gaussian modes as

$$U_{mn}^{(eLG)}(x, y, z; \omega) = \frac{\exp(-in\phi)}{w(z_\zeta)^{n+m+1}} \rho^n L_m^n \left(\frac{\rho^2}{w_0^2(1 + iz_\zeta)}\right)$$
$$\times \exp\left[ikz - \frac{\rho^2}{w_0^2(1 + iz_\zeta)} - i\psi_{eLG}(z_\zeta)\right],$$
$$(2.47)$$

where ϕ is the polar angle, and

$$\psi_{LG}(z_\varsigma) = (n + m + 1)\arctan z_\varsigma. \qquad (2.48)$$

Here index n defines the *topological charge*, i.e. the number of times the phase is changing by 2π, on rotation around the axis of the beam $z = 0$.

Finally we proceed to the standard Laguerre-Gaussian modes, i.e. the modes that are products of the amplitude and phase terms. For this purpose we use the following modification of the angular spectrum for the fundamental Gaussian field

$$a^{(LG)}(k_x, k_y; \omega) = \left[i(k_x + ik_y) + \frac{1}{q}\left(\frac{\partial}{\partial k_x} + i\frac{\partial}{\partial k_y} \right) \right]^m$$

$$\times \left[i(k_x - ik_y) + \frac{1}{q}\left(\frac{\partial}{\partial k_x} - i\frac{\partial}{\partial k_y} \right) \right]^{n+m} \qquad (2.49)$$

$$\times \exp\left[-\frac{w_0^2}{4}(k_x^2 + k_y^2) \right].$$

with q being an arbitrary constant, as before. This leads, after calculations similar to those for the elegant Laguerre-Gaussian modes, to the expression

$$U_{mn}^{(LG)}(x, y, z; \omega) = \frac{\exp(-in\phi)}{w(z_\varsigma)} \frac{\rho^n}{w(z_\varsigma)^n} L_m^n\left[\frac{2\rho^2}{w^2(z_\varsigma)} \right]$$

$$\times \exp\left[ikz - \frac{\rho^2}{w_0^2(1 + iz_\varsigma)} - i\psi_{LG}(z_\varsigma) \right]. \qquad (2.50)$$

Figure 2.4 illustrates the intensity profiles of several first Laguerre-Gaussian modes. In all shown cases one or several intensity rings are well pronounced.

2.2 Superposition of Gaussian beams

Another powerful approach for synthesis of the novel deterministic beams is based on the spatial superposition of basic fields, such as Gaussian beams, for example. Since the propagation laws of basic modes can be readily derived analytically, linear superposition of the results can be immediately used for predicting the behavior of the net field. The same superposition idea can be also used for modeling of the correlation functions, apertures and scattering media, as we will demonstrate in the latter chapters.

2.2.1 Flat-top beams

Optical fields, which possess uniform amplitude or intensity profiles in some part of the transverse cross-section, are required in material thermal process-

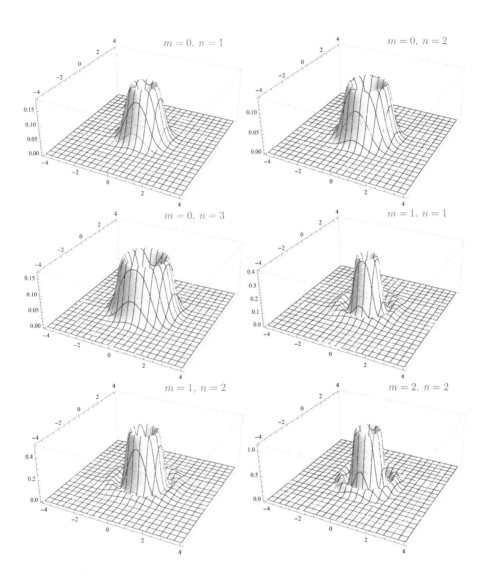

FIGURE 2.4
Intensity distributions of several first beams in the Laguerre-Gaussian family as functions of x/w_0 and y/w_0, at $z = 0$.

ing, inertial confinement fusion and are beneficial for other applications, such as laser communications. For convenient analytical calculations it is preferable to smooth the edge of the profile in order to avoid fringing. In this context several functional forms are used in calculations, for example, the super-Gaussian distribution [14]–[16]

$$U^{(SG)}(\rho, 0) = \exp\left[-\rho^\alpha\right], \tag{2.51}$$

where ρ is the radius in a polar coordinate system and $0 < \alpha < \infty$. In particular, for $\alpha = 2$ the super-Gaussian profile reduces to Gaussian.

A Lorenzian profile is another possibility for describing the flat-top beams [17]

$$U^{(LR)}(\rho, 0) = \frac{1}{1 + \rho^\alpha}, \tag{2.52}$$

with $\alpha \in [1; \infty)$. In spite of being expressed explicitly profiles (2.51) and (2.52) are not particularly convenient for evaluation of field characteristics on propagation.

Flat-top profiles can also be represented by a series of super-Gaussian functions with integral powers as

$$U^{(FG)}(\rho, 0) = \exp(-\rho^2) \sum_{m=0}^{M} \frac{\rho^{2m}}{m!}, \quad M = 0, 1, 2, \dots \tag{2.53}$$

where integer M relates to flatness. For $M = 0$ the series reduces to an ordinary Gaussian profile [18]. Free-space propagation of these beams was studied in [19] and their relation to Hermite-Gaussian and Hermite-Laguerre beams is given in [20]. Comparison of different model fields described above can be found in Ref. [21].

Yet another theoretical model for description of the flat-top profile beams was developed in Refs. [17]–[24] and will be referred here as to the *multi-Gaussian beams*. In this model the flat-topped field is constructed by means of a finite sum of Gaussian functions, namely via the expression

$$U^{(MG)}(\rho, 0) = \sum_{m=1}^{M} \alpha_m \exp\left(-\frac{m\rho^2}{w_0^2}\right), \tag{2.54}$$

where $M = 1, 2, \dots$ and weighting coefficients have the form

$$\alpha_m = \frac{(-1)^{m-1}}{m} \binom{M}{m}, \tag{2.55}$$

where $\binom{M}{m}$ are binomial coefficients and w_0 is the effective waist size. Since all functions in the model (2.54) are ordinary Gaussian it is the optimal model for making the theoretical predictions about of the evolution of such fields in free space and in various media [25]. Even though the right side of (2.54) is the

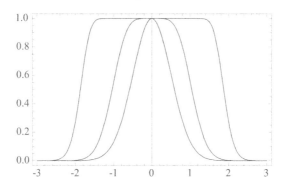

FIGURE 2.5
Intensity distributions of typical multi-Gaussian beams as functions of ρ/w_0, $M = 1, 3, 40$.

sum of sign-alternating terms one can readily prove that it is a non-negative function for any choice of M by noting that

$$1 - (1 - x)^M = \sum_{m=0}^{M} (-1)^m \binom{M}{m} x^m, \qquad (2.56)$$

with $x = \exp(-\rho^2)$ and because of inequality $(1 - x)^M < 1$, which implies that $1 - (1 - x)^M > 0$.

In cases when a normalization is needed for comparison of fields with different indexes M one can divide (2.54) by the factor

$$C_0 = \sum_{m=1}^{M} \alpha_m. \qquad (2.57)$$

Figure 2.5 shows typical one-dimensional multi-Gaussian distributions for $M = 1, 3, 40$.

If flat-top profiles with other than circular symmetry are in need one can readily generalize model (2.54) to two dimensions. In particular, a beam with rectangular symmetry can be obtained if a double sum of one-dimensional multi-Gaussian functions is introduced [23]

$$U^{(MGR)}(x, y, 0) = \sum_{m=1}^{M} \sum_{n=1}^{N} \frac{(-1)^{m+n}}{MN} \binom{M}{m} \binom{N}{n} \exp\left[-\frac{mx^2}{w_{0x}^2} - \frac{ny^2}{w_{0y}^2}\right], \qquad (2.58)$$

where w_{0x} and w_{0y} are the beam waist sizes of a Gaussian beam in the $x-$ and $y-$directions. For $M = N$ and $w_{0x} = w_{0y}$ the beam acquires square shape.

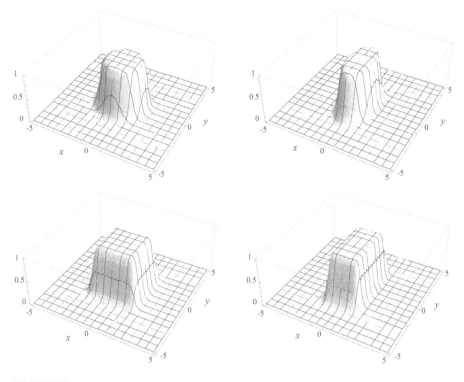

FIGURE 2.6
Two-dimensional intensity distributions of the typical multi-Gaussian beams
as functions of ρ/w_{0x} and ρ/w_{0y}, $N = M = 10$.

An elliptical flat-top beam can be described by the model [24]

$$U^{(MGE)}(x,y,0) = \sum_{n=1}^{N} \frac{(-1)^{n-1}}{N} \binom{N}{n} \exp\left[-n\left(\frac{x^2}{w_{0x}^2} + \frac{2xy}{w_{0xy}^2} + \frac{y^2}{w_{0y}^2}\right)\right],$$
(2.59)

where w_{0x}, w_{0xy} and w_{0y} are the beam waist sizes of an elliptic Gaussian
beam in the $x-$, $y-$ and coupled $xy-$directions.

Figure 2.6 illustrates four flat distributions in two dimensions: circular,
elliptical, square and rectangular flat-top profiles calculated from Eqs. (2.54),
(2.58) and (2.59).

2.2.2 Cusp-Gaussian beams

A counterpart of the flat-top family of beams appears if in the representation
(2.56) one assumes that $0 < M < 1$ [23]. Then the field distribution (2.54) in

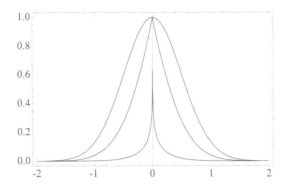

FIGURE 2.7
Intensity distributions of several cusp-Gaussian beams as functions of ρ/w_0, $M = 1, 3, 40$.

the waist plane can be written as

$$U^{(CG)}(\rho, 0) = 1 - \left[1 - \exp\left(-\frac{\rho^2}{w_0^2}\right)\right]^M, \quad 0 < M < 1. \tag{2.60}$$

Such intensity distribution with inherent circular symmetry leads to a conical-like beam. Figure 2.7 shows typical profiles of source distributions (2.60). A natural extension of such a profile is an elliptical sharp beam

$$U^{(CE)}(x, y, 0) = 1 - \left\{1 - \exp\left[-\left(\frac{x^2}{w_{0x}^2} + 2\frac{xy}{w_{0xy}^2} + \frac{y^2}{w_{0y}^2}\right)\right]\right\}^M, \quad 0 < M < 1. \tag{2.61}$$

If the beam profile is modeled in such a manner along the x- and y-directions and such distributions are multiplied, i.e., if

$$U^{(CR)}(x, y, 0) = \left\{1 - \left[1 - \exp\left(-\frac{x^2}{w_{0x}^2}\right)\right]^M\right\}$$

$$\times \left\{1 - \left[1 - \exp\left(-\frac{y^2}{w_{0y}^2}\right)\right]^N\right\}, \tag{2.62}$$

$$0 < M < 1, \quad 0 < N < 1,$$

then the intensity of such a beam acquires a pyramidal structure. Figure 2.8 includes several two-dimensional cusp-Gaussian beams.

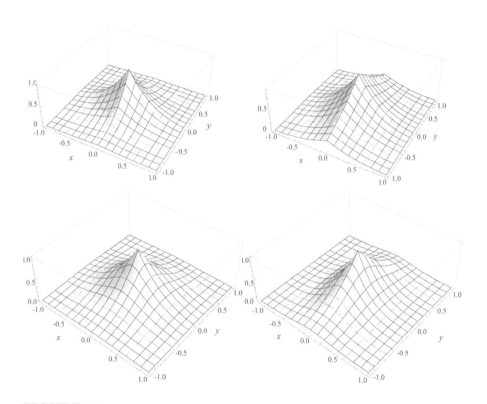

FIGURE 2.8

Typical two-dimensional intensity distributions of cusp-Gaussian beams with different geometries, as functions of x/w_{0x} and y/w_{0y}, $N = M = 10$.

2.2.3 Dark-hollow beams

The *dark-hollow beams* with zero central intensity have important applications in laser optics, atomic optics, binary optics, optical trapping of particles and medical sciences [26], [27]. Up to now, various methods such as the geometrical optical method, mode conversion, optical holography, transverse-mode selection, hollow-fiber method, computer-generated holography, nonlinear optical method and spatial filtering have been proposed to generate dark-hollow beams experimentally.

The simplest possible "doughnut" beam can be modeled as a difference of two Gaussian beams with waist radii w_{0A} and w_{0B}. The resulting beam, which we will refer to as the *annular beam*, has the following field distribution in the source plane [28]

$$U^{(AN)}(\rho,0) = a_A \exp\left[-\frac{\rho^2}{2w_{0A}^2}\right] - a_B \exp\left[-\frac{\rho^2}{2w_{0B}^2}\right], \qquad (2.63)$$

provided $w_{0A} > w_{0B}$. If, in addition, $a_A = a_B$ then the zero intensity spot is generated on the axis of the beam. This dark spot, however, disappears on propagation.

In order to efficiently control the size and the shape of the ring-shaped profiles, one may use, instead of a Gaussian, various multi-Gaussian profiles of the previous section. For instance, the elliptical dark-hollow beams can be modeled as the following single finite sum

$$U^{(DHE)}(x,y,0) = \sum_{n=1}^{N} \frac{(-1)^{n-1}}{N} \binom{N}{n} \left[\exp\left(-\frac{x^2}{w_{nx}^2} - \frac{2xy}{w_{nxy}^2} - \frac{y^2}{w_{ny}^2}\right)\right.$$
$$\left. - \exp\left(-\frac{x^2}{w_{nqx}^2} - \frac{2xy}{w_{nqxy}^2} - \frac{y^2}{w_{nqy}^2}\right)\right], \qquad (2.64)$$

where $w_{nx}^2 = w_{0x}^2/n$, $w_{ny}^2 = w_{0y}^2/n$, $w_{nxy}^2 = w_{0xy}^2/n$, $w_{nqx}^2 = qw_{0x}^2/n$, $w_{nqy}^2 = qw_{0y}^2/n$, $w_{nqxy}^2 = w_{0qxy}^2/n$, $w_{0x}, w_{0y}, w_{0xy}, w_{0qx}, w_{0qy}, w_{0qxy}$ are waist radii as above, $0 < q < 0$ is a scaling factor. When $w_{0x} = w_{0y}$ the elliptical distribution reduces to circular.

Rectangular dark-hollow beams can be modeled as the following double finite sum

$$U^{(DHR)}(x,y,0) = \sum_{m=1}^{M} \sum_{n=1}^{N} \frac{(-1)^{m+n}}{MN} \binom{M}{m} \binom{N}{n}$$
$$\times \left[\exp\left(-\frac{x^2}{w_{mx}^2} - \frac{y^2}{w_{ny}^2}\right) - \exp\left(-\frac{x^2}{w_{mqx}^2} - \frac{y^2}{w_{nqy}^2}\right)\right], \qquad (2.65)$$

where $w_{mx}^2 = w_{0x}^2/m$, $w_{mqx}^2 = qw_{0x}^2/m$, the other parameters are defined the same way as for an elliptical dark-hollow beam. When $w_{0x} = w_{0y}$ the rectangular distribution reduces to a square. Rotation of the rectangular profile

58

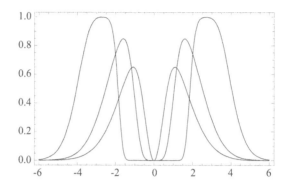

FIGURE 2.9
Intensity distributions of dark-hollow Gaussian beams as functions of ρ/w_0, $N = 1, 3, 40$.

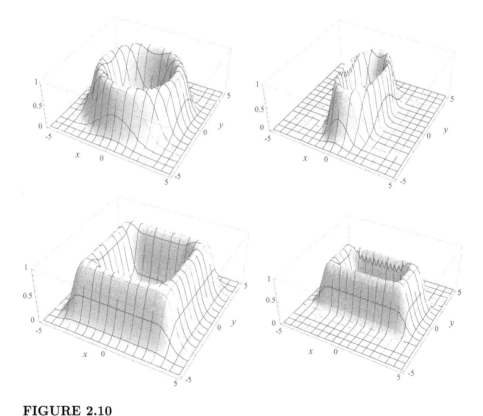

FIGURE 2.10
Intensity distributions of two-dimensional dark-hollow Gaussian beams as functions of x/w_{0x} and y/w_{0y}, $N = M = 10$.

can be achieved in the same way as done for $U^{(DHE)}$, by including the mixed terms.

In Fig. 2.9 typical intensity profiles of dark-hollow beams are shown as a function of the summation index. Figure 2.10 illustrates several two-dimensional dark-hollow beams.

2.3 Other deterministic beams

A variety of other beam profiles have also been introduced including, for instance, the *Ince-Gaussian beams* [10]:

$$U^{(IG)}(\xi, \eta, z; \omega) = \frac{w_0}{w(z)} C_p^m(i\xi, \epsilon) C_p^m(\eta, \epsilon)$$
$$\times \exp\left[ikz - ik\frac{\rho^2}{2q_I(z)} - i(p+1)\psi_G(z)\right], \quad (2.66)$$

where transformation

$$x = \sqrt{\frac{\epsilon}{2}} w(z) \cosh\xi \cos\eta, \quad y = \sqrt{\frac{\epsilon}{2}} w(z) \sinh\xi \sin\eta, \quad (2.67)$$

was used, C_p^m are the even Ince polynomials of order p and degree m, ϵ_I is the ellipticity parameter of Ince-Gaussian beam and

$$q_I(z) = \left[\frac{1}{R(z)} - i\frac{\lambda}{\pi w^2(z)}\right]^{-1}. \quad (2.68)$$

Such fields constitute a family for which the Hermite-Gaussian and the Laguerre-Gaussian modes are two limiting cases, for $\epsilon = 0$ and $\epsilon = \infty$, respectively.

In Ref. [29] all non-diffracting fields, such as plane waves, cosine, Bessel, Mathieu, and parabolic beams modulated with Gaussian functions were shown to be beam-like fields. Perhaps the most significant beams of this kind are the mixtures of the J_0-Bessel beam with a Gaussian beam, also known as J_0-Bessel-Gaussian beam [30] and the J_n-Bessel beam with the Gaussian beams, termed as J_n-Bessel-Gaussian beams (with $n > 0$) [31]. The *spatial* field distribution across the J_0-Bessel-Gaussian source has the form

$$U^{(JBG)}(\rho, 0) = J_0(\beta\rho) \exp\left(-\frac{\rho^2}{w_0^2}\right), \quad (2.69)$$

where w_0 is a positive constant which, being a measure of the width of the Gaussian term, gives the radius of the outer ring, and parameter β determines the size of the inner ring. Moreover, parameter β relates to the angular half-aperture of the cone on which plane waves are superposed for generation of a corresponding non-diffracting Bessel beam (see Chapter 1), i.e., $\beta = k_\perp$.

On propagation in free space the J_0-Bessel-Gaussian beam is given by the expression [30]

$$
U_0^{(JBG)}(\rho, \phi, z; \omega) = -\frac{ikA}{2zq} \exp\left[ik\left(z + \frac{\rho^2}{2z}\right)\right] J_0\left(\frac{i\beta k\rho}{2zq}\right)
$$
$$
\times \exp\left[-\frac{1}{4q}\left(\beta^2 + \frac{k^2\rho^2}{z^2}\right)\right], \tag{2.70}
$$

where $q = 1/w_0^2 - ik/2z$.

Higher-order Bessel-Gaussian beams have in the source plane the form [31]

$$
U_n^{(JBG)}(\rho, \phi, 0) = J_n(\beta\rho) \exp\left[-\frac{\rho^2}{w_0^2}\right] \exp(-in\phi), \quad n = 1, 2, ..., \tag{2.71}
$$

where the last phase term signifies the presence of an optical vertex of order n, since the beam has zero on-axis intensity.

On propagation in free space the J_n-Bessel-Gaussian beams, $n > 0$, evolve as [31]

$$
U_n^{(JBG)}(\rho, \phi, z; \omega) = -\frac{iAk}{2zq} \exp\left[ik\left(z + \frac{\rho^2}{2z}\right)\right] J_n\left(\frac{i\beta k\rho}{2zq}\right)
$$
$$
\times \exp\left[-\frac{1}{4q}(\beta^2 + k^2\rho^2/z^2)\right] \exp(-in\phi). \tag{2.72}
$$

Some other beam-like field profiles and their classifications can be found in Refs. [32]–[38].

Bibliography

[1] H. W. Kogelnik and T. Li, "Laser beams and resonators," *Appl. Opt.* **5**, 1550–1567 (1966).

[2] A. Siegman, *Lasers*, University Science, 1986.

[3] K. Shimoda, *Introduction to Laser Physics*, Springer, 1991.

[4] B. E. A. Saleh and M. C. Teich, *Fundamentals of Photonics*, John Wiley and Sons, 1991.

[5] R. Menzel, *Photonics*, Springer, 2001.

[6] O. Svelto, *Principles of Lasers*, 4th Ed., Plenum Press, 2010.

[7] F. Pampaloni and J. Enderlein, "Gaussian, Hermite-Gaussian, and Laguerre-Gaussian beams: A primer," arXiv:physics/0410021.

[8] E. D. Solomentsev, "Cauchy inequality," M. Hazewinkel, *Encyclopedia of Mathematics*, Springer, 2001.

[9] L. Mandel and E. Wolf, *Optical Coherence and Quantum Optics*, Cambridge University Press, 1995.

[10] M. A. Bandres and J. C. Gutierrez-Vega, "Ince-Gaussian beams," *Opt. Lett.* **29**, 144–146 (2004).

[11] W. H. Carter, "Spot size and divergence for Hermite Gaussian beams of any order," *Appl. Opt.* **19**, 1027–1029 (1980).

[12] W. H. Carter, "Energy carried over the rectangular spot within a Hermite-Gaussian beam," *Appl. Opt.* **21**, 7 (1982).

[13] R. L. Phillips and L. C. Andrews, "Spot size and divergence for Laguerre-Gaussian beams of any order," *Appl. Opt.* **22**, 643–644 (1983).

[14] S. De Silvestri, P. Laporta, V. Magni, and O. Svelto, "Solid-state laser unstable resonators with tapered reflectivity mirrors: the super-Gaussian approach," *IEEE J. Quantum Electron.* **24**, 1172–1177 (1988).

[15] A. Parent, M. Morin, and P. Lavigne, "Propagation of super-Gaussian field distributions," *Opt. Quantum Electron.* **24**, 1071–1079 (1992).

[16] M. S. Bowers, "Diffractive analysis of unstable optical resonator with super-Gaussian mirrors," *Opt. Lett.* **17**, 1319–1321 (1992).

[17] Y. Li, "Light beam with flat-topped profiles," *Opt. Lett.* **27**, 1007–1009 (2002).

[18] F. Gori, "Flattened Gaussian beams," *Opt. Comm.* **107**, 335–341 (1994).

[19] V. Bagini, R. Borghi, F. Gori, A. M. Pacileo, M. Santarsiero, D. Ambrosini, and G. S. Spagnolo, "Propagation of axially symmetric flattened Gaussian beams," *J. Opt. Soc. Am. A* **13**, 1385–1394 (1996).

[20] M. Santarsiero and R. Borghi, "Correspondence between super-Gaussian and flattened Gaussian beams," *J. Opt. Soc. Am. A* **16**, 188–190 (1999).

[21] D. L. Shealy and J. A. Hoffnagle, "Laser beam shaping profiles and propagation," *Appl. Opt.* **45**, 5118–5131 (2006).

[22] Y. Li, "New expressions for flat-topped beams," *Opt. Comm.* **206**, 225–234 (2002).

[23] Y. Li, "Flat-topped beam with non-circular cross-sections," *J. Mod. Opt.* **50**, 1957–1966 (2003).

[24] Y. Cai and Q. Lin, "Light beams with elliptical flat-topped profiles," *J. Opt. Am. A: Pure Appl. Opt.* **6**, 390–395 (2004).

[25] Y. Cai, "Propagation of various flat-topped beams in a turbulent atmosphere," *J. Opt. A: Pure Appl. Opt.* **8**, 537–545 (2006).

[26] T. Kuga, Y. Torii, N. Shiokawa, T. Hirano, Y. Shimizu, and H. Sasada, "Novel optical trap of atoms with a doughnut beam," *Phys. Rev. Lett.* **78**, 4713–4716 (1997).

[27] J. Yin, W. Gao, and Y. Zhu, "Generation of dark hollow beams and their applications," in *Progress in Optics* Vol. **44**, E. Wolf, ed., North-Holland, 2003, pp. 119–204.

[28] L. C. Andrews and R. L. Phillips, *Laser Beam Propagation in the Turbulent Atmosphere*, 2nd edition, SPIE Press, 2005.

[29] J. G. Gutierrez-Vega and M. A. Bandres, "Helmholtz-Gauss waves," *J. Opt. Soc. Am. A* **22**, 289–298 (2005).

[30] F. Gori, G. Guattari, and C. Padovani, "Bessel-Gauss beams," *Opt. Commun.* **64**, 491–495 (1987).

[31] L. Xuan-Hui, C. Xu-Min, Z. Lei and X. Da-Jian, "High-order Bessel-Gaussian beam and its propagation properties," *Chinese Phys. Lett.* **20**, 2155–2157 (2003).

[32] O. E. Gawhary and S. Severini, "Lorentz beams and symmetry properties in paraxial optics," *J. Opt. A, Pure Appl. Opt.* **8**, 409-414 (2006).

[33] E. G. Abramochkin and V. G. Volostnikov, "Generalized Gaussian beams," *J. Opt. A, Pure Appl. Opt.* **6**, S157-S161 (2004).

[34] D. Deng and Q. Guo, "Elegant Hermite-Laguerre-Gaussian beams," *Opt. Lett.* **33**, 1225–1227 (2008).

[35] V. V. Kotlyar, R. V. Skidanov, S. N. Khonina, and V. A. Soifer, "Hypergeometric modes," *Opt. Lett.* **32**, 142–144 (2007).

[36] E. Karimi, G. Zito, B. Piccirillo, L. Marrucci, and E. Santamato, "Hypergeometric-Gaussian modes," *Opt. Lett.* **32**, 3053–3055 (2007).

[37] M. A. Bandres and J. C. Gutierrez-Vega, "Cartesian beams," *Opt. Lett.* **32**, 3459–3461 (2007).

[38] M. A. Bandres and J. C. Gutierrez-Vega, "Circular beams," *Opt. Lett.* **33**, 177–179 (2008).

3

Scalar stochastic beams: theory

CONTENTS

3.1 Statistical description

The complete characterization of a stochastic process describing variation of an optical signal in time and space requires specification of the joint probability density functions of the field of all orders at all time instances and at all spatial positions. While in some cases, such as calculations involving Gaussian distributions, it can be done, this task becomes unattainable in problems involving random sources and media which are governed by non-Gaussian statistics. In such instances specification of the first several moments of the field is usually sufficient at one or two spatial and temporal arguments. For example, for optical signals propagating in random media such as the Earth's atmosphere or oceans, the exact forms of the probability density functions of the fluctuating phase and intensity are not generally known and hence the calculations are usually confined to the first several moments [1].

The vast majority of investigations concerns the second-order moments of a field at one or two spatial arguments, since at optical frequencies they are observable quantities [2] and, at the same rate, often contain satisfactory

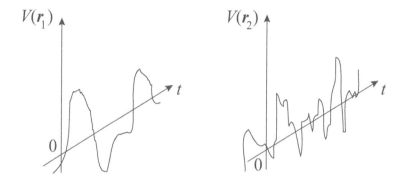

FIGURE 3.1
The mutual coherence function of the beam is obtained on correlating the fields $V(\mathbf{r}_1)$ and $V(\mathbf{r}_2)$ over a long time interval.

amount of information. We will start our analysis with the review of the classical coherence theory which makes it possible to determine the evolution of the second-order statistical moments of random fields propagating in free space or interacting with linear media. We postpone the discussion of higher-order moments of beams governed by Gaussian statistics to Chapter 4, where we will relate them not only to coherence but also to polarization properties.

3.1.1 Mutual coherence function

In the early studies of the stochastic fields the spatio-temporal description of the optical signal was used [3]–[5]. Consider a scalar quasi-monochromatic field ($\Delta\omega << \bar{\omega}$), occupying a finite closed planar domain D, and associated with it complex analytic signal $V(\mathbf{r};t)$ at a point with the three-dimensional position vector $\mathbf{r}$, and at time instant t. The central quantity in the classical coherence theory is the second-order statistical moment of $V(\mathbf{r};t)$ at two spatial positions, say, $\mathbf{r}_1$ and $\mathbf{r}_2$, and two moments in time, t_1 and t_2 (see Fig. 3.1)

$$\Gamma_V(\mathbf{r}_1, \mathbf{r}_2; t_1, t_2) = \langle V^*(\mathbf{r}_1; t_1)V(\mathbf{r}_2; t_2)\rangle_T, \qquad (3.1)$$

where we have used the time average. In the following the subscripts of the correlation functions will be omitted for brevity. Function Γ is called the *mutual coherence function* of the field. Under the assumption that the random process governing the evolution of the optical field is wide-sense statistically stationary (see Chapter 1), we now can write [4]

$$\Gamma(\mathbf{r}_1, \mathbf{r}_2; \tau) = \langle V^*(\mathbf{r}_1; t)V(\mathbf{r}_2; t + \tau)\rangle_T, \qquad (3.2)$$

where τ is the time delay. It follows from Maxwell's equations that the mutual coherence function obeys a pair of wave equations with respect to each of the two spatial arguments, $\mathbf{r}_1$ and $\mathbf{r}_2$ ([3], Section 10.8.1):

$$\nabla_1^2 \Gamma(\mathbf{r}_1, \mathbf{r}_2; \tau) = \frac{1}{c^2} \frac{\partial^2 \Gamma(\mathbf{r}_1, \mathbf{r}_2; \tau)}{\partial \tau^2},$$
$$\nabla_2^2 \Gamma(\mathbf{r}_1, \mathbf{r}_2; \tau) = \frac{1}{c^2} \frac{\partial^2 \Gamma(\mathbf{r}_1, \mathbf{r}_2; \tau)}{\partial \tau^2}, \tag{3.3}$$

where ∇_1^2 and ∇_2^2 are Laplacian operators taken with respect to points $\mathbf{r}_1$ and $\mathbf{r}_2$. The solution of such a system of equations is, in general, very lengthy. Even for free-space propagation and, of course, for propagation in some media, the mutual coherence function exhibits significant time delays at different frequencies [3]. Regardless of these severe limitations the mutual coherence function has played the crucial part in understanding the stochastic fields because of its relatively easy measurability.

3.1.2 Cross-spectral density function

Several decades ago the second-order coherence theory for the wide-sense statistically stationary fields has been also formulated in the space-frequency domain [6]–[8]. The central quantity of this theory is the *cross-spectral density function*, defined at two spatial arguments $\mathbf{r}_1$ and $\mathbf{r}_2$ and at angular frequency ω. By definition, the cross-spectral density function is the Fourier transform of the mutual coherence function with respect to the time delay τ, namely,

$$W(\mathbf{r}_1, \mathbf{r}_2; \omega) = \mathcal{F}[\Gamma(\mathbf{r}_1, \mathbf{r}_2; \tau)]. \tag{3.4}$$

This relation is the natural extension of the Wiener-Khintchine theorem (see Chapter 1). Even though this function is introduced as a Fourier transform of the correlation function, it might actually not be the correlation function by itself. Under the assumption that the random process is *ergodic* the cross-spectral density function has been proven to be the valid correlation function, with averaging process not being carried out with respect to a time interval but rather with respect to an ensemble of monochromatic realizations $U(\mathbf{r}; \omega)$ of the optical field, viz., [7]

$$W(\mathbf{r}_1, \mathbf{r}_2; \omega) = \langle U^*(\mathbf{r}_1; \omega) U(\mathbf{r}_2; \omega) \rangle_\omega. \tag{3.5}$$

The optical field U differs from the field V entering Eqs. (3.1) and (3.2): the former denotes a quasi-monochromatic optical field at some time instant while the latter can be regarded as the *equivalent monochromatic field* at frequency ω.

The cross-spectral density function obeys a pair of Helmholtz equations with respect to spatial arguments $\mathbf{r}_1$ and $\mathbf{r}_2$. Indeed, on taking the Fourier transform of Eqs. (3.3) we find that

$$\nabla_1^2 W(\mathbf{r}_1, \mathbf{r}_2; \omega) + k^2 W(\mathbf{r}_1, \mathbf{r}_2; \omega) = 0,$$
$$\nabla_2^2 W(\mathbf{r}_1, \mathbf{r}_2; \omega) + k^2 W(\mathbf{r}_1, \mathbf{r}_2; \omega) = 0, \tag{3.6}$$

where $k = \omega/c$ is the wave number of light. The solutions of the Helmholtz equations have a much simpler form than those of the wave equations no matter how wide the bandwidth of the signal is. Due to this reason the cross-spectral density function has been recognized as a more efficient tool than the mutual coherence function for solving problems involving interactions between optical fields with arbitrary spectral composition and various media. In the case when the field is quasi-monochromatic the solution for the mutual coherence function is approximately the same as that for the cross-spectral density function.

Not every function of two spatial arguments and a frequency may serve as the cross-spectral density function of a field. A proper cross-spectral density W should satisfy the following conditions (see Ref. [4], Chapter 4).

1. W must be square-integrable with respect to ω, i.e.,

$$\int_0^\infty |W(\mathbf{r}_1, \mathbf{r}_2; \omega)|^2 d\omega < \infty. \tag{3.7}$$

2. If W is a continuous function of its spatial arguments then

$$\int\int |W(\mathbf{r}_1, \mathbf{r}_2; \omega)|^2 d^2 r_1 d^2 r_2 < \infty. \tag{3.8}$$

3. W must be Hermitian, i.e.,

$$W(\mathbf{r}_1, \mathbf{r}_2; \omega) = W^*(\mathbf{r}_1, \mathbf{r}_2; \omega). \tag{3.9}$$

4. For any square-integrable function $f(\mathbf{r})$ the following integral inequality must hold

$$\int\int W(\mathbf{r}_1, \mathbf{r}_2; \omega) f^*(\mathbf{r}_1) f(\mathbf{r}_2) d^2 r_1 d^2 r_2 \geq 0, \tag{3.10}$$

i.e., W must be non-negative definite.

It is generally a very difficult task to verify whether a certain function of two spatial arguments is a *genuine* (valid) correlation function, mostly because of the condition (3.10). A sufficient condition considerably simplifying this problem is given in Ref. [8] (see also [9]) stating that a valid correlation function of a fluctuating field must have the representation of the form

$$W(\mathbf{r}_1, \mathbf{r}_2; \omega) = \int p(\mathbf{s}; \omega) H^*(\mathbf{r}_1, \mathbf{s}; \omega) H(\mathbf{r}_2, \mathbf{s}; \omega) d^2 s, \tag{3.11}$$

where H is an arbitrary function, p is a non-negative Fourier-transformable function and $\mathbf{s}$ is a two-dimensional vector.

As the final remark we note that unlike for many statistical quantities, for the cross-spectral density function the laws of interaction with any linear media are known and exact. In other words, from the knowledge of the cross-spectral density function in one cross-section of the beam (say, perpendicular to direction of propagation) it is possible to uniquely determine it in some other cross-section. This is the direct consequence of the fact that the cross-spectral density function exactly obeys the pair of the Helmholtz equations (1.77). On the contrary, such exact propagation laws do not exist for other important field statistics, for example, the intensity.

3.1.3 Spectral and coherence properties

The key importance of the cross-spectral density function follows from the fact that two readily measurable quantities: the spectral density and the spectral degree of coherence of a random field can be readily calculated from it. Therefore, because the cross-spectral density obeys the exact propagation laws, one can also deduce how these two quantities propagate. The *spectral density* of the field is just the cross-spectral density function calculated at the coinciding spatial arguments, $\mathbf{r} = \mathbf{r}_1 = \mathbf{r}_2$:

$$S(\mathbf{r}; \omega) = W(\mathbf{r}, \mathbf{r}; \omega). \tag{3.12}$$

It provides the information about the distribution of power carried by various frequency components of the spectrum at a certain location in space and can be directly measured by a spectroscope. Because of the simple relation between angular frequency ω and wavelength λ, $\omega = 2\pi c/\lambda$, we can either use $S(\mathbf{r}; \omega)$ or $S(\mathbf{r}; \lambda)$ interchangeably.

For characterization of spectral changes it is useful to introduce the normalized spectral density of the beam

$$S_N(\mathbf{r}; \omega) = \frac{S(\mathbf{r}; \omega)}{\int_0^\infty S(\mathbf{r}; \omega)d\omega}, \quad S_N(\mathbf{r}; \lambda) = \frac{S(\mathbf{r}; \lambda)}{\int_0^\infty S(\mathbf{r}; \lambda)d\lambda}. \tag{3.13}$$

Further, the central frequently (wavelength) of the spectrum can be found from the ratios

$$\omega_1(\mathbf{r}) = \frac{\int_0^\infty \omega S(\mathbf{r}; \omega)d\omega}{\int_0^\infty S(\mathbf{r}; \omega)d\omega}, \quad \lambda_1(\mathbf{r}) = \frac{\int_0^\infty \lambda S(\mathbf{r}; \lambda)d\lambda}{\int_0^\infty S(\mathbf{r}; \lambda)d\lambda}. \tag{3.14}$$

If the initial (position-independent) central wavelength is λ_0, the normalized spectral shift at position $\mathbf{r}$ may be quantified by the expressions

$$\varrho_\omega(\mathbf{r}) = \frac{\omega_1(\mathbf{r}) - \omega_0}{\omega_0}, \quad \varrho_\lambda(\mathbf{r}) = \frac{\lambda_1(\mathbf{r}) - \lambda_0}{\lambda_0}. \tag{3.15}$$

The shift $\varrho_\omega(\mathbf{r})$ in the spectrum is called *blue* if its value is positive and *red* if it is negative. The definition is of course the opposite for $\varrho_\lambda(\mathbf{r})$.

The *spectral degree of coherence* of the optical field at points $\mathbf{r}_1$ and $\mathbf{r}_2$

is the cross-spectral density function normalized by the square roots of the values of the spectral density at these points, i.e., [4]

$$\mu(\mathbf{r}_1, \mathbf{r}_2; \omega) = \frac{W(\mathbf{r}_1, \mathbf{r}_2; \omega)}{\sqrt{S(\mathbf{r}_1; \omega)}\sqrt{S(\mathbf{r}_2; \omega)}}. \tag{3.16}$$

This function is, in general, complex-valued. The physical importance of the spectral degree of coherence follows from the fact that its absolute value, varying between 0 and 1, can be shown to be equal to the *visibility* of fringes produced in the interference experiment (see Fig. 1.11) defined by the expression [4]

$$\mathcal{V}(\omega) = \frac{S_{max}(\omega) - S_{min}(\omega)}{S_{max}(\omega) + S_{min}(\omega)}. \tag{3.17}$$

Indeed, one can show that instead of Eq. (1.141) describing the fringe pattern for monochromatic light, one has a more general expression [4]

$$S(\mathbf{r}; \omega) \approx 2S(\boldsymbol{\rho}; \omega)[1 + |\mu(\boldsymbol{\rho}_1, \boldsymbol{\rho}_2; \omega)| \cos\{\arg[\mu(\boldsymbol{\rho}_1, \boldsymbol{\rho}_2; \omega)] - \Delta R\}], \tag{3.18}$$

where $\Delta R = 2\pi(R_1 - R_2)/\lambda$, while $S(\mathbf{r}; \omega)$ and $S(\boldsymbol{\rho}; \omega) = S(\boldsymbol{\rho}_1; \omega) = S(\boldsymbol{\rho}_2; \omega)$ are the spectral densities at the observation point and at the pinholes.

In the two limiting cases when the modulus of the spectral degree of coherence is either 0 or 1 the field is called *incoherent* or *coherent*. The phase $\arg[\mu(\boldsymbol{\rho}_1, \boldsymbol{\rho}_2; \omega)]$ of the spectral degree of coherence should not be confused with the phase of the field itself. At optical frequencies the former is measurable [10] while the latter is not.

3.1.4   Total, encircled and fractional power

In some cases the spectral density which provides the information about the distribution of energy carried by particular frequency components of the spectrum can be directly measured, for example, with the help of the narrow spectral filters. However, in the majority of practical situations the measurement of the spectral density can be limited or not possible. Instead, several integrated quantities that we list below can be obtained. First of all, the intensity of the beam at point $\mathbf{r}$ can be found by integrating the spectral density over all the frequencies, i.e.,

$$I(\mathbf{r}) = \int_0^\infty S(\mathbf{r}; \omega)d\omega. \tag{3.19}$$

The integration interval $(0; \infty)$ in Eq. (3.19) can be changed to any other interval, say (ω', ω''), to include various filtering effects.

On the other hand, a detector which can exactly measure the signal with prescribed angular frequency ω but has an extended finite area, say Σ_d, returns the *encircled optical energy*, p_E:

$$p_E(\Sigma_d; \omega) = \int_{\Sigma_d} S(\mathbf{r}; \omega)d^2\rho. \tag{3.20}$$

In some calculations relating to signal quality it is useful to estimate the loss of power on transmission, also known as *fractional power* or *power in the bucket*, defined by the ratio of detected energy to transmitted (total) energy,

$$p_F(\Sigma_d, \Sigma_\infty; \omega) = \frac{p_E(\Sigma_d; \omega)}{p_E(\Sigma_\infty; \omega)}, \qquad (3.21)$$

where Σ_d and Σ_∞ are the areas of the detector and the source, respectively, and monochromatic detection is implied.

Fractional power can also play an important role in setting a threshold on whether and where a field initially having beam-like properties (according to the free-space beam conditions) ceases to be such, after interaction with random media, like turbulence or scatterers. Clearly if an appreciable fraction of the radiated power, say 90%, traverses a transverse cross-sectional area at a distance from the source plane, whose linear dimensions subtend an angle of, say, 1° at the source, one may regard the radiated field to have retained its beam-like form up to that distance. The somewhat arbitrary choice of 90% and 1° may, of course, be replaced by other values, depending on the purpose for which the beam is to be used [11]. It is possible to demonstrate that the total transmitted energy is conserved on propagation in free space and in linear media, in the absence of gain or absorption. More general proof of such a statement is included in Chapter 4, where the electromagnetic conservation laws are derived for free fields.

3.1.5 Higher-order statistical properties

In practically all applications including the ones we will consider in Chapters 6–8 measurements or calculations of some statistical moments of the field higher than the second are necessary. This need arises in all situations when the higher-order moments of the field provide additional information to that obtained from the second-order moment. The sole exception from this rule is the stochastic optical field generated by a source obeying complex Gaussian statistics and interacting with either deterministic media (including free-space propagation) or random media whose fluctuations are also Gaussian. Generally the n-th order moment of the optical field can be easily defined [4], but in this study we will restrict ourselves only to the fourth-order moments of the field, i.e., the correlation functions of the form

$$\Gamma^{(4)}(\mathbf{r}_1, \mathbf{r}_2, \mathbf{r}_3, \mathbf{r}_4; t_1, t_2, t_3, t_4) = \langle V^*(\mathbf{r}_1; t_1) V^*(\mathbf{r}_2; t_2) V(\mathbf{r}_3; t_3) V(\mathbf{r}_4; t_4) \rangle_T. \qquad (3.22)$$

For practical purposes the fluctuations are assumed to be stationary, and function $\Gamma^{(4)}$ is then measured at two spatial positions, $\mathbf{r}_1$ and $\mathbf{r}_2$, with time delay τ:

$$\Gamma^{(4)}(\mathbf{r}_1, \mathbf{r}_2; \tau) = \langle V^*(\mathbf{r}_1; t) V^*(\mathbf{r}_2; t+\tau) V(\mathbf{r}_1; t) V(\mathbf{r}_2; t+\tau) \rangle_T \qquad (3.23)$$

In this setting $\Gamma^{(4)}$ approximates the correlation in the intensity of the field, i.e.,

$$\Gamma^{(4)}(\mathbf{r}_1,\mathbf{r}_2;\tau) \approx \langle I(\mathbf{r}_1;t)I(\mathbf{r}_2;t+\tau)\rangle_T. \qquad (3.24)$$

The degree of intensity correlation can be determined with the help of the following normalization of $\Gamma^{(4)}(\mathbf{r}_1,\mathbf{r}_2;\tau)$:

$$b_I(\mathbf{r}_1,\mathbf{r}_2;\tau) = \frac{\Gamma^{(4)}(\mathbf{r}_1,\mathbf{r}_2;\tau) - I(\mathbf{r}_1;t)I(\mathbf{r}_2;t+\tau)}{I(\mathbf{r}_1;t)I(\mathbf{r}_2;t+\tau)}. \qquad (3.25)$$

Further, at coinciding spatial arguments and zero time delay the latter quantity reduces to the *fluctuation contrast* or the *scintillation index*:

$$c_I(\mathbf{r}) = b_I(\mathbf{r},\mathbf{r};0), \qquad (3.26)$$

being, perhaps, the most useful from all the higher-order moments and practically the most accessible. For moments higher than the second the coherence theory in space-frequency domain has not been completely developed.

3.1.6 Coherent mode decomposition

Because of certain difficulties associated with description of interaction of the stochastic optical beams with media it often becomes very convenient to decompose them into simpler structures, which we may call *modes*, carry over the calculations for the modes individually and then "sum the results up" in order to reconstruct the net stochastic field after the interaction. For stochastic beams two such decompositions have gained popularity: the coherent mode decomposition and the angular spectrum decomposition, which we will discuss in this and the next sections, respectively.

For scalar fields the coherent mode decomposition was introduced first by Gamo [6], then developed by Wolf [7] and was further theoretically and experimentally explored in [12]–[19]. A concise account and several applications of this representation belong to the monograph by Ostrovsky [20]. Coherent mode decomposition allows representation of a stochastic field via infinite number of fully coherent, uncorrelated elements, or modes. Such coherent modes are separable in space, i.e., they depend on a single spatial argument, unlike the cross-spectral density function, and are uncorrelated, which makes it very convenient to analyze interaction of fields with media and optical systems using the knowledge of such interaction for individual modes.

The coherent mode representation of a stochastic optical field can be obtained as the solution of a homogeneous Fredholm's integral equation of the second kind [21]

$$\int_D W(\boldsymbol{\rho}_1,\boldsymbol{\rho}_2;\omega)\psi_n(\boldsymbol{\rho}_1;\omega)d\boldsymbol{\rho}_1 = \lambda_n(\omega)\psi_n(\boldsymbol{\rho}_2;\omega), \qquad (3.27)$$

where D is a region occupied by the beam in the source plane, $\lambda_n(\omega)$ and $\psi_n(\boldsymbol{\rho};\omega)$ are the *eigenvalues* and the *eigenfunctions*, respectively. While the form of the eigenvalues depends on the choice of the spectral density, the form of the eigenfunctions is independent from it. It is well known that kernel $W(\boldsymbol{\rho}_1, \boldsymbol{\rho}_2; \omega)$ satisfying Eq. (3.27) can then be represented by the *Mercer's series* [22]

$$W(\boldsymbol{\rho}_1, \boldsymbol{\rho}_2; \omega) = \sum_n \lambda_n(\omega)\psi_n^*(\boldsymbol{\rho}_1; \omega)\psi_n(\boldsymbol{\rho}_2; \omega), \qquad (3.28)$$

where $\psi_n(\boldsymbol{\rho}; \omega)$ are at least mutually orthogonal in D and can be made orthonormal by the Gram-Schmidt procedure. The orthonormality implies that

$$\int_D \psi_n^*(\boldsymbol{\rho}; \omega)\psi_n(\boldsymbol{\rho}; \omega)dr^2 = \delta_{n,m}, \qquad (3.29)$$

$\delta_{n,m}$ being the Kronecker symbol

$$\delta_{n,m} = \begin{cases} 1, & n = m; \\ 0, & n \neq m. \end{cases} \qquad (3.30)$$

In their turn, the eigenvalues are all real and non-negative.

Let us now consider a single mode defined by the formula

$$W_n(\boldsymbol{\rho}_1, \boldsymbol{\rho}_2; \omega) = \psi_n^*(\boldsymbol{\rho}_1; \omega)\psi_n(\boldsymbol{\rho}_2; \omega). \qquad (3.31)$$

It is not hard to show that each of the modes W_n satisfies the same Helmholtz equations (Eqs. (3.6)) as the cross-spectral density function itself. Indeed, on substituting from Eq. (3.31) into the second of the Helmholtz equations (3.6) we find that

$$\sum_n \lambda_n(\omega)\psi_n^*(\boldsymbol{\rho}_1; \omega)\nabla_2^2\psi_n(\boldsymbol{\rho}_2; \omega)$$
$$+ k^2 \sum_n \lambda_n(\omega)\psi_n^*(\boldsymbol{\rho}_1; \omega)\psi_n(\boldsymbol{\rho}_2; \omega) = 0. \qquad (3.32)$$

Further, after multiplying all terms in Eq. (3.32) by $\psi_m(\boldsymbol{\rho}; \omega)$, integrating the result over variable $\boldsymbol{\rho}_1$ and using the orthonormality of the eigenfunctions we arrive at equation

$$\nabla_1^2\psi_n(\boldsymbol{\rho}; \omega) + k^2\psi_n(\boldsymbol{\rho}; \omega) = 0. \qquad (3.33)$$

It then follows from Eq. (3.33) together with Eq. (3.31) that

$$\nabla_1^2 W_n(\boldsymbol{\rho}_1, \boldsymbol{\rho}_2; \omega) + k^2 W_n(\boldsymbol{\rho}_1, \boldsymbol{\rho}_2; \omega) = 0,$$
$$\nabla_2^2 W_n(\boldsymbol{\rho}_1, \boldsymbol{\rho}_2; \omega) + k^2 W_n(\boldsymbol{\rho}_1, \boldsymbol{\rho}_2; \omega) = 0. \qquad (3.34)$$

Thus, the coherent modes can be propagated independently.

On the other hand, the individual coherent modes are completely coherent structures. Indeed the spectral degree of coherence of each mode, given by the formula

$$\mu_n(\boldsymbol{\rho}_1, \boldsymbol{\rho}_2; \omega) = \frac{\psi_n^*(\boldsymbol{\rho}_1; \omega)\psi_n(\boldsymbol{\rho}_2; \omega)}{|\psi_n(\boldsymbol{\rho}_1; \omega)||\psi_n(\boldsymbol{\rho}_2; \omega)|}, \qquad (3.35)$$

is unimodular, i.e., has the absolute value of one.

The eigenvalues $\lambda_n(\omega)$ of the coherent mode decomposition carry important physical meaning: they are the weights of the modes' spectral densities. Indeed, for $\rho_1 = \rho_2 = \rho$, we have

$$S(\rho; \omega) = \sum_n \lambda_n(\omega) |\psi_n(\rho; \omega)|^2. \tag{3.36}$$

Further, the total energy of the optical signal at frequency ω is obtained on integrating the spectral density over the source plane, see Eq. (3.20). It then follows from the expression (3.36) that

$$p_E(\Sigma_\infty; \omega) = \sum_n \lambda_n(\omega), \tag{3.37}$$

where the orthonormality (3.29) of the eigenfunctions was used.

3.1.7 Angular spectrum decomposition

As we have seen in Chapter 1 the efficient test for finding whether or not radiation is beam-like can be derived from its angular spectrum representation. In this section we will show how to extend such representation from monochromatic to stochastic fields and how to employ it in deriving the beam conditions (see also [4]).

We recall that any monochromatic realization $U(x, y, z; \omega)$ in the statistical ensemble of the fluctuating field can be decomposed into the spectrum of plane waves (1.63). On substituting this result into the definition of the cross-spectral density function (3.5) and after changing the order of averaging and integration we find that

$$W(\mathbf{r}_1, \mathbf{r}_2; \omega) = \int_{-\infty}^{\infty} \int_{-\infty}^{\infty} \int_{-\infty}^{\infty} \int_{-\infty}^{\infty} A(k_{1x}, k_{1y}, k_{2x}, k_{2y}; \omega) \exp[i(k_{2z}z_2 - k_{1z}z_1)]$$
$$\times \exp[i(k_{2x}x_2 - k_{1x}x_1 + k_{2y}y_2 - k_{1y}y_1)] dk_{1x} dk_{1y} dk_{2x} dk_{2y}, \tag{3.38}$$

where

$$A(k_{1x}, k_{1y}, k_{2x}, k_{2y}; \omega) = \langle a^*(k_{1x}, k_{1y}; \omega) a(k_{2x}, k_{2y}; \omega) \rangle \tag{3.39}$$

is the *angular correlation function* of the field. Inversion of this formula and evaluation of the result at $z = 0$ leads to the relation

$$A(k_{1x}, k_{1y}, k_{2x}, k_{2y}; \omega) = \frac{1}{(2\pi)^2} \int_{-\infty}^{\infty} \int_{-\infty}^{\infty} \int_{-\infty}^{\infty} \int_{-\infty}^{\infty} W(x_1, y_1, 0, x_2, y_2, 0; \omega)$$
$$\times \exp[-i(k_{2x}x_2 - k_{1x}x_1 + k_{2y}y_2 - k_{1y}y_1)] dx_1 dy_1 dx_2 dy_2. \tag{3.40}$$

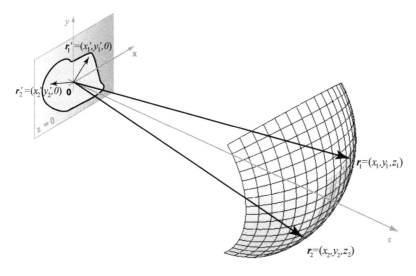

FIGURE 3.2
Notation relating to propagation of stochastic beams.

which represents the four-dimensional Fourier transform with respect to spatial arguments. Similarly to the case of the monochromatic fields, k_{1x}, k_{1y}, k_{2x} and k_{2y} are the spatial frequencies.

In the far zone of the source, for points with position vectors $\mathbf{r}_1 = r_1 \mathbf{u}_1$ and $\mathbf{r}_2 = r_2 \mathbf{u}_2$, where $\mathbf{u}_1$ and $\mathbf{u}_2$ are unit vectors along directions $\mathbf{r}_1$ and $\mathbf{r}_2$, respectively, the expression for the cross-spectral density function can be easily obtained on substituting from Eq. (1.66) into its definition (3.5) and we arrive at the result (see Fig. 3.2)

$$W^{(\infty)}(r_1\mathbf{u}_1, r_2\mathbf{u}_2; \omega) = \frac{(2\pi)^2}{k^2} u_{z1} u_{z2} A(\mathbf{u}_{1\perp}, \mathbf{u}_{2\perp}; \omega) \frac{\exp[ik(r_2 - r_1)]}{r_1 r_2}, \quad (3.41)$$

where $\mathbf{u}_{1\perp}$ and $\mathbf{u}_{2\perp}$ are the two-dimensional projections of vectors $\mathbf{u}_1$ and $\mathbf{u}_2$ onto the source plane $z = 0$.

The *radiant intensity* of the field $J^{(\infty)}(\mathbf{u}; \omega)$ provides the information about the angular distribution of energy. It is defined by the formula [4]

$$J^{(\infty)}(\mathbf{u}; \omega) = \lim_{r \to \infty} [r^2 S^{(\infty)}(r\mathbf{u}; \omega)], \quad (3.42)$$

where $S^{(\infty)}(r\mathbf{u}; \omega) = W^{(\infty)}(r\mathbf{u}, r\mathbf{u}; \omega)$ is the spectral density in the far field. On substituting from Eq. (3.41) into Eq. (3.42) we find that the radiant intensity has the form

$$J^{(\infty)}(r\mathbf{u}; \omega) = \left(\frac{2\pi}{k}\right)^2 A(-\mathbf{u}_\perp, \mathbf{u}_\perp; \omega) u_z^2. \tag{3.43}$$

We will now establish the conditions for the scalar stochastic source under which it generates a beam-like field, following a similar procedure to that for monochromatic fields in Chapter 1. In order to do so we need to assume that the radiant intensity can be appreciable only along directions contained in a region very close to the positive z direction, i.e., when $|\mathbf{u}_\perp|^2 \ll 1$. It is possible to show that this inequality implies that $|\mathbf{u}_{1\perp}|^2 \ll 1$ and $|\mathbf{u}_{2\perp}|^2 \ll 1$ (see the argument in [4], Section 5.6.3). These inequalities imply that the angular correlation function should only contain the components such that

$$|A(-k_x, -k_y, k_x, k_y; \omega)| \approx 0, \quad \text{unless} \quad k_x^2 + k_y^2 \ll k^2. \tag{3.44}$$

Then, we can derive the propagation laws for the stochastic beam-like fields in a manner similar to that for the monochromatic fields. Namely, using the approximations

$$k_{1z} \approx k \left[1 - \frac{1}{2}\left(\frac{k_{1x}^2 + k_{1y}^2}{k^2}\right)\right],$$

$$k_{2z} \approx k \left[1 - \frac{1}{2}\left(\frac{k_{2x}^2 + k_{2y}^2}{k^2}\right)\right]. \tag{3.45}$$

in Eq. (3.38) we arrive at the formula

$$W(\mathbf{r}_1, \mathbf{r}_2; \omega) = \exp[ik(z_2 - z_1)] \int_{-\infty}^{\infty}\int_{-\infty}^{\infty}\int_{-\infty}^{\infty}\int_{-\infty}^{\infty} A(k_{1x}, k_{1y}, k_{2x}, k_{2y}; \omega)$$

$$\times \exp[-i(k_{2x}x_2 - k_{1x}x_1 + k_{2y}y_2 - k_{1y}y_1)]$$

$$\times \exp\left[i\frac{z_1}{2k}(k_{1x}^2 + k_{1y}^2) - i\frac{z_2}{2k}(k_{2x}^2 + k_{2y}^2)\right] dk_{1x}dk_{1y}dk_{2x}dk_{2y}. \tag{3.46}$$

Further, using in Eq. (3.46) the expression for $A(k_{1x}, k_{1y}, k_{2x}, k_{2y}; \omega)$ given by Eq. (3.40) and passing through integration over k_{1x}, k_{1y}, k_{2x}, and k_{2y} one can obtain the free-space propagation law for scalar stochastic beam-like fields:

$$W(\mathbf{r}_1, \mathbf{r}_2; \omega) = \frac{k^2 \exp[ik(z_2 - z_1)]}{(2\pi z)^2} \int_{-\infty}^{\infty}\int_{-\infty}^{\infty}\int_{-\infty}^{\infty}\int_{-\infty}^{\infty} W(x_1', y_1', 0; x_2', y_2', 0; \omega)$$

$$\times \exp\left[\frac{ik}{2z_1}[(x_1 - x_1')^2 + (y_1 - y_1')^2]\right]$$

$$\times \exp\left[\frac{ik}{2z_2}[(x_2 - x_2')^2 + (y_2 - y_2')^2]\right] dx_1'dy_1'dx_2'dy_2'. \tag{3.47}$$

The same propagation formula for the cross-spectral density function of the beam that we have derived using the angular spectrum representation can also be obtained as a solution of the paraxial equation, provided the condition (3.44) is met.

3.2 Mathematical models

Since the cross-spectral density function of the field in the source plane should satisfy several constraints, the class of eligible functions substantially narrows down. Furthermore, even if a certain function can be shown to satisfy all such requirements, it remains a very difficult task to develop analytic expressions for the corresponding coherent modes, angular spectra, paraxial evolution formulas and far-field distributions. We will primarily restrict our attention to the sources for which such derivations have been made.

The model for the cross-spectral density function of planar sources that is most commonly used in the literature is the product of the intensity profile, which can be pertinent to both fully coherent and partially coherent beams and the degree of coherence, which is only the attribute of the random field. The most famous intensity profiles were introduced in Chapter 2. In this section we will be primarily interested in discussing different profiles of the degree of coherence. In particular, will treat in great details several Schell-model sources, including the Gaussian correlated, the J-Bessel correlated and others. We will also illustrate how one can devise novel source correlation functions that lead to desirable far fields, such as those with flat and ring-shaped intensity profiles. Then we will show how one family of sources, known as I-Bessel correlated sources, may be constructed from the coherent modes. Other degrees of freedom that stochastic fields may offer such as the twist phase, anisotropic correlations and non-uniform correlations will also be outlined.

3.2.1 General structure

In the most general setting the cross-spectral density function of the beam in the source plane may be represented by a product

$$W(\boldsymbol{\rho}_1', \boldsymbol{\rho}_2'; \omega) = f[S(\boldsymbol{\rho}_1''; \omega)]g[S(\boldsymbol{\rho}_2''; \omega)]\mu(\boldsymbol{\rho}_1', \boldsymbol{\rho}_2'; \omega), \qquad (3.48)$$

where f and g are some functions, $\boldsymbol{\rho}_1''$ and $\boldsymbol{\rho}_2''$ are linear combinations of $\boldsymbol{\rho}_1'$ and $\boldsymbol{\rho}_2'$, $S(\boldsymbol{\rho}''; \omega)$ is the spectral density at a point in the source plane with position vector $\boldsymbol{\rho}''$ and $\mu(\boldsymbol{\rho}_1', \boldsymbol{\rho}_2'; \omega)$ is the spectral degree of coherence at points $\boldsymbol{\rho}_1'$ and $\boldsymbol{\rho}_2'$.

Two special cases of this form are frequently used. The first is known as the *quasi-homogeneous* model source, for which ($f(x) = x$, $g(x) = 1$, $\boldsymbol{\rho}_1'' =$

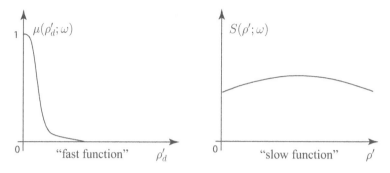

FIGURE 3.3
Illustrating a scalar quasi-homogeneous source.

$(\boldsymbol{\rho}'_1 + \boldsymbol{\rho}'_2)/2$

$$W(\boldsymbol{\rho}'_1, \boldsymbol{\rho}'_2; \omega) \approx S\left(\frac{\boldsymbol{\rho}'_1 + \boldsymbol{\rho}'_2}{2}; \omega\right) \mu(\boldsymbol{\rho}'_2 - \boldsymbol{\rho}'_1; \omega), \qquad (3.49)$$

i.e., for which the spectral density term is approximated by its value at point $\boldsymbol{\rho}''_1 = \boldsymbol{\rho}'_s = (\boldsymbol{\rho}'_1 + \boldsymbol{\rho}'_2)/2$, and the degree of coherence depends on the difference $\boldsymbol{\rho}'_d = \boldsymbol{\rho}'_2 - \boldsymbol{\rho}'_1$ of position-vectors. It is assumed that function μ is much more rapidly varying function of its argument than S (see Fig. 3.3). Scalar quasi-homogeneous sources may be considered as a special case of the electromagnetic quasi-homogeneous sources, which will be discussed in details in Chapter 4.

The other frequently used model for the cross-spectral density function has the structure proposed by Schell [23], [24] ($f(x) = g(x) = \sqrt{x}$, $\boldsymbol{\rho}''_1 = \boldsymbol{\rho}'_1$, $\boldsymbol{\rho}''_2 = \boldsymbol{\rho}'_2$)

$$W(\boldsymbol{\rho}'_1, \boldsymbol{\rho}'_2; \omega) = \sqrt{S(\boldsymbol{\rho}'_1; \omega)}\sqrt{S(\boldsymbol{\rho}'_2; \omega)}\mu(\boldsymbol{\rho}'_2 - \boldsymbol{\rho}'_1; \omega), \qquad (3.50)$$

being the product of three factors: the square roots of the spectral densities S at two spatial arguments $\boldsymbol{\rho}'_1$ and $\boldsymbol{\rho}'_2$; and the degree of coherence μ being a function of the vector difference $\boldsymbol{\rho}'_d = \boldsymbol{\rho}'_2 - \boldsymbol{\rho}'_1$.

For the model (3.49) the following important relation can be rigorously proven [4]:

$$S^{(\infty)}(r\mathbf{u}; \omega) \sim \tilde{\mu}(k\mathbf{u}_\perp; \omega), \qquad (3.51)$$

stating that the distribution of the spectral density in the far field of the source is proportional to the Fourier transform of the spectral degree of coherence. This relation carries an important practical message: in order to control the spectral density distribution of the far field generated by a random source one can select a proper profile of the source degree of coherence. In cases when μ is much narrower than S the same relation can be used for model (3.50). In Chapter 4 we will prove this and another reciprocity relation between the spectral density distribution of the source and the far-field degree of coherence, in the more general form valid for electromagnetic beams.

3.2.2 Gaussian Schell-model sources

Our list of model sources starts with the Gaussian Schell-model source, which acquired the popularity due to its remarkable tractability and generality at the same time. In this model, based on Eq. (3.50) both the spectral density S and the degree of coherence μ are having Gaussian forms:

$$S^{(G)}(\boldsymbol{\rho}';\omega) = A_0^2(\omega) \exp\left[-\frac{\rho'^2}{2\sigma^2(\omega)}\right], \tag{3.52}$$

and

$$\mu^{(G)}(\boldsymbol{\rho}_1',\boldsymbol{\rho}_2';\omega) = \exp\left[-\frac{(\boldsymbol{\rho}_1'-\boldsymbol{\rho}_2')^2}{2\delta^2(\omega)}\right], \tag{3.53}$$

where superscript (G) stands for the Gaussian Schell-model. Hence,

$$W^{(G)}(\boldsymbol{\rho}_1',\boldsymbol{\rho}_2';\omega) = A_0^2(\omega) \exp\left[-\frac{\rho_1'^2 + \rho_2'^2}{4\sigma^2(\omega)}\right] \exp\left[-\frac{(\boldsymbol{\rho}_1'-\boldsymbol{\rho}_2')^2}{2\delta^2(\omega)}\right]. \tag{3.54}$$

Also here A_0^2 is the maximum value of the spectral density (attained on the axis) and the root-mean-square (r.m.s.) widths σ^2 and δ^2 are independent of position but generally depend on frequency. The parameters σ^2 and δ^2 may be associated with characteristic (typical) values of the width of the beam's intensity profile and of the width of the degree of coherence, obtained by ensemble or time averaging. Two limiting cases with respect to the state of coherence are: $\sigma \ll \delta$ fairly coherent beam, $\sigma \gg \delta$ nearly incoherent beam. In Fig. 3.4 the modulus of the spectral degree of coherence $\mu^{(G)}$ is plotted as a function of the separation distance ρ_d' between the source points for three values of parameter δ.

The coherent mode decomposition for the Gaussian Schell-model sources was developed by Gori [25] (see also [26]). Since a two-dimensional Gaussian function can be represented by a product of two one-dimensional Gaussian functions, it is sufficient, in this case, to find the eigenvalues and eigenfunctions for the one-dimensional version of the Fredholm's equation [see Eq. (3.27)], viz. [27]

$$\int\int W^{(G)}(x_1',x_2';\omega)\psi_n^{(G)}(x_1';\omega)dx_1' = \lambda_n^{(G)}(\omega)\psi_n^{(G)}(x_2';\omega), \tag{3.55}$$

It can be verified by direct substitution that the eigenfunctions and the eigenvalues of the Eq. (3.55) are given by the expressions

$$\psi_n^{(G)}(x';\omega) = \sqrt[4]{\frac{2q_3}{\pi}}\frac{1}{\sqrt{2^n n!}}H_n(x'\sqrt{2q_3})\exp(-q_3 x'^2), \tag{3.56}$$

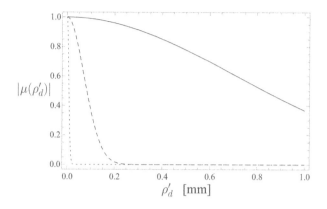

FIGURE 3.4
The degree of coherence of the Gaussian Schell-model source as a function of separation distance $\rho'_d = |\boldsymbol{\rho}'_2 - \boldsymbol{\rho}'_1|$ for several values of parameter δ: $\delta = 0.01$ mm (dotted curve), $\delta = 0.1$ mm (dashed curve) and $\delta = 1$ mm (solid curve).

and

$$\lambda_n^{(G)}(\omega) = A_0 \sqrt{\frac{\pi}{(q_1 + q_2 + q_3)}} \left[\frac{q_2}{q_1 + q_2 + q_3} \right]^n, \qquad (3.57)$$

where $H_n(x)$ are the Hermite polynomials and

$$q_1 = \frac{1}{4\sigma^2}, q_2 = \frac{1}{2\delta^2}, \quad q_3 = \sqrt{q_1^2 + 2q_1q_2}. \qquad (3.58)$$

In the two-dimensional case the cross-spectral density can be represented by the double sum:

$$W^{(G)}(\boldsymbol{\rho}'_1, \boldsymbol{\rho}'_2; \omega) = \sum_m \sum_n \lambda_{mn}^{(G)}(\omega) \psi_{mn}^{(G)*}(\boldsymbol{\rho}'_1; \omega) \psi_{mn}^{(G)}(\boldsymbol{\rho}'_2; \omega), \qquad (3.59)$$

where

$$\lambda_{mn}^{(G)}(\omega) = A_0^2 \left(\frac{\pi}{q_1 + q_2 + q_3} \right) \left(\frac{q_2}{q_1 + q_2 + q_3} \right)^{m+n}, \qquad (3.60)$$

with q_1, q_2 and q_3 defined in Eq. (3.58), and

$$\psi_{mn}^{(G)}(\boldsymbol{\rho}'; \omega) = B_{mn} H_m \left(\sqrt{2q_3} x' \right) H_n \left(\sqrt{2q_3} y' \right) \exp\left[-q_3 \left(x'^2 + y'^2 \right) \right], \qquad (3.61)$$

with

$$B_{mn} = \frac{\sqrt{2q_3}}{\sqrt{\pi 2^{m+n-1} m! n!}}. \qquad (3.62)$$

The distribution of the source eigenvalues is closely related to its coherence properties. The ratio

$$q^{(G)} = \frac{\delta}{\sigma}, \qquad (3.63)$$

sometimes called the *degree of global coherence*, can simply be viewed as the reciprocal of the number of speckle cells within the source area. It was shown to be related to the eigenvalues as

$$\frac{\lambda_{mn}^{(G)}(\omega)}{\lambda_{00}^{(G)}(\omega)} = \left(\frac{1}{q^{(G)2}/2 + 1 + q^{(G)}\sqrt{q^{(G)2}/4 + 1}}\right)^{m+n}. \qquad (3.64)$$

In the limiting case of complete coherence ($q^{(G)} \to \infty$)

$$\frac{\lambda_{mn}^{(G)}(\omega)}{\lambda_{00}^{(G)}(\omega)} = \begin{cases} 1, & m = n = 0, \\ 0, & \text{otherwise.} \end{cases} \qquad (3.65)$$

This implies that only the lowest-order mode is present in this case. In the incoherent limit ($q^{(G)} \to 0$)

$$\frac{\lambda_{mn}^{(G)}(\omega)}{\lambda_{00}^{(G)}(\omega)} = 1, \qquad (3.66)$$

for any m and n, meaning that the infinite number of modes are present and have equal weights.

The angular correlation function of the Gaussian Schell-model beam can be also readily evaluated ([4], Section 5.6.3) using Eqs. (3.40) and (3.54):

$$A^{(G)}(\mathbf{u}_{\perp 1}, \mathbf{u}_{\perp 2}; \omega) = \frac{k^4 A_0^2 \sigma^2}{\pi^2 \left(\frac{4}{\delta^2} + \frac{1}{\sigma^2}\right)}$$

$$\exp\left\{-\frac{k^2}{2}\left[(\mathbf{u}_{\perp 1} - \mathbf{u}_{\perp 2})^2 \sigma^2 + (\mathbf{u}_{\perp 1} + \mathbf{u}_{\perp 2})^2 \frac{1}{\left(\frac{4}{\delta^2} + \frac{1}{\sigma^2}\right)}\right]\right\}. \qquad (3.67)$$

Using the angular spectrum representation (3.67) we can at once derive the conditions that the parameters of the Gaussian Schell-model source should satisfy in order to produce a beam-like field [4]. It turns out that according to Eq. (3.44) it is sufficient for the source to generate a beam if

$$\frac{1}{\delta^2} + \frac{1}{4\sigma^2} \ll \frac{2\pi^2}{\lambda^2}. \qquad (3.68)$$

The cross-spectral density function of a beam generated by a Gaussian Schell-model source and evaluated at points $(\boldsymbol{\rho}_1, z_1)$ and $(\boldsymbol{\rho}_2, z_2)$ can be found

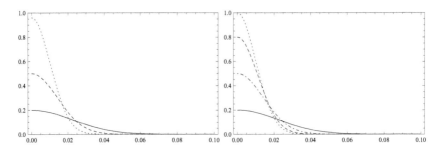

FIGURE 3.5
The spectral density of a Gaussian Schell-model beam with $\sigma = 1$ cm and $\lambda = 628$ μm as a function of ρ [m]. (left) At $z = 100$ m from the source plane for several values of parameter δ: $\delta = 0.5$ mm (dotted curve), $\delta = 0.1$ mm (dashed curve) and $\delta = 0.05$ mm (solid curve). (right) With $\delta = 0.1$ mm for several propagation distances: $z = 0$ m (dotted curve), $z = 50$ m (dashed curve), $z = 100$ m (dot-dashed curve), $z = 150$ m (solid curve).

in the closed form as [28]

$$
W^{(G)}(\boldsymbol{\rho}_1, z_1, \boldsymbol{\rho}_2, z_2; \omega) = \frac{A_0^2}{\Delta^{(G)^2}(z_1, z_2)} \exp\left\{ -\frac{(\boldsymbol{\rho}_1 - \boldsymbol{\rho}_2)^2}{2\Delta^{(G)^2}(z_1, z_2)} \left[\frac{1}{4\sigma^2} + \frac{1}{\delta^2} \right] \right\}
$$

$$
\times \exp\left\{ -\left[\frac{1}{8\sigma^2} + i\frac{z_2 - z_1}{8k\sigma^2} \left(\frac{1}{4\sigma^2} + \frac{1}{\delta^2} \right) \right] \frac{(\boldsymbol{\rho}_1 + \boldsymbol{\rho}_2)^2}{\Delta^{(G)^2}(z_1, z_2)} \right\}
$$

$$
\times \exp\left[-i\frac{k(\rho_1^2 - \rho_2^2)}{2R^{(G)}(z_1, z_2)} \right],
$$

$$(3.69)$$

where

$$
\Delta^{(G)^2}(z_1, z_2) = 1 + \frac{z_1 z_2}{k^2 \sigma^2} \left[\frac{1}{4\sigma^2} + \frac{1}{\delta^2} \right] + i\frac{z_2 - z_1}{k} \left[\frac{1}{2\sigma^2} + \frac{1}{\delta^2} \right], \quad (3.70)
$$

and

$$
R^{(G)}(z_1, z_2) = \sqrt{z_1 z_2} \left[1 + \frac{k^2 \sigma^2}{z_1 z_2} \left(\frac{1}{4\sigma^2} + \frac{1}{\delta^2} \right) \right] \quad (3.71)
$$

are sometimes called the *beam spreading coefficient* and the *generalized curvature parameter*. We omit the derivation here since the similar calculation will be included in Chapter 4 for the electromagnetic version of this model source.

From Eq. (3.69) the expressions for the spectral density and the spectral degree of coherence can be readily deduced. Figure 3.5 illustrates the changes in the spectral density of the Gaussian Schell-model beam as a function of the propagation distance from the source. In Fig. 3.6 the modulus of the transverse

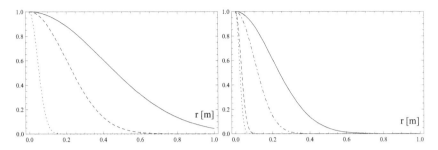

FIGURE 3.6
The modulus of the spectral degree coherence of a Gaussian Schell-model beam with $\lambda = 628$ μm and $\sigma = 1$ cm as a function of the half-distance $\rho_d/2$ [m] between two points $\boldsymbol{\rho}_1 = -\boldsymbol{\rho}_2$. (left) At $z = 100$ m from the source plane for several values of parameter δ: $\delta = 0.1$ mm (dotted curve), $\delta = 0.05$ mm (dashed curve) and $\delta = 0.01$ mm (solid curve). (right) With $\delta = 0.1$ mm for several propagation distances: $z = 0$ m (dotted curve), $z = 10$ m (dashed curve), $z = 50$ m (dot-dashed curve), $z = 100$ m (solid curve).

degree of coherence is shown varying with a function of the separation distance between two positions.

In Fig. 3.7 the fractional power $p_F(\Sigma_d, \Sigma_\infty; \omega)$ of a typical Gaussian Schell-model beam is shown as a function of propagation distance and the detector size [11]. The detector is assumed to be circular, with radius $\bar{\rho}$. It is evident that the decrease in the degree of coherence of the source leads to the substantial loss of power captured by a finite aperture.

The Gaussian Schell-model beam can also be used to illustrate the possible changes in the spectrum. On assuming that the spectral profile of the source is position-independent and is a Gaussian function of wavelength:

$$I_0(\lambda) = A_0^2(\lambda) = \exp\left[-\frac{(\lambda - \lambda_{0I})^2}{2\Lambda_I^2}\right] \tag{3.72}$$

where λ_{0I} is the central wavelength and Λ_I is the r.m.s. spectral width, it is possible to predict the spectral changes on propagation in free space, on substituting from Eq. (3.72) into Eq. (3.69). Figure 3.8 illustrates the fact that typically on paraxial propagation along the optical axis the blue shift occurs.

In addition, parameters σ and δ may also depend on the wavelength, say, as Gaussian functions:

$$\sigma(\lambda) = \sigma_0 \exp\left[-\frac{(\lambda - \lambda_{0\sigma})^2}{2\Lambda_\sigma^2}\right], \quad \delta(\lambda) = \delta_0 \exp\left[-\frac{(\lambda - \lambda_{0\delta})^2}{2\Lambda_\delta^2}\right],$$

causing drastic shifts in the spectral density of the propagating beam.

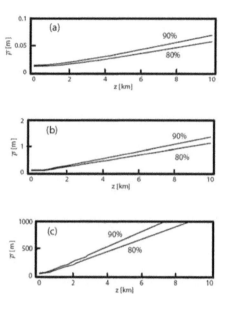

FIGURE 3.7

The contours of the fractional power in transverse cross-sections of a Gaussian Schell-model beam with $\lambda = 0.632$ μ m, $\sigma = 1$ cm. (top) $\delta \gg \sigma$ (almost coherent source); (middle) $\delta = 0.1\sigma$ (partially coherent source); (bottom) $\delta \ll \sigma$ (nearly incoherent source). The beam is captured by a circular aperture with radius $\bar{\rho}$. From Ref. [11].

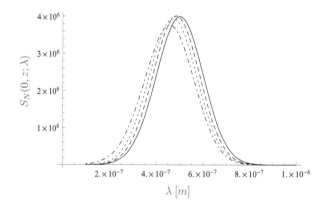

FIGURE 3.8

Spectral shift of a typical Gaussian Schell-model beam with a Gaussian spectral profile propagating in free space; $\lambda_{0I} = 0.5$ μm, $\Lambda_I = 0.1$ μm, $\sigma = 1$ mm, $\delta = 0.1$ mm, $z = 0$ (solid curve), $z = 1$ m (dashed curve), $z = 2$ m (dotted curve) and $z = 100$ m (dash-dotted curve).

Figure 3.9 illustrates the possibility of having spectral changes due to the spectral dependence of the degree of coherence through $\delta(\lambda)$, while $\sigma = const.$ Here two sources are compared with different dependence of the r.m.s. coherence width δ on the wavelength (compare solid and dashed curves on top left figure). The initial spectral profiles for these sources are assumed to be the same (shown on top right figure). From the rest of the figures we see that the two spectra evolve differently.

Several other extensions of the basic scalar Gaussian Schell-model sources have been introduced to include the quadratic wave-front curvature, the twist phase and the anisotropic features relating to the r.m.s width of the beam's intensity and of its correlation function [29]–[35]. In particular, the twist phase factor was first considered by Simon and Mukunda in [31] (see also [32]) theoretically and experimentally demonstrated in [33]. The twist phase, unlike the usual quadratic phase curvature, is absent in a coherent Gaussian beam, is bounded in strength (as the consequence of the non-negative-definiteness of the cross-spectral density matrix) and has an intrinsic chiral property, being responsible for the rotation of the beam spot on propagation. The coherent-mode decomposition of the twisted Gaussian Schell-model beams, the analysis of their transfer of radiance, and the dependence of the orbital angular momentum of a partially coherent beam on its twist phase belong to Refs. [32], [35].

The cross-spectral density function of the most general Gaussian-Schell model source that incudes the aforementioned properties can be at best expressed with the help of the 2×2 matrices, viz. [36]:

$$W^{(G)}(\boldsymbol{\rho}_1', \boldsymbol{\rho}_2'; \omega) = A_0^2 \exp\left\{-\frac{1}{4}[\boldsymbol{\rho}_1'^T(\boldsymbol{\sigma}^2)^{-1}\boldsymbol{\rho}_1' + \boldsymbol{\rho}_2'^T(\boldsymbol{\sigma}^2)^{-1}\boldsymbol{\rho}_2']\right\}$$
$$\times \exp\left\{-\frac{1}{2}(\boldsymbol{\rho}_1' - \boldsymbol{\rho}_2')^T(\boldsymbol{\delta}^2)^{-1}(\boldsymbol{\rho}_1' - \boldsymbol{\rho}_2')\right\} \quad (3.73)$$
$$\times \exp\left\{-\frac{ik}{2}(\boldsymbol{\rho}_1' - \boldsymbol{\rho}_2')^T(\mathbf{R}^{-1} + \tau_\delta \mathbf{J})(\boldsymbol{\rho}_1' + \boldsymbol{\rho}_2')\right\},$$

where vectors $\boldsymbol{\rho}_1'$ and $\boldsymbol{\rho}_2'$ specify two positions in the source plane, superscript T stands for matrix transposition and $\mathbf{J}$ is the transpose antisymmetric matrix,

$$\mathbf{J} = \begin{bmatrix} 0 & 1 \\ -1 & 0 \end{bmatrix}. \quad (3.74)$$

Further, $\boldsymbol{\sigma}^2$, $\boldsymbol{\delta}^2$, $\mathbf{R}^{-1}$ are the transverse spot width matrix, the transverse correlation width matrix and the wave-front curvature matrix, respectively. They are defined as:

$$(\boldsymbol{\sigma}^2)^{-1} = \begin{bmatrix} \sigma_{11}^{-2} & \sigma_{12}^{-2} \\ \sigma_{12}^{-2} & \sigma_{22}^{-2} \end{bmatrix}, \quad (3.75)$$

$$(\boldsymbol{\delta}^2)^{-1} = \begin{bmatrix} \delta_{11}^{-2} & \delta_{12}^{-2} \\ \delta_{12}^{-2} & \delta_{22}^{-2} \end{bmatrix}, \quad (3.76)$$

86

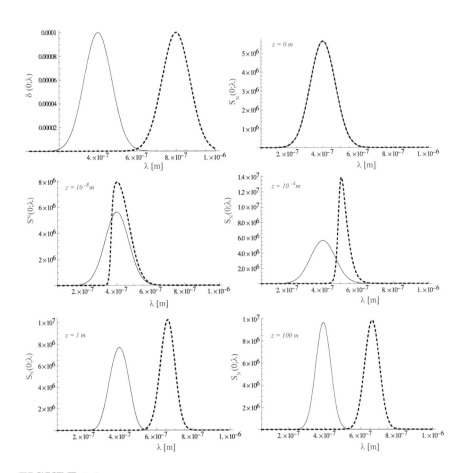

FIGURE 3.9
Spectral shifts of two Gaussian Schell-model beams with different r.m.s. widths
of correlation coefficients, $\delta(\lambda)$, with a Gaussian spectral profile propagating
in free space. Left top figure: $\delta(\lambda)$: $\lambda_{0\delta} = 0.4$ μm (solid curve); $\lambda_{0\delta} = 0.8$ μm
(dashed curve). The rest of the figures: evolution of beams with two $\delta(\lambda)$ in
left top figure in space. The other parameters of two beams are $\lambda_{0I} = 0.4$ μm,
$\Lambda_I = \Lambda_\delta = \lambda_{0I}/4$, $\sigma_0 = 1$ cm, $\delta_0 = 1$ mm.

and

$$\mathbf{R}^{-1} = \begin{bmatrix} R_{11}^{-1} & R_{12}^{-1} \\ R_{12}^{-1} & R_{22}^{-1} \end{bmatrix}, \tag{3.77}$$

all possessing transpose symmetry. The factor τ_δ is a scalar real-valued *twist phase* with the dimension of an inverse distance, limited by the double inequality [31]

$$0 \le \tau_\delta^2 \le \frac{1}{k^2 \mathrm{Det}\boldsymbol{\delta}^2}. \tag{3.78}$$

Written explicitly the twist term takes the form

$$(\boldsymbol{\rho}_1' - \boldsymbol{\rho}_2')^T \mathbf{J}(\boldsymbol{\rho}_1' + \boldsymbol{\rho}_2') = x_1' y_2' - x_2' y_1', \tag{3.79}$$

implying that the two-dimensional cross-spectral density function (3.73) cannot be split into a product of two one-dimensional cross-spectral-density functions, as in the case of the basic Gaussian Schell-model source.

3.2.3 J_0-Bessel correlated sources

The other mathematical model for the cross-spectral density function of a planar Schell-type source is the J_0-Bessel-correlated source introduced by Gori et al. [37]. The analysis of the fields produced by such sources in free-space and the beam conditions can be found in Refs. [38] and [39], respectively. A J_0-Bessel-correlated source has the degree of coherence of the form

$$\mu^{(J)}(\boldsymbol{\rho}_1', \boldsymbol{\rho}_2'; \omega) = J_0(\beta(\omega)|\boldsymbol{\rho}_1' - \boldsymbol{\rho}_2'|), \tag{3.80}$$

J_0 is the 0-th order Bessel function of the first kind and $\beta(\omega)$ is an arbitrary function of angular frequency. We note that $\mu^{(J)}$ depends only on the magnitude of the difference vector between the two position vectors and is independent from its direction. In Fig. 3.10 we plot $\mu^{(J)}$ as a function of $\rho_d' = |\boldsymbol{\rho}_1' - \boldsymbol{\rho}_2'|$ for several values of parameter β.

Also, in the polar coordinates for two-dimensional vectors $\boldsymbol{\rho}_1 = (\rho_1', \phi_1')$ and $\boldsymbol{\rho}_2 = (\rho_2', \phi_2')$ we have:

$$|\boldsymbol{\rho}_1' - \boldsymbol{\rho}_2'| = \sqrt{\rho_1'^2 + \rho_2'^2 - 2\rho_1'\rho_2' \cos(\phi_1' - \phi_2')}. \tag{3.81}$$

Hence, generally, the cross-spectral density function of the J_0-Bessel-correlated source is given by the expression

$$W^{(J)}(\boldsymbol{\rho}_1', \boldsymbol{\rho}_2'; \omega) = \sqrt{S(\boldsymbol{\rho}_1'; \omega)}\sqrt{S(\boldsymbol{\rho}_2'; \omega)} J_0(\beta(\omega)|\boldsymbol{\rho}_1' - \boldsymbol{\rho}_2'|), \tag{3.82}$$

where spectral density $S(\boldsymbol{\rho}'; \omega)$ is assumed to be a circularly symmetric function of $\boldsymbol{\rho}'$. Just like the Gaussian Schell-model sources J_0-correlated sources also have the coherent mode decomposition in the closed form [37]. Indeed, recall that the general form of the coherent mode expansion is the infinite sum

$$W^{(J)}(\boldsymbol{\rho}_1', \boldsymbol{\rho}_2'; \omega) = \sum_{n=0}^{\infty} \lambda_n^{(J)}(\omega)\psi_n^{*(J)}(\boldsymbol{\rho}_1'; \omega)\psi_n^{(J)}(\boldsymbol{\rho}_2'; \omega), \tag{3.83}$$

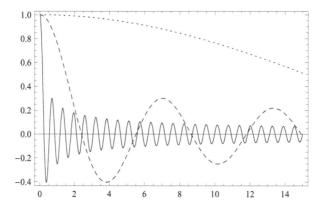

FIGURE 3.10
The spectral degree of coherence of the J_0-Bessel correlated source μm as a function of separation distance $\rho'_d = |\boldsymbol{\rho}'_1 - \boldsymbol{\rho}'_2|$ [mm] for several values of parameter β: $\beta = 0.1$ (dotted curve), $\beta = 1$ (dashed curve) and $\beta = 10$ (solid curve).

where $\lambda_n^{(J)}(\omega)$ are the eigenvalues and $\psi_n^{(J)}(\boldsymbol{\rho}'_1;\omega)$ are the eigenfunctions of the integral equation

$$\int W^{(J)}(\boldsymbol{\rho}'_1, \boldsymbol{\rho}'_2; \omega)\psi_n^{(J)}(\boldsymbol{\rho}'_1;\omega)d^2\rho'_1 = \lambda_n^{(J)}(\omega)\psi_n^{(J)}(\boldsymbol{\rho}'_2;\omega). \qquad (3.84)$$

Here $n = 0, 1, 2, ...$ and the integration is performed over the source plane. On substituting from Eq. (3.82) into Eq. (3.84) we find that, in the polar coordinate system,

$$\sqrt{S(\boldsymbol{\rho}'_2;\omega)}\int\limits_0^\infty\int\limits_0^{2\pi}\sqrt{S(\boldsymbol{\rho}'_1;\omega)}J_0\left\{\beta(\omega)\sqrt{\rho'^2_1 + \rho'^2_2 - 2\rho_1\rho_2\cos(\phi'_1 - \phi'_2)}\right\}$$

$$\times\,\psi_n^{(J)}(\rho'_1, \phi'_1;\omega)\rho'_1 d\rho'_1 d\theta'_1 = \lambda_n^{(J)}(\omega)\psi_n^{(J)}(\rho'_2, \phi'_2;\omega). \qquad (3.85)$$

Using Neumann's addition theorem ([40], p. 358) we find that

$$J_0[\beta(\omega)\sqrt{\rho'^2_1 + \rho'^2_2 - 2\rho'_1\rho'_2\cos(\phi'_1 - \phi'_2)}]$$

$$= \sum_{m=-\infty}^{\infty} J_m(\beta(\omega)\rho'_1)J_m(\beta(\omega)\rho'_2)\exp[im(\phi'_1 - \phi'_2)], \qquad (3.86)$$

where J_m is the Bessel function of order m. Let us now prove that the eigenfunctions $\psi_n^{(J)}(\rho', \phi'; \omega)$ are of the form

$$
\psi_n^{(J)}(\rho', \phi'; \omega) = c_n(\omega)\sqrt{S(\rho'; \omega)}\Big[a_n(\omega)J_n(\beta(\omega)\rho')\exp(-in\phi')
$$
$$
+ b_n(\omega)J_{-n}(\beta(\omega)\rho')\exp(in\phi')\Big],
$$
(3.87)

where ratios a_n/b_n are arbitrary and coefficients c_n are to be found. Indeed, on substituting from Eqs. (3.86) and (3.87) into Eq. (3.85) we find that

$$
c_n(\omega)\sqrt{S(\rho_2'; \omega)} \sum_{m=-\infty}^{\infty} J_m(\beta(\omega)\rho_2')\exp(-im\phi_2') \int_0^\infty \int_0^{2\pi} S(\rho_1'; \omega)J_m[\beta(\omega)\rho_1']
$$
$$
\times [a_n(\omega)J_n[\beta(\omega)\rho_1']\exp(-in\phi_1') + b_n(\omega)J_{-n}[\beta(\omega)\rho_1']\exp(in\phi_1')]
$$
$$
\times \exp(im\phi_1')\rho_1' d\rho_1' d\phi_1'
$$
$$
= \lambda_n^{(J)}(\omega)c_n(\omega)[a_n(\omega)\sqrt{S(\rho_2'; \omega)}J_n[\beta(\omega)\rho_2']\exp(-in\phi_2')
$$
$$
+ b_n(\omega)\sqrt{S(\rho_2'; \omega)}J_{-n}[\beta(\omega)\rho_2']\exp(in\phi_2')], \quad (n = 0, 1, 2, ...).
$$
(3.88)

The equation above is the identity because integrating over variable ϕ_1' we find that

$$
\int_0^{2\pi} \exp[i(m \mp n)\phi_1']d\phi_1 = \begin{cases} 2\pi, & m = \pm n, \\ 0, & m \neq \pm n, \end{cases}
$$
(3.89)

where the upper and the lower signs must be taken simultaneously. Hence, only the terms with $m = \pm n$ survive on the left side of Eq. (3.88) and the identity is proven.

As for the eigenvalues, they can be found on the basis of the fact that functions J_n have the same parity as their index, and we have

$$
\lambda_n^{(J)}(\omega) = 2\pi \int_0^\infty S(\rho'; \omega)J_n^2[\beta(\omega)\rho']\rho' d\rho', \quad (n = 0, 1, 2, ...).
$$
(3.90)

Further, factor $c_n(\omega)$ can be found from the fact that the eigenfunctions are orthonormal, i.e.,

$$
\int |\psi_n(\rho', \phi'; \omega)|^2 d^2\rho' = 1, \quad (n = 0, 1, 2, ...).
$$
(3.91)

It then follows from Eqs. (3.87), (3.90) and (3.91) that

$$
c_n(\omega) = \lambda^{-1/2}(\omega),
$$
(3.92)

i.e., can be determined if the spectral density of the source is specified.

For instance, when the spectral density is a Gaussian function with unit amplitude, i.e.,

$$S(\rho'; \omega) = \exp\left[-\frac{\rho'^2}{w_0^2}\right] \tag{3.93}$$

the eigenvalues take the form [37]

$$\lambda_n^{(J)}(\omega) = \frac{\pi w_0^2}{2} \exp\left[-\frac{\beta^2(\omega)w_0^2}{4}\right] I_n\left[\frac{\beta^2(\omega)w_0^2}{4}\right]. \tag{3.94}$$

The beam condition for the J_0-Bessel correlated source was recently derived in [39] and has the form of inequality

$$\frac{1}{w_0^2 + w_0^4\beta^2(\omega)} \ll k^2. \tag{3.95}$$

We will now derive the formula for the cross-spectral density function of a paraxial beam generated by a J_0-Bessel-correlated source with the Gaussian spectral density. Since the coherent modes for this source are Bessel-Gaussian beams, as was shown above, it will be sufficient to propagate the modes independently and then to sum the obtained results up. For this purpose, without loss of generality, we will use the following equation for the source modes:

$$\psi_n^{(J)}(\rho', \phi', 0; \omega) = A_0 \exp\left(-\frac{\rho'^2}{w_0^2}\right) J_n\left[\beta(\omega)\rho'\right] \exp(-in\phi'). \tag{3.96}$$

Using the Huygens-Fresnel diffraction integral we find that at distance z from the source the modes are given by the expression

$$\psi_n^{(J)}(\rho, \phi, z; \omega) = -\frac{ik}{2\pi z} \exp\left[i\left(kz + \frac{k\rho^2}{2z}\right)\right] \int_0^\infty \int_0^{2\pi} \psi_n^{(J)}(\rho', \phi', 0; \omega)$$

$$\times \exp\left\{i\frac{k}{2z}[\rho'^2 - 2\rho\rho'\cos(\phi - \phi')]\right\} \rho' d\rho' d\phi'$$

$$= -\frac{ikA_0}{2\pi z} \exp\left[i\left(kz + \frac{k\rho^2}{2z}\right)\right] \tag{3.97}$$

$$\times \int_0^\infty \int_0^{2\pi} J_n[\beta(\omega)\rho'] \exp\left[-\left(\frac{1}{w_0^2} - \frac{ik}{2z}\right)\rho'^2\right]$$

$$\times \exp\left\{-i\left[n\phi' + \frac{k\rho\rho'}{z}\cos(\phi - \phi')\right]\right\} \rho' d\rho' d\phi'.$$

Since

$$\int_0^{2\pi} \exp\left\{-i\left[n\phi' + \frac{k\rho\rho'}{z}\cos(\phi - \phi')\right]\right\} = (-i)^n 2\pi J_n\left(\frac{k\rho\rho'}{z}\right) \exp(-in\phi'),$$

$$\tag{3.98}$$

Equation (3.97) becomes

$$\psi_n^{(J)}(\rho,\phi,z;\omega) = -\frac{ikA_0}{z}\exp\left[ik\left(z+\frac{\rho^2}{2z}\right)\right](-i)^n\exp(-in\phi')$$

$$\times\int_0^\infty\exp\left[-\left(\frac{1}{w_0^2}-\frac{ik}{2z}\right)\rho'^2\right]J_n[\beta(\omega)\rho']J_n\left(\frac{k\rho\rho'}{z}\right)\rho'd\rho'.$$

$$(3.99)$$

This integral is tabulated as [41]:

$$\int_0^\infty\exp\left(-q_1x\right)J_n(2q_2\sqrt{x})J_n(2q_3\sqrt{x})dx = \frac{1}{q_1}\exp\left(-\frac{q_2^2+q_3^2}{q_1}\right)I_n\left(\frac{2q_2q_3}{q_1}\right),$$

$$(3.100)$$

where I_n is the modified Bessel function of the first kind and n-th order, for which the following relation holds

$$I_n(x) = (-i)^n J_n(ix). \tag{3.101}$$

Employing these results we finally obtain the formula

$$\psi_n^{(J)}(\rho,\phi,z;\omega) = A_0\frac{w_0}{w(z)}\exp\left\{i\left[\left(k-\frac{\beta^2(\omega)}{2k}\right)z-\psi_G(z)\right]\right\}$$

$$\times J_n\left\{\frac{w_0}{w(z)}\beta(\omega)\rho\exp[-i\psi_G(z)]\right\}$$

$$\times\exp\left[\left(-\frac{1}{w^2(z)}+\frac{ik}{2R(z)}\right)\left(\rho^2+\frac{\beta^2(\omega)z^2}{k^2}\right)\right]\exp[-in\phi],$$

$$(3.102)$$

where

$$w(z) = w_0\sqrt{1+(z/z_R)^2} \tag{3.103}$$

$$R(z) = z+z_R^2/z, \tag{3.104}$$

$$\psi_G(z) = \arctan(z/z_R), \tag{3.105}$$

$z_R = kw_0^2/2$ being the Rayleigh distance, as before. These formulas are the same as in the case of a deterministic Gaussian beam (see Chapter 2).

In order to determine the cross-spectral density of the propagating beam at distance z from the source, it suffices to substitute the propagating modes (3.102) into the coherent mode decomposition of the form

$$W^{(J)}(\mathbf{r}_1,\mathbf{r}_2;\omega) = \sum_{n=0}^\infty\lambda_n^{(J)}(\omega)\psi_n^{(J)*}(\mathbf{r}_1;\omega)\psi_n^{(J)}(\mathbf{r}_2;\omega). \tag{3.106}$$

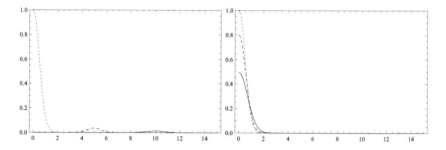

FIGURE 3.11

Spectral density profiles of the J_0-correlated Schell-model beam as a function of the normalized radius ρ/w_0 at several normalized distances: $z/z_R = 0$ (dotted curve), $z/z_R = 0.5$ (dashed curve), $z/z_R = 1$ (solid curve). (left) $\beta w_0/2 = 10$; (right) $\beta w_0/2 = 0.1$.

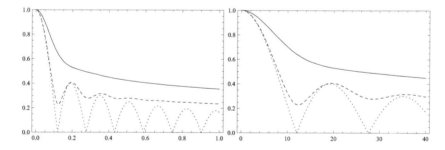

FIGURE 3.12

The modulus of the spectral degree of coherence of the J_0-correlated Schell-model beam as a function of the separation distance ρ'_d at several normalized distances: at several normalized distances: $z/z_R = 0$ (dotted curve), $z/z_R = 0.2$ (dashed curve), $z/z_R = 1$ (solid curve). (left) $\beta w_0/2 = 10$; (right) $\beta w_0/2 = 0.1$.

Hence, we obtain the following representation

$$W^{(J)}(\mathbf{r}_1, \mathbf{r}_2; \omega) = \frac{A_0^2 w_0^2}{w^2(z)} \exp\left[\left(-\frac{1}{w^2(z)} - \frac{ik}{2R(z)}\right)\left(\rho_1^2 + \frac{\beta^2(\omega)z^2}{k^2}\right)\right]$$

$$\times \exp\left[\left(-\frac{1}{w^2(z)} + \frac{ik}{2R(z)}\right)\left(\rho_2^2 + \frac{\beta^2(\omega)z^2}{k^2}\right)\right]$$

$$\times \sum_{n=-\infty}^{\infty} J_n\left\{\frac{w_0\beta(\omega)\rho_1}{w(z)}\exp[i\psi_G(z)]\right\} J_n\left\{\frac{w_0\beta(\omega)\rho_2}{w(z)}\exp[-i\psi_G(z)]\right\}$$

$$\times \exp[in(\phi_1 - \phi_2)].$$

$$(3.107)$$

In the same manner as in the source plane we find, on the basis of Neumann's theorem [40], that in the closed form the expression (3.107) can be written as

$$W^{(J)}(\mathbf{r}_1, \mathbf{r}_2; \omega)$$

$$= A_0^2 \frac{w_0^2}{w^2(z)} \exp\left[-\frac{\rho_1^2 + \rho_2^2}{w^2(z)} + \frac{ik}{2R(z)}(\rho_2^2 - \rho_1^2)\right] \exp\left[-\frac{2\beta^2 z^2}{k^2 w^2(z)}\right]$$

$$\times J_0\left[\frac{w_0\beta(\omega)}{w(z)}\sqrt{\rho_1^2 \exp[2i\psi_G(z)] + \rho_2^2 \exp[-2i\psi_G(z)] - 2\rho_1\rho_2\cos(\phi_1 - \phi_2)}\right],$$

$$(3.108)$$

$\psi_G(z)$ being defined in Eq. (3.105). Since the cross-spectral density depends on the angular difference $\phi_1 - \phi_2$ the beam preserves its circular symmetry on propagation.

In Figs. 3.11 and 3.12 the evolution of the spectral density and the spectral degree of coherence of the J_0-Bessel correlated beam is plotted as a function of the normalized radius ρ/w_0.

3.2.4 Multi-Gaussian correlated sources

In this section we introduce a generalization of the Gaussian Schell-model sources whose spectral degree of coherence is modeled by the multi-Gaussian family of functions. Such functions have previously been employed for modeling of optical fields, apertures [42] and scattering potentials [43]. Generally, the multi-Gaussian functions make it possible to control the width of the flat center of the profile and the slope of its edge by the choice of two parameters. But being employed for modeling of the correlation, rather than the field, the multi-Gaussian functions can serve as an important tool for generating far fields with flat intensity profiles.

Let us set the spectral degree of coherence $\mu(\boldsymbol{\rho}_1', \boldsymbol{\rho}_2'; \omega)$, at a pair of points in the source plane with position vectors $\boldsymbol{\rho}_1'$ and $\boldsymbol{\rho}_2'$ and at frequency ω in the

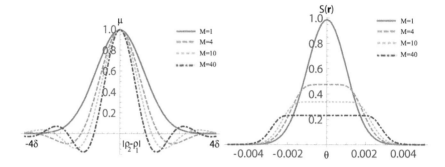

FIGURE 3.13

(left) The degree of coherence $\mu^{(MG)}$ for several values of M; (right) Far-field spectral density of the field generated by the multi-Gaussian Schell-model source vs. $\theta = \arcsin(u_\perp)$ (radians) for several values of M. From Ref. [44].

following form [44]:

$$\mu^{(MG)}(\boldsymbol{\rho}_1', \boldsymbol{\rho}_2'; \omega)$$

$$= \frac{1}{C_0} \sum_{m=1}^{M} \binom{M}{m} \frac{(-1)^{m-1}}{m} \exp\left[-\frac{|\boldsymbol{\rho}_2' - \boldsymbol{\rho}_1'|^2}{2m\delta^2}\right], \qquad (3.109)$$

where $C_0 = \sum_{m=1}^{M} \frac{(-1)^{m-1}}{m}\binom{M}{m}$ is the normalization factor, $\binom{M}{m}$ stands for binomial coefficients and δ is the r.m.s. correlation width of the term with $m = 1$. As is illustrated in Fig. 3.13 (left), the profile function defined by Eq. (3.109) visually resembles a Bessel-correlated source or a Lambertian source (see [4], Figs. 5.9 and 5.10), even though it has a different functional form. As in the case of a Bessel function whose Taylor's expansion is a sum of sign-alternating terms, the multi-Gaussian function is also represented by a sum of positive and negative exponentials.

Since not any form of the spectral degree of coherence is suitable for defining a physically realizable random source [8], one must first establish that the sum (3.109) is such. Recall that in order for the cross-spectral density function to be physically realizable, or genuine, the integral representation (3.11) must hold for some function H and non-negative, Fourier-transformable function p. Assume that function $H(\boldsymbol{\rho}, \mathbf{s})$ has the form

$$H(\boldsymbol{\rho}, \mathbf{s}) = \tau(\boldsymbol{\rho}) \exp[-i\mathbf{s} \cdot \boldsymbol{\rho}], \qquad (3.110)$$

leading to the cross-spectral density of the form

$$W(\boldsymbol{\rho}_1, \boldsymbol{\rho}_2; \omega) = \tau^*(\boldsymbol{\rho}_1; \omega)\tau(\boldsymbol{\rho}_2; \omega)\tilde{p}[(\boldsymbol{\rho}_1 - \boldsymbol{\rho}_2); \omega], \qquad (3.111)$$

where $\tau(\rho; \omega)$ is a (possibly complex) profile intensity function, and the tilde denotes the Fourier transform.

The form of function $p(\mathbf{s})$ determines a family of sources with different correlation functions. The Fourier transform of Eq. (3.109) results in the expression:

$$p(\mathbf{s}; \omega) = \frac{\delta^2}{C_0} \sum_{m=1}^{M} (-1)^{m-1} \binom{M}{m} \exp\left[-\frac{m\delta^2 |\mathbf{s}|^2}{2}\right], \qquad (3.112)$$

representing a family of flat-top profiles. It has the Fourier transform because it is a finite sum of Gaussian functions. Also its non-negativity has been proven in Section 2.3.1 where the deterministic multi-Gaussian sources were discussed.

We will also chose the Gaussian profile for function τ:

$$\tau(\rho'; \omega) = \exp\left[-\frac{\rho'^2}{4\sigma^2}\right]. \qquad (3.113)$$

Then, together with the weighting function $p(s)$ given by Eq. (3.112), one obtains, on substituting them into Eq. (3.111), the cross-spectral density function of the form

$$\begin{aligned} W^{(MG)}(\rho'_1, \rho'_2; \omega) &= \frac{1}{C_0} \exp\left[-\frac{\rho'^2_1 + \rho'^2_2}{4\sigma^2}\right] \\ &\times \sum_{m=1}^{M} \frac{(-1)^{m-1}}{m} \binom{M}{m} \exp\left[-\frac{|\rho'_2 - \rho'_1|^2}{2m\delta^2}\right], \end{aligned} \qquad (3.114)$$

which has been termed the *multi-Gaussian Schell-model source* [44].

We will now derive the cross-spectral density function of the far field radiated by the source (3.114), at two points specified by position vectors $\mathbf{r}_1 = r_1 \mathbf{u}_1$ and $\mathbf{r}_2 = r_2 \mathbf{u}_2$, with $\mathbf{u}_1^2 = \mathbf{u}_2^2 = 1$. On substituting from Eq. (3.114) into Eq. (3.41), one finds that the cross-spectral density $W_{(\infty)}^{(MG)}(r_1 \mathbf{u}_1, r_2 \mathbf{u}_2; \omega)$ in the far field becomes

$$\begin{aligned} W_{(\infty)}^{(MG)}(\mathbf{r}_1, \mathbf{r}_2; \omega) &= \frac{1}{C_0} k^2 \cos\theta_1 \cos\theta_2 \frac{\exp[ik(r_2 - r_1)]}{r_1 r_2} \\ &\times \sum_{m=1}^{M} \frac{(-1)^{m-1}}{m} \binom{M}{m} \frac{1}{(a_m^2 - b_m^2)} \\ &\times \exp[-k^2(\alpha_m \mathbf{u}_{1\perp}^2 + \alpha_m \mathbf{u}_{2\perp}^2 - 2\beta_m \mathbf{u}_{1\perp} \cdot \mathbf{u}_{2\perp})], \end{aligned} \qquad (3.115)$$

where $\cos\theta_1 = u_{z1}$, $\cos\theta_2 = u_{z2}$, and

$$a_m = \frac{1}{2}\left(\frac{1}{2\sigma^2} + \frac{1}{m\delta^2}\right), \qquad b_m = \frac{1}{2m\delta^2}, \qquad (3.116)$$

$$\alpha_m = \frac{a_m}{4(a_m^2 - b_m^2)}, \qquad \beta_m = \frac{b_m}{4(a_m^2 - b_m^2)}. \qquad (3.117)$$

The spectral density in the far zone can be now found by the formula $S^{(MG)}_{(\infty)}(\mathbf{r};\omega) = W^{(MG)}_{(\infty)}(\mathbf{r},\mathbf{r};\omega)$:

$$
\begin{aligned}
S^{(MG)}_{(\infty)}(\mathbf{r};\omega) = &\frac{k^2 \cos^2\theta}{C_0 r^2} \\
&\times \sum_{m=1}^{M} \frac{(-1)^{m-1}}{m} \binom{M}{m} \frac{\exp[-2k^2 \mathbf{u}^2_\perp (\alpha_m - \beta_m)]}{(a_m^2 - b_m^2)}.
\end{aligned}
\tag{3.118}
$$

In order for the source (3.114) to generate a beam, the spectral density in Eq. (3.118) must be negligible except for directions within a narrow solid angle about the z-axis. This is so if

$$
\exp[-2k^2 \mathbf{u}^2_\perp \theta^2 (\alpha_m - \beta_m)] \approx 0,
\tag{3.119}
$$

for any $m = 1, ..., M$, unless $|\mathbf{u}_\perp|^2 \ll 1$, implying that

$$
2k^2 (\alpha_m - \beta_m) \ll 1,
\tag{3.120}
$$

or, in terms of the source parameters,

$$
\frac{1}{4\sigma^2} + \frac{1}{m}\frac{1}{\delta^2} \ll \frac{2\pi^2}{\lambda^2}, \quad m = 1, ..., M.
\tag{3.121}
$$

If this inequality is valid for $m = 1$, it is also so $m = 2, ..., M$. It implies that the beam condition for the multi-Gaussian Schell-model sources is the same as that for the Gaussian Schell-model sources ([4], Eq. 5.6-73):

$$
\frac{1}{4\sigma^2} + \frac{1}{\delta^2} \ll \frac{2\pi^2}{\lambda^2}.
\tag{3.122}
$$

Figure 3.13 (right) shows several typical far fields radiated by the source (3.114) with the following parameters: $\lambda = 632$ nm, $\sigma = 1$ mm, $\delta = 0.1$ mm. It is clearly seen that a beam with the Gaussian degree of coherence in the source plane decreases with the increase of angle $\theta = \arcsin(u_\perp)$, while the beams with multi-Gaussian source correlations have flat profiles with different heights and sharpness of the edges.

Let us now investigate the behavior of both the spectral density and the degree of coherence for these beams at arbitrary distance z from the source. Note that Eq. (3.69) implies that the "m"-th term in the sum (3.114) can be treated in the same manner if one sets:

$$
\delta_m = \sqrt{m}\delta.
\tag{3.123}
$$

The sum of all M such terms results in the expression

$$W^{(MG)}(\boldsymbol{\rho}_1, z_1, \boldsymbol{\rho}_2, z_2; \omega) = \frac{A_0^2}{C_0} \sum_{m=1}^{M} \binom{M}{m} \frac{(-1)^{m-1}}{m} \frac{1}{\Delta_m^{(MG)^2}(z_1, z_2)}$$

$$\times \exp\left\{-\frac{(\boldsymbol{\rho}_1 - \boldsymbol{\rho}_2)^2}{2\Delta_m^{(MG)^2}(z_1, z_2)}\left[\frac{1}{4\sigma^2} + \frac{1}{\delta_m^2}\right]\right\}$$

$$\times \exp\left\{-\left[\frac{1}{8\sigma^2} + i\frac{z_2 - z_1}{8k\sigma^2}\left(\frac{1}{4\sigma^2} + \frac{1}{\delta_m^2}\right)\right]\frac{(\boldsymbol{\rho}_1 + \boldsymbol{\rho}_2)^2}{\Delta_m^{(MG)^2}(z_1, z_2)}\right\}$$

$$\times \exp\left[-i\frac{k(\rho_1^2 - \rho_2^2)}{2R_m^{(MG)}(z_1, z_2)}\right],$$

$$(3.124)$$

where

$$\Delta_m^{(MG)^2}(z_1, z_2) = 1 + \frac{z_1 z_2}{k^2\sigma^2}\left[\frac{1}{4\sigma^2} + \frac{1}{\delta_m^2}\right] + i\frac{z_2 - z_1}{k}\left[\frac{1}{2\sigma^2} + \frac{1}{\delta_m^2}\right], \quad (3.125)$$

and

$$R_m^{(MG)}(z_1, z_2) = \sqrt{z_1 z_2}\left[1 + \frac{k^2\sigma^2}{z_1 z_2}\left(\frac{1}{4\sigma^2} + \frac{1}{\delta_m^2}\right)\right]. \quad (3.126)$$

Hence the spectral density at any point $(\boldsymbol{\rho}, z)$ within the cross-section of the beam assumes the form:

$$S^{(MG)}(\boldsymbol{\rho}, z; \omega) = W^{(MG)}(\boldsymbol{\rho}, \boldsymbol{\rho}, z; \omega)$$

$$= \frac{A_0^2}{C_0} \sum_{m=1}^{M} \binom{M}{m} \frac{(-1)^{m-1}}{m\Delta_m^{(MG)^2}(z)} \exp\left[-\frac{\rho^2}{2\sigma^2\Delta_m^{(MG)^2}(z)}\right]. \quad (3.127)$$

Figure 3.14 shows the contours of the transverse cross-sections of a multi-Gaussian Schell-model beam at several fixed distances from the source and for several values of summation index M. While at sufficiently small distances from the source all curves preserve Gaussian shape [Fig. 3.14 (a)], as the propagation distance grows the transverse intensity profiles begin to depend on M. Namely, the larger the value of M is, the smaller its height at the beam axis and the flatter the profile becomes [Fig. 3.14 (b)]. For sufficiently large propagation distances, in the far zone of the source [see Fig. 3.14 (c)], all contours with $M > 1$ take on shapes profiles with flat plateaus around the optical axis. Such flat regions are structurally preserved but grow in width as the distance from the source is increased even further [compare Figs. 3.14 (c) and 3.14 (d)].

In the case when the points are located symmetrically about the optical axis, i.e. if $\boldsymbol{\rho}_d = 2\boldsymbol{\rho}_1 = -2\boldsymbol{\rho}_2$, $z_1 = z_2 = z$ the modulus of the transverse

98

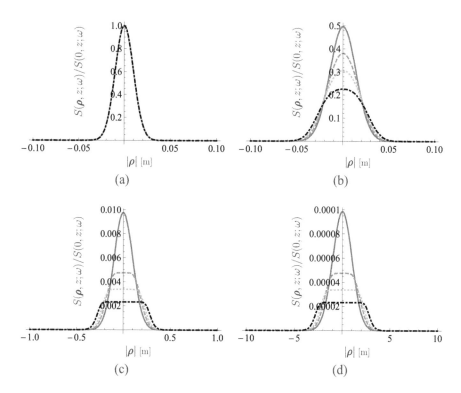

(a)

(b)

(c)

(d)

FIGURE 3.14

Transverse cross-section of the spectral density of the multi-Gaussian Schell-model beam propagating in free space vs. $|\boldsymbol{\rho}|$, at several distances from the source plane: (a) 1 m; (b) 100 m; (c) 1 km; and (d) 10 km; $M = 1$ solid curve, $M = 4$ dashed curve, $M = 10$ dotted curve and $M = 40$ dash-dotted curve. Other beam parameters are: $\lambda = 632$ μ m, $\sigma = 1$ cm, $\delta = 1$ mm. From Ref. [45].

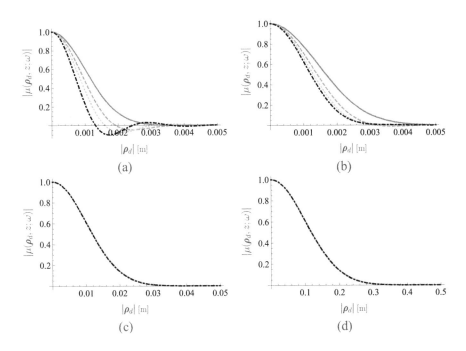

FIGURE 3.15

The modulus of the spectral degree of coherence of the multi-Gaussian Schell-model beam propagating in free space vs. ρ_d, at the same distances from the source, values of M and source parameters as in Fig. 3.14. From Ref [45].

spectral degree of coherence reduces to expression:

$$|\mu^{(MG)}(\boldsymbol{\rho}_d, z; \omega)| = \frac{|W^{(MG)}(\boldsymbol{\rho}_1, \boldsymbol{\rho}_2, z; \omega)|}{S^{(MG)}(0, z; \omega)}$$

$$= \frac{\displaystyle\sum_{m=1}^{M} \binom{M}{m} \frac{(-1)^{m-1}}{m} \frac{1}{\Delta_m^{(MG)^2}(z)} \exp\left[-\frac{\rho_d^2\left(4\sigma^2 + \frac{1}{\delta_m^2}\right)}{2\Delta_m^{(MG)^2}(z)}\right]}{\displaystyle\sum_{m=1}^{M} \binom{M}{m} \frac{(-1)^{m-1}}{m} \frac{1}{\Delta_m^{(MG)^2}(z)}}. \qquad (3.128)$$

In Fig. 3.15 the evolution of $|\mu^{(MG)}|$ on propagation in free space as a function of transverse difference variable ρ_d for several values of z and M is illustrated. For small distances z from the source the shapes of $|\mu^{(MG)}|$ resemble that in the source plane. As the range grows, however, the profiles become Gaussian-like while the dependence on M gradually disappears [see Figures 3.15 (a) – 3.15 (c)]. In the far zone of the source, all the curves behave in the same way: they remain Gaussian with monotonically increasing variance, due to the free-space diffraction [compare Figures 3.15 (c) and 3.15 (d)].

3.2.5 Bessel-Gaussian-correlated and Laguerre-Gaussian-correlated Schell-model sources

In this section we will introduce two more families of random sources producing beam-like fields, which may be viewed as a generalization of the Gaussian Schell-model sources. The reason for combining the two families into one discussion is the similarity of their far fields: both generate ring-shaped intensity distributions [46]. Optical beams with ring-shaped intensity profiles are of particular interest for applications involving particle manipulation. However, if a ring is generated by a deterministic source then its shape is not invariant on free-space propagation. For instance, an annular beam transforms to a Gaussian beam. On the other hand if an intensity profile is formed in the far field it remains shape-invariant, while its features spatially expand. In order to determine a source that produces a ring-shaped far field it suffices to select a correlation function whose Fourier transform is rotationally symmetric, vanishes on the optical axis, has a maximum at some off-axis location and decreases to zero with growing distance from the axis.

For the analysis of the two families of sources of interest we first need to discuss some general issues about the valid mathematical models for rotationally symmetric random sources. A position vector $\boldsymbol{\rho}'$ and the corresponding vector in Fourier space $\mathbf{s}$ can be expressed in the polar coordinate system, i.e., in terms of radii and orientation angles: $\boldsymbol{\rho}' = (\rho', \phi')$ and $\mathbf{s} = (s, \vartheta)$. The function $p(\mathbf{s})$, used in the proof of the non-negativity of cross-spectral density function, can be set as

$$p(\mathbf{s}) = (-i)^n p(s) \exp(in\vartheta), \qquad (3.129)$$

where $p(s)$ is a scalar function, n is the azimuthal mode index, $(-i)^n$ is a transform coefficient, independent on the coordinate variables s and ϑ. On substituting from Eq. (3.129) into Eq. (3.111) one finds that the cross-spectral density function takes the form

$$W(\boldsymbol{\rho}_1', \boldsymbol{\rho}_2'; \omega) = 2\pi \exp(in\phi_{12})\tau^*(\boldsymbol{\rho}_1'; \omega)\tau(\boldsymbol{\rho}_2'; \omega)$$

$$\times \int_0^\infty p(s) J_n(2\pi s |\boldsymbol{\rho}_1' - \boldsymbol{\rho}_2'|) s ds, \qquad (3.130)$$

where ϕ_{12} is the phase coordinate of vector $\boldsymbol{\rho}_1' - \boldsymbol{\rho}_2'$, $J_n(x)$ is the n-th order Bessel function of the first kind. Equation (3.130) implies that the source plane spectral density vanishes, i.e., $S(\boldsymbol{\rho}'; \omega) = 0$, for $n \neq 0$, leading to the restriction $n = 0$. Then, substitution of Eq. (3.130) into function

$$Q(f) = \iint W(\boldsymbol{\rho}_1', \boldsymbol{\rho}_2'; \omega) f^*(\boldsymbol{\rho}_1'; \omega) f(\boldsymbol{\rho}_2'; \omega) d^2 \rho_1' d^2 \rho_2' \qquad (3.131)$$

results in

$$Q(f) = 2\pi \int_0^\infty p(s) \left| \int \tau(\boldsymbol{\rho}) f(\boldsymbol{\rho}) d^2 \rho \right|^2 s ds. \qquad (3.132)$$

Function $Q(f) \geq 0$ for any f, therefore function $p(s)$ is non-negative for all values of s. It is to be noted that the integral in Eq. (3.130) (with $n = 0$) represents the Hankel transform.

We will now choose the forms for $p(s)$ to obtain the two ring shaped families of beams. For the Bessel-Gaussian-correlated beams we set

$$p(s) = 2\pi \delta_b^2 I_0(2\pi \beta_b \delta_b s) \exp(-\beta_b/2 - 2\pi^2 \delta_b^2 s^2), \qquad (3.133)$$

where δ_b and β_b are real constants and $I_0(x)$ is the 0-th order modified Bessel function of the first kind, which is always greater than or equal to unity for any real argument x. Hence, function $p(s)$ is non-negative for any values of β_b and δ_b. The Fourier transform of $p(s)$ in Eq. (3.133) leads to the following form of the source spectral degree of coherence

$$\mu^{(BG)}(\boldsymbol{\rho}_2', \boldsymbol{\rho}_1'; \omega) = J_0\left(\frac{\beta_b}{\delta_b}|\boldsymbol{\rho}_2' - \boldsymbol{\rho}_1'|\right) \exp\left(-\frac{|\boldsymbol{\rho}_2' - \boldsymbol{\rho}_1'|^2}{2\delta_b^2}\right). \qquad (3.134)$$

The sources with correlation (3.134) have been called the *Bessel-Gaussian-correlated Schell-model sources* [46]. In particular, on setting the Gaussian

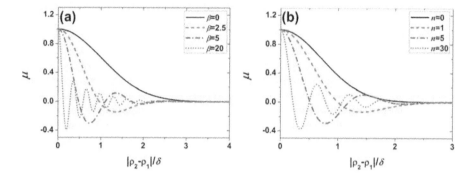

FIGURE 3.16

The spectral degree of coherence of (a) the Bessel-Gaussian-correlated sources, (b) the Laguerre-Gaussian-correlated sources. From Ref. [46].

profile for function τ, i.e., $\tau(\boldsymbol{\rho}; \omega) = \exp(-\rho'^2/2\sigma^2)$ the following cross-spectral density function can be obtained

$$
W^{(BG)}(\boldsymbol{\rho}_1', \boldsymbol{\rho}_2'; \omega) = \exp\left(-\frac{\rho_1'^2 + \rho_2'^2}{2\sigma^2} - \frac{|\boldsymbol{\rho}_2' - \boldsymbol{\rho}_1'|}{2\delta_b^2}\right) J_0\left(\frac{\beta_b}{\delta_b}|\boldsymbol{\rho}_2' - \boldsymbol{\rho}_1'|\right).
$$

(3.135)

Figure 3.16 (a) shows the spectral degree of coherence of the Bessel-Gaussian-correlated Schell-model source calculated from Eq. (3.134) for several values of parameter β_b. This family of model sources has two interesting limiting cases: it reduces to the Gaussian Schell-model source if $\beta_b = 0$ and to J_0-correlated source if $\beta_b \to \infty$.

The beam conditions for the parameters of the Bessel-Gaussian-correlated source can now be derived from its cross-spectral density function in the far field, at two points specified by position vectors $\mathbf{r}_1 = r_1\mathbf{u}_1$ and $\mathbf{r}_2 = r_2\mathbf{u}_2$, with $u_1^2 = u_2^2 = 1$. The field in the far zone of the Bessel-Gaussian-correlated source can be found by substituting Eq. (3.135) into (3.40), (3.41):

$$
W_{(\infty)}^{(BG)}(r\mathbf{u}_1, r\mathbf{u}_2; \omega) = \frac{1}{2}\frac{\sigma^2 k^2}{b_b} \cos\theta_1 \cos\theta_2 \frac{\exp\left[ik(r_2 - r_1)\right]}{r_1 r_2}
$$

$$
\times I_0\left[\frac{\beta_b}{4b_b\delta_b}|k(\mathbf{u}_{\perp 1} + \mathbf{u}_{\perp 2})|\right] \exp\left[-\frac{\beta_b^2}{4b_b\delta_b^2}\right]
$$

(3.136)

$$
\times \exp\left[-\frac{\sigma^2}{2}|k(\mathbf{u}_{\perp 1} - \mathbf{u}_{\perp 2})|^2\right] \exp\left[-\frac{1}{16b_b}|k(\mathbf{u}_{\perp 1} + \mathbf{u}_{\perp 2})|^2\right],
$$

where $b_b = 1/(8\sigma^2) + 1/(2\delta_b^2)$. The condition for the Bessel-Gaussian-correlated source to generate a beam can be readily verified to be just the same as for the Gaussian Schell-model sources (3.44), due to the fact that $I_0 \geq 1$ for all values of its argument. Further, on substituting from Eq. (3.136)

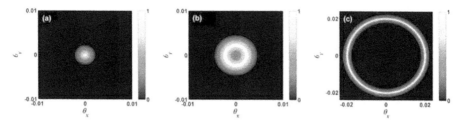

FIGURE 3.17
Far field generated by a typical Bessel-Gaussian-correlated source for several
values of order n: (a) $\beta_b = 0$, (b) $\beta_b = 2.5$, (c) $\beta_b = 20$. From Ref. [46].

into the definition of the spectral density, one finds that in the far zone the
Bessel-Gaussian Schell-model beam has profile

$$S_{(\infty)}^{(BG)}(r\mathbf{u},\omega) = \frac{k^2\sigma^2}{2b_br^2}\cos^2\theta I_0\left[\frac{\beta_b k}{2b_b\delta_b}u_\perp\right]\exp\left[-\frac{\beta_b^2}{4b_b\delta_b^2} - \frac{k^2}{4b_b}u_\perp^2\right]. \quad (3.137)$$

The angle between the direction of the global maximum of the intensity
profile and the optical axis, $\theta_{max}^{(BG)}$, can be readily determined from Eq. (3.137)
to be:

$$\theta_{max}^{(BG)} = \frac{\sqrt{16\beta_b^2 - 56b_b\beta_b^2\delta_b^2 + b_b^2\delta_b^4 + 4\beta_b^2} - b_b\delta_b^2}{8k\beta_b\delta_b}. \quad (3.138)$$

Figure 3.17 illustrates the far field of a Bessel-Gaussian Schell-model beam,
for several fixed values of parameter β_b. While the Gaussian Schell-model
beam ($\beta_b = 0$) in the source plane has the Gaussian intensity distribution
in the far field, with increasing β_b, the central intensity relative to the total
light intensity gradually decreases and the profile start to resemble that of the
dark-hollow beam. Also, the area of the central dark region in the far field is
directly proportional to the value of β_b.

The other class of sources that radiate ring-shaped intensity profiles in the
far zone can be introduced if, instead of choosing function $p(s)$ to be in the
form (3.133), one sets

$$p(s) = \frac{\pi^{2n+1}\delta_l^{2n+2}2^{n+1}}{n!}s^{2n}\exp(-2\pi^2\delta_l^2 s^2), \quad n = 0, 1, ... \quad (3.139)$$

The Fourier transform of this function leads to the following degree of source
coherence:

$$\mu^{(LG)}(\boldsymbol{\rho}_1',\boldsymbol{\rho}_2';\omega) = L_n\left(\frac{|\boldsymbol{\rho}_2' - \boldsymbol{\rho}_1'|}{2\delta_l^2}\right)\exp\left(-\frac{|\boldsymbol{\rho}_2' - \boldsymbol{\rho}_1'|}{2\delta_l^2}\right). \quad (3.140)$$

Since the right side of Eq. (3.139) is non-negative, for any values of δ_l, the
mathematical model (3.140) describes a physically realizable source. The

sources with correlation (3.140) have been named the *Laguerre-Gaussian-correlated Schell-model sources* [46]. In Fig. 3.16 (b) the variation of the degree of coherence of this with the separation distance ρ_d between two points, for several values of index n is illustrated.

The angular correlation function of the Laguerre-Gaussian-correlated Schell-model sources takes the form

$$A^{(LG)}(r\mathbf{u}_1, r\mathbf{u}_2; \omega) = \frac{\sigma^2 (2b_b - 1/\delta_l^2)^n}{(2\pi)^2 (2b_b)^{n+1}} L_n \left(-\frac{\sigma^2 |r\mathbf{u}_1 - r\mathbf{u}_2|^2}{4b_b \delta_l^2} \right)$$

$$\times \exp\left(-\frac{\delta_l^2}{2} |r\mathbf{u}_1 + r\mathbf{u}_2|^2 \right) \exp\left(-\frac{1}{16 b_b} |r\mathbf{u}_1 - r\mathbf{u}_2|^2 \right).$$

$$(3.141)$$

It is readily seen from Eq. (3.141) that the argument of the Laguerre function is always negative or equal to zero. Since for all values of n, $L_n(x) \geq 1$ when $x \leq 0$, the same inequality as for the Gaussian Schell-model source is obtained for the beam conditions, see Eq. (3.68). The far-field spectral density of the Laguerre-Gaussian Schell-model beams then reduces to the expression:

$$S_{(\infty)}^{(LG)}(r\mathbf{u}, \omega) = \frac{k^2 \sigma^2}{(2b_b)^{n+1} r^2} \cos^2 \theta L_n \left[-\frac{k^2 \sigma^2}{b_b \delta_l^2} u_\perp^2 \right] \exp\left[-\frac{k^2}{4b_b} u_\perp^2 \right]. \quad (3.142)$$

The angle $\theta_{max}^{(LG)}$ corresponding to the the maximum ring's intensity maximum depends on the source parameters as

$$\theta_{max}^{(LG)} = \frac{4b_l^2 - c_l + \sqrt{16 b_l^2 - 56 b_l^2 c_l + c_l^2}}{16 b_l c_l}, \quad (3.143)$$

where $b_l = 2k\sigma\sqrt{n/b_b\delta_l}$, $c_l = k^2\sigma^2/(2b_b\delta_l^2) + k^2/(4b_b)$.

Figure 3.18 shows the variation of the far-zone spectral density radiated by the Laguerre-Gaussian Schell-model sources for different values of index n. Similarly to the case of the Bessel-Gaussian Schell-model beams, the intensity profiles are rings centered about the optical axis, except for the case $n = 0$.

we stress again that unlike the dark-hollow beams, the Bessel-Gaussian and Laguerre-Gaussian Schell-model beams are formed not at the source plane but in the far field, preserving their profiles there. This feature makes them particularly suitable for applications involving manipulation of particles in cases when the presence of the propagation path between the source and the particle cannot be avoided.

3.2.6 Non-uniformly correlated sources

So far we have only considered a variety of mathematical models for random beams that are based on Schell's assumption, i.e., implying that the degree of coherence of the field at two spatial arguments depends only on their separation. The Schell model leads to optical fields that have uniform

FIGURE 3.18
The far fields generated by typical Laguerre-Gaussian-correlated sources for several values of n: (a) $n = 0$; (b) $n = 1$; (c) $n = 30$. From Ref. [46].

(position-independent) correlations at any plane transverse to the direction of propagation. However, the general properties of the cross-spectral density function do not preclude from prescribing more general, position-dependent correlations [47].

Without loss of generality we limit our attention to the one-dimensional case. Hence, the sufficient condition for the genuine cross-spectral density function [see Eq. (3.11)] reduces to the form

$$W(x_1, x_2; \omega) = \int p(s; \omega) H^*(x_1, s; \omega) H(x_2, s; \omega) ds, \qquad (3.144)$$

where H is an arbitrary kernel and p is an arbitrary non-negative, Fourier-transformable function. In particular, for the choice [8]

$$H(x, s; \omega) = \tau(x) \exp[-if(x)s], \qquad (3.145)$$

$\tau(x)$ being a complex function, $f(x)$ being any real function, the cross-spectral density function (3.144) becomes

$$W(x_1, x_2; \omega) = \tau^*(x_1) \tau(x_2) \tilde{p}[f(x_1) - f(x_2)], \qquad (3.146)$$

the tilde standing for the Fourier transform. If $f(x) = x$ the Schell model is deduced, otherwise a variety of fields with non-uniform correlations emerge. For instance, on taking [47]

$$p(s; \omega) = (\pi w_b^2)^{-1/2} \exp\left(-\frac{s^2}{w_b^2}\right), \qquad (3.147)$$

for some positive real w_b possibly depending on ω, and

$$H(x, s; \omega) = \exp\left(-\frac{x^2}{2w_0^2}\right) \exp[-ik(x - \gamma_0)^2 s], \qquad (3.148)$$

where w_0 and γ_0 are real constants, we find that the cross-spectral density function takes on the form

$$W^{(NUC)}(x_1, x_2; \omega) = \exp\left(-\frac{x_1^2 + x_2^2}{2w_0}\right) \mu^{(NUC)}(x_1, x_2; \omega). \qquad (3.149)$$

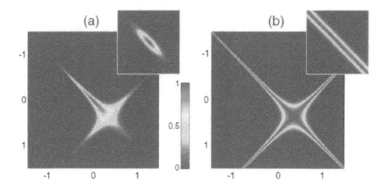

FIGURE 3.19
(a) Absolute value of the cross-spectral density and (b) the degree of coherence
of a typical beam with non-uniform correlations. The axes correspond to x_1
and x_2 in millimeters and their scale in the insets is the same as in the larger
figures. From Ref. [47].

Here the spectral degree of coherence has the form

$$\mu^{(NUC)}(x_1, x_2; \omega) = \exp\left[-\frac{[(x_1 - \gamma_0)^2 - (x_1 - \gamma_0)^2]^2}{w_c^4}\right], \qquad (3.150)$$

where $w_c = \sqrt{2/(kw_b)}$. Just like the classic Schell-model beams the beam
defined in (3.149) has the Gaussian intensity profile. Its degree of coherence
(3.150), however, differs substantially from Schell's degree of coherence, depending on $x_1^2 - x_2^2$ rather than on separation $x_1 - x_2$. Figure 3.19 illustrates
the cross-spectral density function and the spectral degree of coherence in a
typical beam with non-uniform correlations.

Free-space propagation of random fields with non-uniform correlations can
be analyzed either numerically, with the help of the Huygens-Fresnel integral
for the cross-spectral density function or via the mode decomposition (3.144).
We will follow the latter approach while restricting our attention to the propagation of the spectral density of the field. On substituting the modes $H(x, s; \omega)$
into the Huygens-Fresnel formula we find that at any plane $z > 0$

$$H(x, s, z; \omega) = \sqrt{\frac{k}{2\pi z}} \int H(x', s; \omega) \exp\left[\frac{ik(x - x')^2}{2z}\right] dx', \qquad (3.151)$$

and hence, the mode's intensity becomes

$$|H(x, s, z; \omega)|^2 = \frac{w_0}{w(z, s)} \exp\left[-\frac{(x - 2sz\gamma_0)^2}{w^2(z, s)}\right], \qquad (3.152)$$

with

$$w^2(z, s) = w_0^2(1 - 2sz)^2 + \left(\frac{z}{kw_0}\right)^2. \qquad (3.153)$$

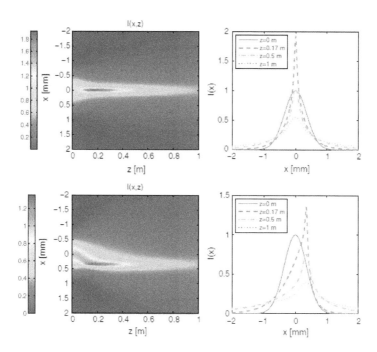

FIGURE 3.20

Propagation of a typical beam with non-uniform correlations. Figures on the left side show the evolution of the intensity at the $(x - z)$ plane. On the right side, the lateral intensity distribution is shown at selected propagation distances. The upper row: $\gamma_0 = 0$, the bottom row: $\gamma_0 \neq 0$. From Ref. [47].

The spectral density of the field in a plane $z > 0$ can be found from expression

$$S^{(NUC)}(x, z; \omega) = \int p(s)|H(x, s, z; \omega)|^2 ds. \tag{3.154}$$

Figure 3.20 compares the propagation of the spectral density of beams with non-uniform correlations, for no off-axis shift, $\gamma_0 = 0$, and with the shift, $\gamma_0 \neq 0$. In the latter case there exists a lateral shift of the maximum intensity from the optical axis. This example illustrates the possibility of directing the beam's power by tuning the spectral degree of coherence in the source plane. Another interesting feature of beams with non-uniform correlations is an enhanced concentration of the intensity in some region in space at some propagation distance from the source, with the maximum value being greater than the initial intensity maximum.

3.2.7 I_0-Bessel correlated sources

Another class of stochastic sources and beams was introduced by Ponomarenko [48], that is capable of carrying separable optical vortices. We will refer to such sources and the beams they generate as I_0-*Bessel correlated* because their cross-spectral density function can be constructed, with the help of the coherent mode decomposition (see Section 2.3), from the coherent Laguerre-Gaussian modes. The cross-spectral density matrix of an I_0-Bessel correlated source has the form

$$W^{(IB)}(\boldsymbol{\rho}_1', \boldsymbol{\rho}_2'; \omega) = \frac{A_0^2 \xi^{-n/2}}{1 - \xi} \exp\left[-\frac{(1+\xi)}{(1-\xi)} \frac{(\rho_1'^2 + \rho_2'^2)}{w_I^2}\right]$$
$$\times I_n\left(\frac{4\sqrt{\xi}}{1-\xi} \frac{\rho_1' \rho_2'}{w_I^2}\right) \exp[-in(\phi_1' - \phi_2')], \tag{3.155}$$

where $I_n(x)$ is a modified Bessel function of order n, ξ is a real-valued parameter with $0 < \xi < 1$, $\rho_1' = |\boldsymbol{\rho}_1'|$, $\rho_2' = |\boldsymbol{\rho}_2'|$, $\phi_1' = \arg \boldsymbol{\rho}_1'$, $\phi_2' = \arg \boldsymbol{\rho}_2'$. This formula can be derived from the expression for the Laguerre-Gaussian laser modes in the source plane $z = 0$ [see Eq. (2.50)]:

$$U_{mn}^{(LG)}(\boldsymbol{\rho}'; \omega) = \left(\frac{\sqrt{2}\rho'}{w_I}\right)^n L_m^n\left(\frac{2\rho'^2}{w_I^2}\right) \exp(-in\phi') \exp\left(-\frac{\rho'^2}{w_I^2}\right), \tag{3.156}$$

where $\rho' = |\boldsymbol{\rho}'|$, $\phi' = \arg \boldsymbol{\rho}'$, and $L_m^n(x)$ is the Laguerre polynomial of order m with azimuthal mode index n, applied in the coherent mode representation of a scalar stochastic beam:

$$W^{(IB)}(\boldsymbol{\rho}_1', \boldsymbol{\rho}_2'; \omega) = \sum_{m,n} \lambda_{mn} U_{mn}^{(LG)*}(\boldsymbol{\rho}_1'; \omega) U_{mn}^{(LG)}(\boldsymbol{\rho}_2'; \omega). \tag{3.157}$$

In fact, on substituting from Eq. (3.156) into Eq. (3.157) and using the summation formula for Laguerre polynomials [41]

$$
\sum_{n=0}^{\infty} \frac{m!}{(m+n)!} q_3^m L_m^n(q_1) L_m^n(q_2)
$$

$$
= \frac{(q_1 q_2 q_3)^{-n/2}}{1-q_3} \exp\left[-\frac{q_3(q_1+q_2)}{1-q_3}\right] I_n\left(\frac{\sqrt{4q_1 q_2 q_3}}{1-q_3}\right),
\tag{3.158}
$$

with $q_1 = 2\rho_1'^2/w^2$, $q_2 = 2\rho_2'^2/w^2$, $q_3 = \xi$, where $|q_3| < 1$ and on multiplying both sides of the last formula by $A_0(\rho_1'/w)^n (\rho_2'/w)^n \exp[-(\rho_1'^2 + \rho_2'^2)/w^2]$, A_0 being a positive factor, we arrive at the formula (3.155). We note that the coherent mode decomposition for the I_0-Bessel correlated sources has the form of the double sum

$$
W^{(IB)}(\boldsymbol{\rho}_1', \boldsymbol{\rho}_2'; \omega) = A_0^2 \sum_{l=-\infty}^{\infty} \sum_{m=0}^{\infty} \lambda_{ml} U_{ml}^{(LG)*}(\boldsymbol{\rho}_1'; \omega) U_{ml}^{(LG)}(\boldsymbol{\rho}_2'; \omega),
\tag{3.159}
$$

where

$$
\lambda_{ml} = \frac{m!}{(m+l)!} \xi^l \delta_{n,l},
\tag{3.160}
$$

$\delta_{n,l}$ denoting the Kronecker symbol. Since the eigenvalues λ_{ml} should be all non-negative, parameter ξ must have values in the interval $0 < \xi < 1$. The spectral density and the spectral degree of coherence of an I_0-Bessel correlated source are given by the formulas

$$
S^{(IB)}(\rho'; \omega) = \frac{A_0^2 \xi^{-n/2}}{1-\xi} \exp\left[-\frac{2\rho'^2(1+\xi)}{w_I^2(1-\xi)}\right] I_n\left(\frac{4\sqrt{\xi}}{1-\xi}\frac{\rho'^2}{w_I^2}\right)
\tag{3.161}
$$

and

$$
\mu^{(IB)}(\boldsymbol{\rho}_1', \boldsymbol{\rho}_2'; \omega) = \frac{I_n(2\rho_1'\rho_2'/\delta_I^2)}{\sqrt{I_n(2\rho_1'^2/\delta_I^2)}\sqrt{I_n(2\rho_2'^2/\delta_I^2)}} \exp[-in(\phi_1' - \phi_2')],
\tag{3.162}
$$

where

$$
\delta_I^2 = \frac{(1-\xi)w_0^2}{2\sqrt{\xi}}
\tag{3.163}
$$

can be regarded as the spatial coherence length. On expressing parameter ξ in terms of w_I and δ_I we find that

$$
\xi = \frac{\delta_I^4}{w_I^4}\left[\sqrt{1+\frac{w_I^4}{\delta_I^4}} - 1\right]^2.
\tag{3.164}
$$

Therefore, as $\delta_I \to \infty$, in the spatially coherent limit, $\xi \to 0$, and as $\delta_I \to 0$, in the spatially incoherent case, $\xi \to 1$.

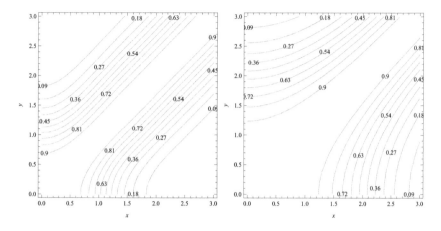

FIGURE 3.21
Contours of the modulus of the spectral degree of coherence of the I_0-Bessel correlated source, as function of $x = \rho'_1/\delta_I$ and $y = \rho'_2/\delta_I$. (left) $n = 0$; (right) $n = 10$.

In Fig. 3.21 the modules of the spectral degree of coherence of two I_0-Bessel-correlated sources are presented as a function of the normalized variables ρ'_1/δ_I, ρ'_2/δ_I for two values of azimuthal mode index n.

The expression for the radiant intensity distribution of the far field produced by a I_0-Bessel-correlated field and the conditions for the source to generate a beam-like field can be obtained, as before, with the help of the angular spectrum representation for random fields. In the case under interest it will be more convenient to obtain the Fourier transforms of the individual coherent modes $U_{ml}^{(LG)}(\boldsymbol{\rho};\omega)$ in the far field and then to sum the contributions of the coherent modes to arrive at

$$\widetilde{W}(\mathbf{f}_1,\mathbf{f}_2;\omega) = \sum_{m,l} \lambda_{ml}\widetilde{U}_{ml}^{(LG)*}(\mathbf{f}_1;\omega)\widetilde{U}_{ml}^{(LG)}(\mathbf{f}_2;\omega), \qquad (3.165)$$

where $\mathbf{f}_1 = k\mathbf{u}_{1\perp}$, $\mathbf{f}_2 = k\mathbf{u}_{2\perp}$ are the spatial frequency vectors, $\widetilde{U}_{ml}^{(LG)}(\mathbf{f};\omega)$ the Fourier transform of a single mode. By representing the Bessel function $J_n(x)$ by the integral of the form [41]

$$J_n(x) = \frac{1}{2\pi} \int_0^{2\pi} \exp[in\phi - ix\cos\phi]d\phi \qquad (3.166)$$

and using the formula [49]

$$
\int_0^\infty x^{q_1/2} \exp(-q_2 x) J_{q_1}(q_3\sqrt{x}) L_m^{q_1}(q_4 x)dx
$$

$$
= \left(\frac{q_3}{2}\right)^{q_1} \frac{(q_2-q_4)^m}{q_2^{m+q_1+1}} \exp\left(-\frac{q_3^2}{4q_2}\right) L_m^{q_1}\left(\frac{q_3^2 q_4/4q_2}{q_4-q_2}\right) \tag{3.167}
$$

with $q_2 = 1/2$, $q_3 = fw_I/2$ and $q_4 = 1$, we determine the mode in the far field. It is important to note that the structure of each mode is the same as in the source plane [compare with Eq. (3.156)]. Finally, on substituting from Eq. (3.167) into Eq. (3.165) and applying summation formula (3.158) we find that

$$
\widetilde{W}^{(IB)}(\mathbf{f}_1, \mathbf{f}_2; \omega) = \frac{A_0^2(w_I^2/2)\xi^{-n/2}}{1-\xi} \exp\left[-\frac{(1+\xi)}{(1-\xi)}\frac{(f_1^2+f_2^2)w_I^2}{4}\right]
$$

$$
\times I_n\left(\frac{\sqrt{\xi}}{1-\xi}f_1 f_2 w_I^2\right) \exp[-in(\phi_1 - \phi_2)], \tag{3.168}
$$

where $f_1 = |\mathbf{f}_1|$, $f_2 = |\mathbf{f}_2|$, $\phi_1 = \arg(\mathbf{f}_1)$ and $\phi_2 = \arg(\mathbf{f}_2)$. It follows from Eq. (3.168) that the radiant intensity of the I_0-Bessel-correlated field is given by the formula

$$
J^{(IB)}(\mathbf{u}; \omega) = \frac{A_0^2 w_I^4 \xi^{-n/2}}{4(1-\xi)} \exp\left[-\frac{(1+\xi)}{(1-\xi)}\frac{k^2 w_I^2|\mathbf{u}_\perp|^2}{2}\right]
$$

$$
\times I_n\left(\frac{\sqrt{\xi}}{1-\xi}k^2 w_I^2|\mathbf{u}_\perp|^2\right) \exp[-in(\theta_1 - \theta_2)]. \tag{3.169}
$$

Equation (3.169) implies that for values of parameter ξ close to zero (almost coherent beam) the radiant intensity tends to Gaussian shape. Further, conditions on the source parameters may be obtained from Eq. (3.44). In analytic form the beam conditions may be obtained only in the special cases of a fairly coherent ($\xi \to 0$) and almost incoherent ($\xi \to 1$) sources. Indeed, from Eq. (3.169) in the former case one finds [48]

$$
\frac{1}{w_I} << k \tag{3.170}
$$

while in the latter

$$
\frac{(1+\sqrt{\xi})^2}{2\sqrt{\xi}}\frac{1}{\delta_I^2} << k^2. \tag{3.171}
$$

The paraxial propagation law for the cross-spectral density function of the I_0-Bessel-correlated beams can also be obtained with the help of the coherent mode decomposition. It was shown in Chapter 2 that at distance z from the source the Laguerre-Gaussian modes obey formula (2.50). On substituting

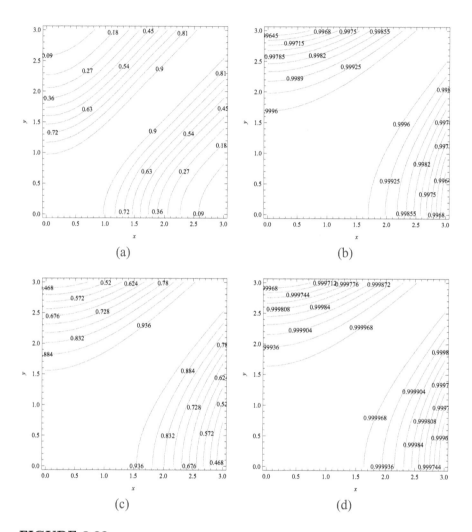

FIGURE 3.22

Contours of the modulus of the spectral degree of coherence of the stochastic I_0-Bessel correlated beam, as a function of $x = \rho_1/\delta_I$ and $y = \rho_2/\delta_I$. (a) $n = 0$, $z/z_R = 1$; (b) $n = 0$, $z/z_R = 10$; (c) $m = 10$, $z/z_R = 1$; (d) $m = 10$, $z/z_R = 10$.

from Eq. (2.50) into the coherent mode decomposition of the propagating beam

$$W^{(IB)}(\boldsymbol{\rho}_1, \boldsymbol{\rho}_2, z_1, z_2; \omega) = \sum_m \sum_n \lambda_{mn} U_{mn}^{(LG)*}(\boldsymbol{\rho}_1, z_1; \omega) U_{mn}^{(LG)}(\boldsymbol{\rho}_2, z_2; \omega),$$

(3.172)

we find, after using the summation formula (3.158), that the cross-spectral density function of the stochastic I_0-Bessel correlated beam has the form

$$W^{(IB)}(\boldsymbol{\rho}_1, \boldsymbol{\rho}_2, z_1, z_2; \omega) = \frac{A_0^2 \xi^{-n/2}}{1 - \xi} \frac{w_I^2}{w(z_1) w(z_2)}$$

$$\times \exp\left[-\frac{(1+\xi)}{(1-\xi)} \left(\frac{\rho_1^2}{w(z_1)^2} + \frac{\rho_2^2}{w(z_2)^2}\right)\right]$$

$$\times I_n\left(\frac{4\sqrt{\xi}}{1-\xi} \frac{\rho_1 \rho_2}{w(z_1)w(z_2)}\right) \exp\left[i\left(\frac{k\rho^2}{2R(z_1)} - \frac{k\rho^2}{2R(z_1)}\right)\right]$$

$$\times \exp i[k(z_1 - z_2) - (n+1)(\Phi(z_1) - \Phi(z_2))] \exp[in(\phi_1 - \phi_2)].$$

(3.173)

It follows from Eq. (3.173) that on paraxial propagation the spectral density of the I_0-Bessel-correlated beam preserves its initial structure, i.e., remains the product of the Gaussian function and the modified Bessel function of order n and its phase carries a vortex with a topological charge equal to the azimuthal index n, just like a coherent Laguerre-Gaussian mode. Moreover, the phase of such a beam is separable, i.e., its cross-spectral density function in a plane, transverse to the direction of propagation of the beam, can be represented by a product of a phase term and term depending on the radial variable.

In Fig. 3.22 the contours for the modulus of the transverse degree of coherence ($z_1 = z_2 = z$) of the propagating beam are plotted for several values of the normalized propagation distance $z/z_R = 2z/kw_I^2$, as a function of the normalized radial distances $x = \rho_1/\delta_I$ and $y = \rho_2/\delta_I$.

3.3   Methods of generation

Not long after their theoretical introduction, scalar random beams with adjustable properties have been successfully realized in the laboratory [50, 51]. Among the traditional techniques for synthesis of a random beam-like field we distinguish the passage of a laser beam through a rotating diffuser as the most practical. A typical diffuser is the grounded glass fixed between two optically transparent plates and is usually employed in the transmission mode. The rotation of the diffuser guaranties the generation of statistical ensemble of realizations at any position on its surface. The operational principle of the

FIGURE 3.23
Spatial Light Modulator.

diffuser relies on the spatial and temporal beam's phase modulation, while the amplitude modulation is assumed to be mild and uniform.

Another important technique for generation of random beams is based on the van Cittert-Zernike theorem [4], implying that the coherence of light generated by a spatially incoherent source increases on free-space propagation. The desired state of coherence for the generated partially coherent beam can be directly related to the propagation distance from the incoherent source. Among less familiar methods are scattering of a field from rough surfaces, or its passage through turbulent medium.

One of the newer methods for the wave randomization is either its transmission through or reflection from (depending on the device) a Spatial Light Modulator [SLM], controlled by a computer [52]–[56], see Fig. 3.23. We will only discuss the method based on the reflection of a laser beam from the phase-only, nematic Spatial Light Modulator (SLM). The surface of the SLM device acts as a phase modulator (phase screen): the resulting phase can be made different for different pixels within one frame. The SLM is capable of rapid change of frames. Regardless of the discrete nature of such phase modulation, both spatial and temporal, the resulting field can be regarded as partially coherent provided the detector's rate is much slower. The other convenient feature of the SLM is the capability of digitally adjusting the statistics of phase at will. Here we confine our attention to fields governing by Gaussian random processes with zero mean and having Schell-like second-order correlation functions.

The operational principle of an SLM is based on the fact that the electric signal (induced voltage) given to a pixel is proportional to the magnitude of the phase modulation of the reflected wave. The assignment of the signals (phases) at all the pixels on the SLM can be done via a simple computer program, which returns a two-dimensional array of phase values. We note that the sequence of images is uncorrelated for any pixel and, hence, can be used as an ensemble of realizations. More specifically, we assume that for each pixel on the SLM surface with position vector $\rho(x, y)$ the real random process $\varpi(\rho)$ describing the phase distribution on the SLM (called below the "SLM phase") is correlated as

$$\langle \varpi(\rho_1) \varpi(\rho_2) \rangle_s = \varpi_0^2 \exp\left[-\frac{(\rho_1 - \rho_2)^2}{\delta_\varpi^2}\right], \qquad (3.174)$$

where $\varpi_0^2 = \sqrt{\langle |\varpi(\rho)|^2 \rangle}$, δ_ϖ^2 is the correlation width, and the angular brackets with subscript s stand for the ensemble of realizations of the SLM phase distributions. In order to simulate such a process we first generate a two-dimensional array $R_\varpi(\rho)$ of independent random variables that obey Gaussian statistics with zero mean. It is well known that the independent Gaussian variables are necessarily uncorrelated, i.e.,

$$\langle R_\varpi(\rho_1) R_\varpi(\rho_2) \rangle_s = \delta^{(2)}(\rho_2 - \rho_1), \qquad (3.175)$$

i.e., delta-correlated in the two-dimensional space. Next, we calculate convolution of array $R_\varpi(\rho)$ with a window function $f_\varpi(\rho)$, which in general has a form of the desired degree of coherence. In particular, for the Gaussian Schell-model source we set

$$f_\varpi(\rho) = \exp\left[-\frac{\rho^2}{\gamma_\varpi^2}\right], \qquad (3.176)$$

to obtain the Gaussian-correlated array $g_f(\rho)$:

$$g_f(\rho) = \int f_\varpi(\rho - \rho') R_\varpi(\rho') d^2\rho'. \qquad (3.177)$$

On taking correlation function of $g_f(\rho)$ at positions ρ_1 and ρ_2 and performing integration we arrive at the expression

$$\langle g_f(\rho_1) g_f(\rho_2) \rangle_s = \frac{\pi \gamma_\varpi^2}{2} \exp\left[-\frac{(\rho_2 - \rho_1)^2}{2\gamma_\varpi^2}\right]. \qquad (3.178)$$

116

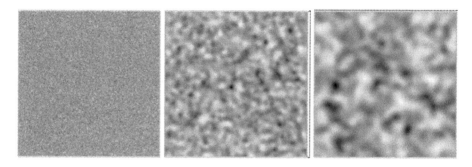

FIGURE 3.24
Typical realizations of the SLM phase screens for the Gaussian Schell-model beams with different correlation widths.

Finally, on associating

$$\varpi_0 = \sqrt{\frac{\pi \gamma_\varpi}{2}}, \quad \delta_\varpi = \gamma_\varpi \qquad (3.179)$$

we obtain Eq. (3.174). By repeating the procedure the necessary sequence of the SLM phase images can be produced. A sample MATLAB® program generating a sequence of phase screen realizations is given below.

```
clear all
jj = 1;
for n=1:3000
I = normrnd(0,1,512,512);
I_min = min(min(I));I_max = max(max(I));
[y,x] = meshgrid(0:length(I),0:length(I));
r = length(x)/2;  c = length(x)/2;
rho = sqrt((x-r).^2 + (y-c).^2);
Corr_width_2 = 1;  window = exp(-rho.^2/Corr_width_2);
GSB = conv2(window,I);
figure(2)
GSB_512 = GSB_512./max(max(GSB_512)); imshow(GSB_512,[]);
jj = jj+1;
figname = printf('GSM.bmp',jj,round(Corr_width_2));
imwrite(GSB_512,figname, 'bmp'); end
```

Figure 3.24 shows several phase screens simulated by this procedure for several values of γ_ϖ^2: 1, 100, 300, from left to right, respectively. If the resolution and physical dimensions of the SLM are known then the equivalent values with the units of millimeters or microns can be determined.

The change in the form of the correlation function (3.176) leads to differently structured phase screens. Figure 3.25 illustrates a computer simulation creating one realization of the two-dimensional phase distribution for

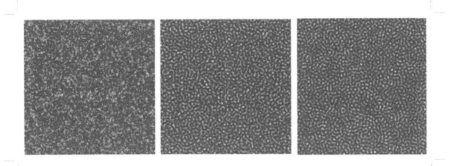

FIGURE 3.25
Comparison of the SLM phase screen realizations for Schell-model sources: (left) Gaussian; (middle) Bessel-Gaussian; (right) Laguerre-Gaussian. From Ref. [46].

the Gaussian (left), the Bessel-Gaussian (middle) and the Laguerre-Gaussian (right) Schell-model sources [46]. The structure of the correlations for the Gaussian Schell-model source is different from the other two implying similarities and differences in intensity profiles of the far fields they generate.

Even though there are several other ways of producing a random beam, such as scattering of a laser beam from a random surface or its propagation in a turbulent medium they are relatively less practical because of the difficulties associated with statistical characterization of such media.

Bibliography

[1] L. C. Andrews and R. L. Phillips, *Laser Beam Propagation in the Turbulent Atmosphere*, 2nd edition, SPIE Press, 2005.

[2] E. Wolf, "Optics in terms of observable quantities," *Nuovo Cimento* **12**, 884–888 (1954).

[3] M. Born and E. Wolf, *Principles of Optics*, Cambridge University Press, 7th Edition, 1999.

[4] L. Mandel and E. Wolf, *Optical Coherence and Quantum Optics*, Cambridge University Press, 1995.

[5] G. W. Goodman, *Statistical Optics*, Wiley, 1985.

[6] H. Gamo, "Matrix treatment of artial coherence," in *Progress in Optics* Vol. **3**, Ed. E. Wolf, North-Holland 1964.

[7] E. Wolf, "New theory of partial coherence in the space-frequency domain. Part 1. Spectra and cross-spectra of steady-state sources," *J. Opt. Soc. Am.* **72**, 343–351 (1982).

[8] F. Gori and M. Santarsiero, "Devising genuine spatial correlation functions," *Opt. Lett.* **32**, 3531–3533 (2007).

[9] R. Martnez-Herrero, P. M. Mejas, and F. Gori, "Genuine cross-spectral densities and pseudo-modal expansions," *Opt. Lett.* **34**, 1399–1401 (2009).

[10] A. Dogariu and G. Popescu, "Measuring the phase of spatially coherent polychromatic fields," *Phys. Rev. Lett.* **89**, 243902 (2002).

[11] O. Korotkova and E. Wolf, "Beam criterion for atmospheric propagation," *Opt. Lett.* **32**, 2137–2139 (2007).

[12] G. Iaconis and I. A. Walmsley, "Direct measurement of the two-point field correlation function," *Opt. Lett.* **21**, 1783-1785 (1996).

[13] E. Tervonen, J. Turunen, and A. T. Friberg, "Transverse laser-mode structure determination from spatial coherence measurements: experimental results," *Appl. Phys. B* **49**, 409-414 (1989).

[14] B. Lu, B. Zhang, B. Cai, and C. Yang, "A simple method for estimating the number of effectively oscillating modes and weighting factors of mixed mode laser beams behaving like Gaussian Schell-model beams," *Opt. Commun.* **101**, 49-52 (1993).

[15] A. Cutolo, T. Isernia, I. Izzo, R. Pierri, and L. Zeni, "Transverse mode analysis of a laser beam by near- and far-field intensity measurements," *Appl. Opt.* **34**, 7974-7978 (1995).

[16] M. Santarsiero, F. Gori, R. Borghi, and G. Guattari, "Evaluation of the modal structure for light beams with Hermite Gaussian modes," *Appl. Opt.* **38**, 5272-5281 (1999).

[17] R. Borghi and M. Santarsiero, "Modal structure analysis for a class of axially symmetric flat-topped laser beams," *IEEE J. Quantum Electron.* **35**, 745-750 (1999).

[18] M. Santarsiero, F. Gori, and R. Borghi, "Modal-weight determination for a class of multimode beams," in *Laser Beam and Optics Characterization*, H. Weber and H. Laabs, Eds., Optisches Institut, Technische Universitat Berlin, Berlin, 2000, pp. 161-170.

[19] X. Xue, H. Wei, and A. G. Kirk, "Intensity-based modal decomposition of optical beams in terms of HermiteGaussian functions," *J. Opt. Soc. Am. A* **17**, 1086-1091 (2000).

[20] A. S. Ostrovsky, *Coherent Mode Representations in Optics*, SPIE Press, 2006.

[21] R. Courant and D. Hilbert, "Methods of Mathematical Physics," Interscience, 1953.

[22] J. Mercer, "Functions of positive and negative type and their connection with the theory of integral equations," *Philosophical Transactions of the Royal Society A* **209**, 441–458 (1909).

[23] A. C. Schell, *The Multiple Plate Antenna,* Doctoral Dissertation, MIT, Sect. 7.5, 1961.

[24] A. Schell, "A technique for the determination of the radiation pattern of a partially coherent aperture," *IEEE Trans. Antennas Propag.* **15**, 187–188 (1967).

[25] F. Gori, "Collett–Wolf sources and multimode lasers," *Opt. Comm.* **34**, 301–305 (1980).

[26] A. Starikov and E. Wolf, "Coherent-mode representation of Gaussian Schell–model sources and their radiation fields," *J. Opt. Soc. Am.* **72**, 923–928 (1982).

[27] F. Gori, "Mode propagation of the field generated by Collett-Wolf Schell–model sources," *Opt. Comm.* **46**, 149–154 (1983).

[28] S. Sahin, O. Korotkova, G. Zhang, and J. Pu, "Free-space propagation of the spectral degree of cross–polarization of stochastic electromagnetic beams," *J. Opt. A: Pure Appl. Opt.* **11**, 085703 (2009).

[29] Y. Li and E. Wolf, "Radiation from anisotropic Gaussian Schell-model sources," *Opt. Lett.* **7**, 256–258 (1982).

[30] R. Simon and N. Mukunda, "Shape-invariant anisotropic Gaussian Schell-model beams: a complete characterization," *J. Opt. Soc. A* **15**, 1361–1370 (1998).

[31] R. Simon, and N. Mukunda, "Twisted Gaussian Schell-model beams," *J. Opt. Soc. Am. A* **10**, 95-109 (1993).

[32] D. Ambrosini, V. Bagini, F. Gori, and M. Santarsiero, "Twisted Gaussian Schell-model beams: a superposition model," *J. Mod. Opt.* **41**, 1391–1399 (1994).

[33] A. T. Friberg, E. Tervonen, and J. Turunen, "Interpretation and experimental demonstration of twisted Gaussian Schell-model beams," *J. Opt. Soc. Am. A* **11**, 1818–1826 (1994).

[34] R. Simon, A. T. Friberg, and E. Wolf, "Transfer of radiance by twisted Gaussian Schell-model beams in paraxial systems," *Pure Appl. Opt.* **5**, 331-343 (1996).

[35] J. Serna and J. M. Movilla, "Orbital angular momentum of partially coherent beams," *Opt. Lett.* **26**, 405–407 (2001).

[36] Q. Lin and Y. Cai, "Tensor ABCD law for partially coherent twisted anisotropic GaussianSchell model beams," *Opt. Lett.* **27**, 216–218 (2002).

[37] F. Gori, G. Guattari, and C. Padovani, "Modal expansion for J_0-correlated Schell-model sources," *Opt. Comm.* **64**, 311–316 (1987).

[38] C. Palma, R. Borghi, and G. Cincotti, "Beams originated by J_0-correlated Schell-model planar sources," *Opt. Commun.* **125** 113–121 (1996).

[39] G. Wu, Q. Lou, J. Zhou, H. Guo, H. Zhao, and Z. Yuan, "Beam conditions for radiation by an electromagnetic J_0-correlated Schell-model source," *Opt. Lett.* **33**, 2677–2679 (2008).

[40] G. N. Watson, *A Treatise on the Theory of Bessel Functions*, Cambridge University Press, Cambridge, 1922.

[41] I. S. Gradstein and I. M. Ryzhik, *Tables of Integrals, Series and Products*, Academic, 1980.

[42] F. Gori, "Flattened Gaussian beams," *Opt. Comm.* **107** 335–341 (1994).

[43] S. Sahin, G. Gbur, and O. Korotkova, "Scattering of light from particles with semi-soft boundaries," *Opt. Lett.* **36** 3957–3959 (2011).

[44] S. Sahin and O. Korotkova, "Light sources generating far fields with tunable flat profiles," *Opt. Lett.* **37**, 2970–2972 (2012).

[45] O. Korotkova, S. Sahin, and E. Shchepakina, "Multi-Gaussian Schell-model beams," *J. Opt. Soc. Am. A* **29**, 2159–2164 (2012).

[46] Z. Mei and O. Korotkova, "Random sources generating ring-shaped beams," *Opt. Lett.* **38**, 91–93 (2013).

[47] H. Lajunen and T. Saastamoinen, "Propagation characteristics of partially coherent beams with spatially varying correlations," *Opt. Lett.* **36**, 4104–4106 (2011).

[48] S. A. Ponomarenko, "A class of partially coherent beams carrying optical vortices," *J. Opt. Soc. Am. A* **18**, 150–156 (2001).

[49] A. P. Prudnikov, Y. A. Brichkov, and O. I. Marichev, *Integrals and Series*, Gordon, 1992.

[50] P. De Santis, F. Gori, G. Guattari, and C. Palma, "Synthesis of partially coherent fields," *J. Opt. Soc. Am. A* **3**, 1258–1262 (1986).

[51] D. Mendlovic, G. Shabtay, and A. W. Lohmann, "Synthesis of spatial coherence," *Opt. Lett.* **24**, 361–363 (1999).

[52] T. Shirai and E. Wolf, "Coherence and polarization of electromagnetic beams modulated by random phase screens and their changes on propagation in free space," *J. Opt. Soc. Am. A* **21**, 1907–1916 (2004).

[53] R. Betancur and R Castaneda, "Spatial coherence modulation," *J. Opt. Soc. Am. A* **26**, 147–155 (2009).

[54] A. S. Ostrovsky, G. Martinez-Niconoff, V. Arrizon, P. Martinez-Vara, M. A. Olivera-Santamaria, and C. Rickenstorff-Parrao, "Modulation of coherence and polarization using liquid crystal spatial light modulators," *Opt. Express* **17**, 5257–5264 (2009).

[55] C. Macias-Romero, R. Lim, M. R. Foreman, and P. Torok, "Synthesis of structured partially spatially coherent beams," *Opt. Lett.* **36**, 1638– (2011).

[56] O. Korotkova, S. Avramov-Zamurovic, C. Nelson, and R. Malek-Madani, "Probability Density Function (PDF) of intensity of a stochastic light beam propagating in the turbulent atmosphere," *Proc. SPIE.* **8238**, 82380J (2012).

4

Electromagnetic stochastic beams: theory

CONTENTS

4.1 Statistical description

4.1.1 Beam coherence polarization matrix

The extension of the second-order coherence theory from scalar to electromagnetic beams has been made in the early 1990s. It was first shown by James [1]

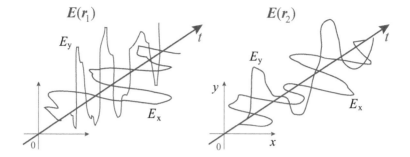

FIGURE 4.1
Fluctuating in time electric vector field at two spatial positions $\mathbf{r}_1$ and $\mathbf{r}_2$.

via a numerical example that the state of coherence of the source generating a stochastic beam may influence its degree of polarization when the beam propagates in free space. Later the beam coherence-polarization matrix was introduced by Gori [2], which can be regarded as a generalization of the mutual coherence function (3.1) of a scalar beam-like field to the electromagnetic domain.

If a stochastic beam is of an electromagnetic nature then its space-time evolution can be characterized by the electric field vector $\mathbf{E}(\mathbf{r};t) = [E_x(\mathbf{r};t), E_y(\mathbf{r};t)]$, where E_x and E_y are the two mutually orthogonal components fluctuating in the plane perpendicular to the direction of propagation, z. In order to take into account all the second-order correlation properties of the fluctuating in time electric vector-field at two points in space, say $\mathbf{r}_1$ and $\mathbf{r}_2$, and moments t_1 and t_2 (see Fig. 4.1) we now employ the beam coherence-polarization matrix denoted here as $\boldsymbol{\Gamma}$ [2],

$$\boldsymbol{\Gamma}(\mathbf{r}_1,\mathbf{r}_2;t_1,t_2) = [\Gamma_{\alpha\beta}(\mathbf{r}_1,\mathbf{r};t_1,t_2)]$$
$$= \begin{bmatrix} \langle E_x^*(\mathbf{r}_1;t_1)E_x(\mathbf{r}_2;t_2)\rangle_T & \langle E_x^*(\mathbf{r}_1;t_1)E_y(\mathbf{r}_2;t_2)\rangle_T \\ \langle E_y^*(\mathbf{r}_1;t_1)E_x(\mathbf{r}_2;t_2)\rangle_T & \langle E_y^*(\mathbf{r}_1;t_1)E_y(\mathbf{r}_2;t_2)\rangle_T \end{bmatrix}. \quad (4.1)$$

Further, the beam coherence-polarization matrix of a wide-sense statistically stationary field reduces to the matrix of the form

$$\boldsymbol{\Gamma}(\mathbf{r}_1,\mathbf{r}_2;\tau) = [\Gamma_{\alpha\beta}(\mathbf{r}_1,\mathbf{r}_2;\tau)]$$
$$= \begin{bmatrix} \langle E_x^*(\mathbf{r}_1;t)E_x(\mathbf{r}_2;t+\tau)\rangle_T & \langle E_x^*(\mathbf{r}_1;t)E_y(\mathbf{r}_2;t+\tau)\rangle_T \\ \langle E_y^*(\mathbf{r}_1;t)E_x(\mathbf{r}_2;t+\tau)\rangle_T & \langle E_y^*(\mathbf{r}_1;t)E_y(\mathbf{r}_2;t+\tau)\rangle_T \end{bmatrix}, \quad (4.2)$$

where $\tau = t_2 - t_1$ is the time delay between two time instants.

4.1.2 Cross-spectral density matrix

In a similar manner to how it was done for scalar fields one can take the Fourier transform of each of the elements of the beam coherence-polarization matrix with respect to variable τ and obtain the *cross-spectral density matrix* characterizing the beam in the space-frequency domain,

$$W_{\alpha\beta}(\mathbf{r}_1, \mathbf{r}_2; \omega) = \mathcal{F}[\Gamma^{(2)}_{\alpha\beta}(\mathbf{r}_1, \mathbf{r}_2; \tau)]. \tag{4.3}$$

This matrix was first introduced by Wolf [3] and its properties were summarized in [4]. The experimental determination of the cross-spectral dencity matrix elements was performed in [5]. In a similar way as was done with the cross-spectral density function of a scalar beam it can be proven that matrix (4.3) is the correlation matrix by itself [6] (see also [7]):

$$\mathbf{W}(\mathbf{r}_1, \mathbf{r}_2; \omega) = [W_{\alpha\beta}(\mathbf{r}_1, \mathbf{r}_2; \omega)]$$
$$= \begin{bmatrix} \langle E_x^*(\mathbf{r}_1; \omega) E_x(\mathbf{r}_2; \omega) \rangle_\omega & \langle E_x^*(\mathbf{r}_1; \omega) E_y(\mathbf{r}_2; \omega) \rangle_\omega \\ \langle E_y^*(\mathbf{r}_1; \omega) E_x(\mathbf{r}_2; \omega) \rangle_\omega & \langle E_y^*(\mathbf{r}_1; \omega) E_y(\mathbf{r}_2; \omega) \rangle_\omega \end{bmatrix}, \tag{4.4}$$

where, as for the spectral density function, the angular brackets $\langle \cdot \rangle_\omega$ denote the average over the ensemble of monochromatic realizations of the electric field components at frequency ω.

We stress here that unlike the elements of the beam coherence-polarization matrix that can be shown to satisfy a pair of wave equations with respect to $\mathbf{r}_1$ and $\mathbf{r}_2$, the elements of the cross-spectral density matrix satisfy a pair of Helmholtz equations with respect to these arguments. This fact results in the discrepancy in the two solutions that describe the propagation of beams and their interaction with media.

As in the case of the scalar cross-spectral density function, not every 2×2 matrix with elements depending on two spatial variables and a frequency can be an eligible cross-spectral density matrix. A proper, or *genuine*, cross-spectral density matrix should satisfy the following conditions [8]:

1. Each of the four elements should be square-integrable with respect to ω, i.e.,

$$\int\limits_0^\infty |W_{\alpha\beta}(\mathbf{r}_1, \mathbf{r}_2; \omega)|^2 d\omega < \infty. \tag{4.5}$$

2. If each matrix element is a continuous function of its spatial arguments then

$$\int \int |W_{\alpha\beta}(\mathbf{r}_1, \mathbf{r}_2; \omega)|^2 d^2 r_1 d^2 r_2 < \infty. \tag{4.6}$$

3. It must be quasi-Hermitian

$$W_{\alpha\beta}(\mathbf{r}_1, \mathbf{r}_2; \omega) = W_{\alpha\beta}^*(\mathbf{r}_2, \mathbf{r}_1; \omega), \tag{4.7}$$

i.e., in addition to ordinary Hermiticity condition the order of arguments should be changed.

126

4. It must be non-negative definite, i.e., the following integral inequality must hold

$$\int \int [W_{xx}(\mathbf{r}_1,\mathbf{r}_2;\omega)f_x^*(\mathbf{r}_1)f_x(\mathbf{r}_2)$$
$$+ W_{xy}(\mathbf{r}_1,\mathbf{r}_2;\omega)f_x^*(\mathbf{r}_1)f_y(\mathbf{r}_2)$$
$$+ W_{yx}(\mathbf{r}_1,\mathbf{r}_2;\omega)f_y^*(\mathbf{r}_1)f_x(\mathbf{r}_2)$$
$$+ W_{yy}(\mathbf{r}_1,\mathbf{r}_2;\omega)f_y^*(\mathbf{r}_1)f_y(\mathbf{r}_2)] \, d^2r_1d^2r_2 \geq 0,$$
(4.8)

where $f_x(\mathbf{r})$ and $f_y(\mathbf{r})$ are two arbitrary square-integrable, in general complex-valued functions [Ref. [8], Eq. (6.6-8)].

The electromagnetic generalization of the sufficient condition expressed by Eq. (3.11) has been derived by Gori [9] (see also [10]) stating that in order for a cross-spectral density matrix to be genuine its elements must have representation of the form

$$W_{\alpha\beta}(\mathbf{r}_1,\mathbf{r}_2;\omega) = \int p_{\alpha\beta}(\mathbf{s})H_\alpha^*(\mathbf{r}_1,\mathbf{s})H_\beta(\mathbf{r}_2,\mathbf{s})d^2s,$$
(4.9)

where s is a two-dimensional vector, H_α ($\alpha = x, y$) are arbitrary functions and $p_{\alpha\beta}$ obey the inequalities

$$p_{\alpha\alpha}(\mathbf{s}) \geq 0,$$
(4.10)

and

$$p_{xx}(\mathbf{s})p_{yy}(\mathbf{s}) - |p_{xy}(\mathbf{s})|^2 \geq 0$$
(4.11)

for all values of $\mathbf{s}$.

4.1.3 Spectral, coherence and polarization properties

Because of the fact that a fluctuating electromagnetic beam-like field is represented by a two-dimensional vector the set of its properties is much richer than that of its scalar counterpart. In what follows we will only be concerned with the properties that carry a certain physical interpretation or can be directly measured, i.e., for determination of which the knowledge of the correlation matrix is not required.

We begin by introducing two quantities that are natural extensions of those in scalar theory. The *spectral density* of an electromagnetic stochastic beam at a point specified by position vector $\mathbf{r}$ is defined by the formula [8]

$$S(\mathbf{r};\omega) = Tr\mathbf{W}(\mathbf{r},\mathbf{r};\omega),$$
(4.12)

where Tr stands for the trace of the matrix. The spectral shifts can be examined with the same expressions as in the scalar case, Eqs. (3.13)–(3.15), but using definition (4.12).

The *spectral degree of coherence* of the electromagnetic stochastic beam, the quantity which can be regarded as the generalization of that for a scalar beam [see Eq. (3.16)], can be shown to be given by the formula [4]

$$\eta(\mathbf{r}_1, \mathbf{r}_2; \omega) = \frac{Tr\mathbf{W}(\mathbf{r}_1, \mathbf{r}_2; \omega)}{\sqrt{S(\mathbf{r}_1; \omega)}\sqrt{S(\mathbf{r}_2; \omega)}}, \tag{4.13}$$

where spectral densities $S(\mathbf{r}_1; \omega)$ and $S(\mathbf{r}_2; \omega)$ are given by Eq. (4.12). The absolute value of the spectral degree of coherence can be measured directly, since it is equal to the visibility of interference fringes in the Young's double-slit experiment [11] (see also [4]).

In order to account for the vectorial nature of the field at any given point in the beam's cross-section one must specify the set of its polarization properties. In this section we will discuss the polarization ellipse representation and in the following section we will proceed with an alternative description, via the Stokes parameters.

If evaluated at coinciding spatial arguments, i.e., at $\mathbf{r}_1 = \mathbf{r}_2 = \mathbf{r}$ the cross-spectral density matrix of the fluctuating field can be represented by a sum of two matrices, say, $\mathbf{W}^{(p)}$ and $\mathbf{W}^{(u)}$ [12] (see also [13])

$$\mathbf{W}(\mathbf{r}, \mathbf{r}; \omega) = \mathbf{W}^{(p)}(\mathbf{r}, \mathbf{r}; \omega) + \mathbf{W}^{(u)}(\mathbf{r}, \mathbf{r}; \omega), \tag{4.14}$$

where

$$\mathbf{W}^{(p)}(\mathbf{r}, \mathbf{r}; \omega) = \begin{bmatrix} B(\mathbf{r}; \omega) & D(\mathbf{r}; \omega) \\ D^*(\mathbf{r}; \omega) & C(\mathbf{r}; \omega) \end{bmatrix}, \tag{4.15}$$

is the singular matrix, i.e., with

$$Det[\mathbf{W}^{(p)}(\mathbf{r}, \mathbf{r}; \omega)] \equiv 0. \tag{4.16}$$

Here Det stands for determinant, and

$$\mathbf{W}^{(u)}(\mathbf{r}, \mathbf{r}; \omega) = \begin{bmatrix} A(\mathbf{r}; \omega) & 0 \\ 0 & A(\mathbf{r}; \omega) \end{bmatrix}, \tag{4.17}$$

is the multiple of the 2×2 identity matrix. Matrices $\mathbf{W}^{(p)}$ and $\mathbf{W}^{(u)}$ are associated with *polarized* and *unpolarized* portions of the beam, for the reasons discussed below. Moreover, it is possible to show ([13], p. 627) that decomposition (4.14) is unique.

On expressing the elements of matrices $\mathbf{W}^{(p)}$ and $\mathbf{W}^{(u)}$ via the elements of the cross-spectral density matrix $\mathbf{W}$, we obtain the following expressions [12]

$$A(\mathbf{r}; \omega) = \frac{1}{2}\left[W_{xx} + W_{yy} - \sqrt{(W_{xx} + W_{yy})^2 + 4|W_{xy}|^2}\right], \tag{4.18}$$

$$B(\mathbf{r}; \omega) = \frac{1}{2}\left[W_{xx} - W_{yy} + \sqrt{(W_{xx} - W_{yy})^2 + 4|W_{xy}|^2}\right], \tag{4.19}$$

$$C(\mathbf{r};\omega) = \frac{1}{2}\left[W_{yy} - W_{xx} + \sqrt{(W_{xx} - W_{yy})^2 + 4|W_{xy}|^2}\right], \qquad (4.20)$$

$$D(\mathbf{r};\omega) = W_{xy}, \qquad D^*(\mathbf{r};\omega) = W_{yx}, \qquad (4.21)$$

where the arguments of all matrices on the right have been omitted.

Further, since matrix $\mathbf{W}^{(p)}$ (4.15) is singular it may be represented in a factorized form ([13], Chapter 4)

$$\mathbf{W}^{(p)}(\mathbf{r},\mathbf{r};\omega) = \begin{bmatrix} \varepsilon_x^*\varepsilon_x & \varepsilon_x^*\varepsilon_y \\ \varepsilon_y^*\varepsilon_x & \varepsilon_y^*\varepsilon_y \end{bmatrix}, \qquad (4.22)$$

where $\varepsilon_x(\mathbf{r};\omega)$ and $\varepsilon_y(\mathbf{r};\omega)$ are the time-independent components of the *equivalent monochromatic electric field* $[\varepsilon_x(\mathbf{r};\omega)e^{-i\omega t}, \varepsilon_y(\mathbf{r};\omega)e^{-i\omega t}]$ at position $\mathbf{r}$ oscillating at frequency ω. The real parts of the components of such equivalent field, denoted here by $\varepsilon_x^{(r)}$ and $\varepsilon_y^{(r)}$ are related to the $\mathbf{W}$-matrix elements by the formulas

$$\begin{aligned} \varepsilon_x^{(r)}(\mathbf{r};\omega) &= \sqrt{B(\mathbf{r};\omega)}\cos[\omega t + \phi_x] \\ \varepsilon_y^{(r)}(\mathbf{r};\omega) &= \sqrt{C(\mathbf{r};\omega)}\cos[\omega t + \phi_y]. \end{aligned} \qquad (4.23)$$

where

$$\phi(\mathbf{r};\omega) = \phi_y(\mathbf{r};\omega) - \phi_x(\mathbf{r};\omega) = arg[D(\mathbf{r};\omega)], \qquad (4.24)$$

is the phase difference between x- and y-field components, since $D = \varepsilon_x^*\varepsilon_y$. On eliminating from Eq. (4.23) the time dependence one can obtain the elliptic quadratic form (see Appendix A of Ref. [12])

$$\begin{aligned} C(\mathbf{r};\omega)[\varepsilon_x^{(r)}(\mathbf{r};\omega)]^2 &- 2Re[D(\mathbf{r};\omega)]\varepsilon_x^{(r)}(\mathbf{r};\omega)\varepsilon_y^{(r)}(\mathbf{r};\omega) + B(\mathbf{r};\omega)[\varepsilon_y^{(r)}(\mathbf{r};\omega)]^2 \\ &= (Im[D(\mathbf{r};\omega)])^2. \end{aligned}$$

$$(4.25)$$

This ellipse is called the *spectral polarization ellipse* of the stochastic field at position $\mathbf{r}$. Thus, at each point within the stochastic beam it is possible to determine the spectral polarization ellipse which coincides with the one associated with the equivalent monochromatic electric field at a given frequency.

It is often convenient to relate some of the characteristics of the polarization ellipse (see Fig. 1.9) to the elements of the cross-spectral density matrix $\mathbf{W}$. For instance, the magnitudes of the semi-axes of the ellipse can be written as

$$\begin{aligned} \varsigma_{1,2}(\mathbf{r};\omega) = \frac{1}{\sqrt{2}}\Bigg[&\sqrt{(W_{xx} - W_{yy})^2 + 4|W_{xy}|^2} \\ &\pm \sqrt{(W_{xx} - W_{yy})^2 + 4[ReW_{xy}]^2}\Bigg]^{1/2}, \end{aligned} \qquad (4.26)$$

where $+$ and $-$ signs on the right side of the formula correspond to *major semi-axis* ς_1 and *minor semi-axis* ς_2, respectively. As in the case of a monochromatic field, the shape of ellipse can be then defined by the ratio

$$\epsilon(\mathbf{r};\omega) = \frac{\varsigma_2(\mathbf{r};\omega)}{\varsigma_1(\mathbf{r};\omega)}, \tag{4.27}$$

which we will refer to as the *spectral degree of ellipticity*. As for monochromatic beam, if $\epsilon = 0$ the elliptic polarization state degenerates into linear, while if $\epsilon = 1$ the state of polarization becomes circular.

The orientation of the ellipse can be determined with the help of the smallest angle formed by the positive x–direction and the direction of the major semi-axis of the ellipse. Such an angle, say ψ, which we will refer to as a *spectral orientation angle*, can be shown to be given by the expression ([12], Appendix B)

$$\psi(\mathbf{r};\omega) = \frac{1}{2} \arctan\left(\frac{2Re[W_{xy}(\mathbf{r};\omega)]}{W_{xx}(\mathbf{r};\omega) - W_{yy}(\mathbf{r};\omega)}\right). \tag{4.28}$$

The angle χ in Fig. 1.9, the *spectral ellipticity angle*, can be shown to be related to the degree of ellipticity by the formula [14]

$$\chi(\mathbf{r};\omega) = \arctan[\epsilon(\mathbf{r};\omega)]. \tag{4.29}$$

The matrix $\mathbf{W}^{(u)}$ in decomposition (4.14) being a multiple of an identity matrix represents the completely random portion of the beam, or, its *unpolarized* portion, because the field it describes does not possess any preferential direction of oscillations.

Frequently it is of importance to find what portion of the total beam is completely polarized at a given point in space. The quantitative measure for this, called the *spectral degree of polarization*, $\wp$, can then be defined as the ratio of the spectral densities

$$\wp(\mathbf{r};\omega) = \frac{S^{(p)}(\mathbf{r};\omega)}{S(\mathbf{r};\omega)}, \tag{4.30}$$

where, evidently, $S^{(p)}(\mathbf{r};\omega) = Tr\mathbf{W}^{(p)}(\mathbf{r};\omega)$. Generally, $0 \leq \wp \leq 1$, with limits $\wp = 0$ for unpolarized light and $\wp = 1$ for completely polarized light. After some algebraic manipulations it is possible to express the degree of polarization via the Tr and Det of the cross-spectral density matrix as [8]

$$\wp(\mathbf{r};\omega) = \sqrt{1 - \frac{4Det[\mathbf{W}(\mathbf{r},\mathbf{r};\omega)]}{[Tr\mathbf{W}(\mathbf{r},\mathbf{r};\omega)]^2}}. \tag{4.31}$$

It is crucial to point out that unlike some other polarization properties, the degree of polarization that we have discussed earlier is invariant under rotations of the coordinate axes, since it is just an algebraic function of two matrix invariants, the trace and the determinant.

We also note that the decomposition (4.14) relies purely on mathematics.

No physical device has been invented so far which would have directly acted as a "polarization beam splitter," separating polarized and unpolarized portions of the random beam. Nevertheless, there exist several theoretical studies that explore individual behavior of the polarized and unpolarized portions of the beam, for instance, on free-space propagation (cf. [15]).

A natural generalization of the degree of polarization to two spatial arguments has been recently developed and is called the *degree of cross-polarization* [16], [17]

$$P(\mathbf{r}_1, \mathbf{r}_2; \omega) = \sqrt{1 - \frac{4Det[\mathbf{W}(\mathbf{r}_1, \mathbf{r}_2; \omega)]}{[Tr\mathbf{W}(\mathbf{r}_1, \mathbf{r}_2; \omega)]^2}}, \tag{4.32}$$

being the measure of similarity between classic degrees of polarization of the beam at two points. The degree of cross-polarization plays an important role in determining the intensity-intensity correlations in the light fields governed by Gaussian statistics.

4.1.4   Classic and generalized Stokes parameters

Among several possible ways of describing the polarization properties at a given point within a light beam the oldest and most frequently used for measurements is based on the set of four quantities, called the *Stokes parameters*, after G. Stokes [18]. The Stokes parameters of deterministic light beams have been reviewed in Chapter 1. In space-frequency treatment the four spectral Stokes parameters at point $\mathbf{r}$ and angular frequency ω have the following definitions

$$
\begin{aligned}
s_0(\mathbf{r}; \omega) &= \langle E_x^*(\mathbf{r}; \omega)E_x(\mathbf{r}; \omega)\rangle + \langle E_y^*(\mathbf{r}; \omega)E_y(\mathbf{r}; \omega)\rangle, \\
s_1(\mathbf{r}; \omega) &= \langle E_x^*(\mathbf{r}; \omega)E_x(\mathbf{r}; \omega)\rangle - \langle E_y^*(\mathbf{r}; \omega)E_y(\mathbf{r}; \omega)\rangle, \\
s_2(\mathbf{r}; \omega) &= \langle E_x^*(\mathbf{r}; \omega)E_y(\mathbf{r}; \omega)\rangle + \langle E_y^*(\mathbf{r}; \omega)E_x(\mathbf{r}; \omega)\rangle, \\
s_3(\mathbf{r}; \omega) &= i[\langle E_y^*(\mathbf{r}; \omega)E_x(\mathbf{r}; \omega)\rangle - \langle E_x^*(\mathbf{r}; \omega)E_y(\mathbf{r}; \omega)\rangle].
\end{aligned}
\tag{4.33}
$$

Hence, all four Stokes parameters are the linear combinations of the cross-spectral density matrix elements, viz.,

$$
\begin{aligned}
s_0(\mathbf{r}; \omega) &= W_{xx}(\mathbf{r}, \mathbf{r}; \omega) + W_{yy}(\mathbf{r}, \mathbf{r}; \omega), \\
s_1(\mathbf{r}; \omega) &= W_{xx}(\mathbf{r}, \mathbf{r}; \omega) - W_{yy}(\mathbf{r}, \mathbf{r}; \omega), \\
s_2(\mathbf{r}; \omega) &= W_{xy}(\mathbf{r}, \mathbf{r}; \omega) + W_{yx}(\mathbf{r}, \mathbf{r}; \omega), \\
s_3(\mathbf{r}; \omega) &= i[W_{yx}(\mathbf{r}, \mathbf{r}; \omega) - W_{xy}(\mathbf{r}, \mathbf{r}; \omega)].
\end{aligned}
\tag{4.34}
$$

We note that the definition of the Stokes parameter $s_0(\mathbf{r}; \omega)$ coincides with that for the spectral density of the stochastic beam [see Eq. (4.12)], i.e., $s_0(\mathbf{r}; \omega) = S(\mathbf{r}; \omega)$. Hence, often Stokes parameters s_1, s_2 and s_3 normalized by s_0 are used for the purpose of treating the polarimetric features of the beam independently of its level of intensity. The degree of polarization and the parameters of the polarization ellipse can be expressed via the Stokes parameters by the formulas

$$\wp(\mathbf{r};\omega) = \frac{\sqrt{s_1^2(\mathbf{r};\omega) + s_2^2(\mathbf{r};\omega) + s_3^2(\mathbf{r};\omega)}}{s_0^2(\mathbf{r};\omega)}, \qquad (4.35)$$

$$\varsigma_{1,2}(\mathbf{r};\omega) = \left[\frac{s_0(\mathbf{r}) \pm \sqrt{s_1^2(\mathbf{r}) + s_2^2(\mathbf{r})}}{s_3(\mathbf{r})}\right]^{1/2}, \qquad (4.36)$$

$$\psi(\mathbf{r};\omega) = \frac{1}{2}\arctan\left[\frac{s_2(\mathbf{r})}{s_1(\mathbf{r})}\right]. \qquad (4.37)$$

The reader can consult [13] and [14] for other ellipsometric relations.

Just like for monochromatic fields, there exists an alternative useful geometrical representation of the Stokes parameters of random beams by means of the Poincaré sphere [19] (see Fig. 1.10). The set of all possible polarization states of a random electromagnetic wave at a given position in space belongs to the surface and the inner region of the Poincaré sphere, i.e., satisfies the inequality

$$s_1^2(\mathbf{r};\omega) + s_2^2(\mathbf{r};\omega) + s_3^2(\mathbf{r};\omega) \le s_0^2(\mathbf{r};\omega). \qquad (4.38)$$

Here s_1, s_2 and s_3 may be viewed as the Cartesian coordinates of the vector which characterizes the state of polarization, while its magnitude, relative to s_0, is equal to the degree of polarization. When the wave is completely polarized, the inequality (4.38) reduces to the equality and the set of possible states of polarization is then confined to the surface of the sphere. Hence, the Poincaré sphere, which we introduced in Chapter 1 (see Fig. 1.10) becomes an extremely useful visual tool in problems involving complex (and variable) distributions of polarization states of electromagnetic fields.

A generalization of conventional Stokes parameters from one to two spatial arguments has been made in Ref. [20], by means of the expressions

$$\begin{aligned}
S_0(\mathbf{r}_1,\mathbf{r}_2;\omega) &= \langle E_x^*(\mathbf{r}_1;\omega)E_x(\mathbf{r}_2;\omega)\rangle + \langle E_y^*(\mathbf{r}_1;\omega)E_y(\mathbf{r}_2;\omega)\rangle, \\
S_1(\mathbf{r}_1,\mathbf{r}_2;\omega) &= \langle E_x^*(\mathbf{r}_1;\omega)E_x(\mathbf{r}_2;\omega)\rangle - \langle E_y^*(\mathbf{r}_1;\omega)E_y(\mathbf{r}_2;\omega)\rangle, \\
S_2(\mathbf{r}_1,\mathbf{r}_2;\omega) &= \langle E_x^*(\mathbf{r}_1;\omega)E_y(\mathbf{r}_2;\omega)\rangle + \langle E_y^*(\mathbf{r}_1;\omega)E_x(\mathbf{r}_2;\omega)\rangle, \\
S_3(\mathbf{r}_1,\mathbf{r}_2;\omega) &= i[\langle E_y^*(\mathbf{r}_1;\omega)E_x(\mathbf{r}_2;\omega)\rangle - \langle E_x^*(\mathbf{r}_1;\omega)E_y(\mathbf{r}_2;\omega)\rangle].
\end{aligned} \qquad (4.39)$$

Hence, the generalized Stokes parameters can be expressed via the elements of the cross-spectral density matrix as

$$\begin{aligned}
S_0(\mathbf{r}_1,\mathbf{r}_2;\omega) &= W_{xx}(\mathbf{r}_1,\mathbf{r}_2;\omega) + W_{yy}(\mathbf{r}_1,\mathbf{r}_2;\omega), \\
S_1(\mathbf{r}_1,\mathbf{r}_2;\omega) &= W_{xx}(\mathbf{r}_1,\mathbf{r}_2;\omega) - W_{yy}(\mathbf{r}_1,\mathbf{r}_2;\omega), \\
S_2(\mathbf{r}_1,\mathbf{r}_2;\omega) &= W_{xy}(\mathbf{r}_1,\mathbf{r}_2;\omega) + W_{yx}(\mathbf{r}_1,\mathbf{r}_2;\omega), \\
S_3(\mathbf{r}_1,\mathbf{r}_2;\omega) &= i[W_{yx}(\mathbf{r}_1,\mathbf{r}_2;\omega) - W_{xy}(\mathbf{r}_1,\mathbf{r}_2;\omega)].
\end{aligned} \qquad (4.40)$$

The two-point Stokes parameters may be used for propagation of classic Stokes parameters and have a useful physical interpretation [21] as well as a number of interesting relations with other statistical properties, for example, with the spectral degree of coherence [20]:

$$\eta(\mathbf{r}_1, \mathbf{r}_2; \omega) = \frac{S_0(\mathbf{r}_1, \mathbf{r}_2; \omega)}{\sqrt{S_0(\mathbf{r}_1, \mathbf{r}_1; \omega)}\sqrt{S_0(\mathbf{r}_2, \mathbf{r}_2; \omega)}}. \qquad (4.41)$$

4.1.5 Coherent mode decomposition

In Chapter 3 we have discussed the representation for the cross-spectral density functions of sources generating scalar beam-like stochastic fields as an infinite series of coherent modes. As was shown in [6], [22]–[24], such representation can also be generalized to the electromagnetic domain. In Refs. [6] and [22] it is suggested that two coupled integral equations are simultaneously solved for determining the series form of the four elements of the cross-spectral density matrix. In the general setting, solving of the coupled integral equations presents a daunting task. On the other hand, in Ref. [24] a simpler approach is taken, in which two uncoupled integral equations are solved for the diagonal matrix elements and the obtained solutions are used for construction of modes for the off-diagonal matrix elements. We will now outline some details of this method. Since the diagonal elements of the cross-spectral density matrix are Hermitian, they can be expressed in terms of a scalar coherent-mode representation, i.e., as:

$$W_{xx}(\boldsymbol{\rho}_1, \boldsymbol{\rho}_2; \omega) = \sum_{n=0}^{\infty} \lambda^{(x)}(\omega)\phi_n^*(\boldsymbol{\rho}_1; \omega)\phi_n(\boldsymbol{\rho}_2; \omega), \qquad (4.42)$$

and

$$W_{yy}(\boldsymbol{\rho}_1, \boldsymbol{\rho}_2; \omega) = \sum_{n=0}^{\infty} \lambda^{(y)}(\omega)\psi_n^*(\boldsymbol{\rho}_1; \omega)\psi_n(\boldsymbol{\rho}_2; \omega), \qquad (4.43)$$

where the eigenfunctions $\phi_n(\boldsymbol{\rho}; \omega)$ and $\psi_n(\boldsymbol{\rho}; \omega)$ as well as the eigenvalues $\lambda^{(x)}(\omega)$ and $\lambda^{(y)}(\omega)$ are found by solving the integral equations

$$\int_D W_{xx}(\boldsymbol{\rho}_1, \boldsymbol{\rho}_2; \omega)\phi_n(\boldsymbol{\rho}_1; \omega) = \lambda^{(x)}(\omega)\phi_n(\boldsymbol{\rho}_2; \omega), \qquad (4.44)$$

and

$$\int_D W_{yy}(\boldsymbol{\rho}_1, \boldsymbol{\rho}_2; \omega)\psi_n(\boldsymbol{\rho}_1; \omega) = \lambda^{(y)}(\omega)\psi_n(\boldsymbol{\rho}_2; \omega), \qquad (4.45)$$

respectively, where the integration is performed over the domain D occupied by the source. On the other hand, the off-diagonal elements of the correlation matrix are not Hermitian and, hence, they do not possess the scalar coherent mode representation. However, it is still possible to expand each of the off-diagonal elements in a double series in terms of the modes $\phi_n(\boldsymbol{\rho};\omega)$ and $\psi_n(\boldsymbol{\rho};\omega)$, $n = 1...\infty$ of the diagonal elements. We first expand the components $E_x(\boldsymbol{\rho};\omega)$ and $E_y(\boldsymbol{\rho};\omega)$ of the electric vector-field as

$$E_x(\boldsymbol{\rho};\omega) = \sum_{n=0}^{\infty} a_n(\omega)\phi_n(\boldsymbol{\rho};\omega), \tag{4.46}$$

and

$$E_y(\boldsymbol{\rho};\omega) = \sum_{n=0}^{\infty} b_n(\omega)\psi_n(\boldsymbol{\rho};\omega), \tag{4.47}$$

with coefficients a_n and b_n being random numbers. Hence, on taking cross-correlation of Eqs. (4.46) and (4.47) we find that

$$W_{xy}(\boldsymbol{\rho}_1,\boldsymbol{\rho}_2;\omega) = \langle E_x^*(\boldsymbol{\rho}_1;\omega)E_y^*(\boldsymbol{\rho}_2;\omega)\rangle =$$
$$= \sum_{n=0}^{\infty}\sum_{m=0}^{\infty} \langle a_n^*(\omega)b_m(\omega)\rangle \phi_n^*(\boldsymbol{\rho}_1;\omega)\psi_m(\boldsymbol{\rho}_2;\omega). \tag{4.48}$$

The coefficients in the double sum can be determined from the orthogonality of the eigenfunctions. Indeed, after multiplying each side of Eq. (4.48) by the product $\phi_{m'}^*(\boldsymbol{\rho}_1;\omega)\psi_{m'}(\boldsymbol{\rho}_2;\omega)$, integrating over domain D and using the orthogonality relations

$$\int_D \phi_n^*(\boldsymbol{\rho};\omega)\phi_m(\boldsymbol{\rho};\omega)d^3r = \delta_{nm}, \tag{4.49}$$

$$\int_D \psi_n^*(\boldsymbol{\rho};\omega)\psi_m(\boldsymbol{\rho};\omega)d^3r = \delta_{nm}, \tag{4.50}$$

with δ_{nm} being the Kronecker symbol, we find that

$$\Lambda_{nm}(\omega) \equiv \langle a_n^*(\omega)b_m(\omega)\rangle = \int_D\int_D \phi_n(\boldsymbol{\rho}_1;\omega)\psi_m^*(\boldsymbol{\rho}_2;\omega)d^2\rho_1 d^2\rho_2. \tag{4.51}$$

It then follows from Eqs. (4.48) and (4.51) that

$$W_{xy}(\boldsymbol{\rho}_1,\boldsymbol{\rho}_2;\omega) = \sum_{n=0}^{\infty}\sum_{m=0}^{\infty} \Lambda_{nm}(\omega)\phi_n^*(\boldsymbol{\rho}_1;\omega)\psi_m(\boldsymbol{\rho}_2;\omega). \tag{4.52}$$

By following the same procedure the representation of the other off-diagonal element can be established as:

$$W_{xy}(\boldsymbol{\rho}_1,\boldsymbol{\rho}_2;\omega) = \sum_{n=0}^{\infty}\sum_{m=0}^{\infty} \Lambda_{nm}^*(\omega)\phi_n^*(\boldsymbol{\rho}_2;\omega)\psi_m(\boldsymbol{\rho}_1;\omega), \tag{4.53}$$

where the anti-Hermitian property of the cross-spectral density function, $W_{yx}^*(\boldsymbol{\rho}_1,\boldsymbol{\rho}_2;\omega) = W_{xy}(\boldsymbol{\rho}_2,\boldsymbol{\rho}_1;\omega)$ was employed. The representations of the off-diagonal elements as the double sums can be viewed as the *bimodal* expansions.

4.1.6 Angular-spectrum decomposition

Angular spectrum decomposition electromagnetic fields, both deterministic and random, is another important mathematical representation that substantially simplifies the analysis of the properties of sources and fields they generate. It has played the major role in radiometry, scattering theory and wave propagation in random media [8]. As was already stated in the previous chapters, the possibility of decomposing of arbitrary wave fronts into as simple forms as plane waves leads, in many instances, to the significant reduction in the problem's complexity.

In this section we will first briefly review the angular spectrum representation for deterministic electromagnetic waves [8] and then will turn to the extension of this classic development to the case of the electromagnetic stochastic fields [25].

The two Cartesian components x and y transverse to the direction of propagation, z, of a monochromatic electric field vector $\mathbf{E}(\mathbf{r}; \omega)$ at a point with position vector $\mathbf{r} = (x, y, z)$ and frequency ω may be represented as the double integral:

$$E_\alpha(x, y, z; \omega) = \int\limits_{-\infty}^{\infty} \int\limits_{-\infty}^{\infty} a_\alpha(k_x, k_y; \omega) \exp[i(k_x x + k_y y + k_z z)] dk_x dk_y, \quad (4.54)$$

where $\alpha = (x, y)$ and $a_\alpha(u, v; \omega)$ is the angular spectrum component of the field given by the expression

$$a_\alpha(k_x, k_y; \omega) = \frac{1}{(2\pi)^2} \int\limits_{-\infty}^{\infty} \int\limits_{-\infty}^{\infty} E_\alpha(x', y', 0; \omega) \exp[-i(k_x x' + k_y y')] dx' dy',$$

$$(4.55)$$

where k_z was defined in (1.64). For the sake of completeness we would like to note that one can also decompose the z-component $E_z(\mathbf{r}; \omega)$ of the electric field in a similar fashion. Namely, it follows from the Maxwell's divergence equation $\nabla \cdot \mathbf{E}(\mathbf{r}; \omega) = 0$ that the z component of the angular spectrum may be related to the x and y components as

$$a_z(k_x, k_y, k_z) = -\frac{1}{k_z}[k_x a_x(k_x, k_y, k_z) + k_y a_y(k_x, k_y, k_z)]. \quad (4.56)$$

Similarly, the angular spectrum representation may be developed for the magnetic field vector [13]. Thus, Eqs. (4.54)–(4.56) represent a monochromatic electromagnetic field vector at position $\mathbf{r}$ in the half-space $z > 0$ in terms of the x and y components of the electric field in the plane $z = 0$, via the modes which possess the plane-wave structure.

As for the scalar waves, one of the major advantages of the electromagnetic angular spectrum decomposition stems from the fact that the components of the field along a particular direction in the far zone of the source plane $z = 0$

are simply related to the corresponding components of the electromagnetic angular spectrum evaluated at certain spatial frequencies. Namely, the following set of relations may be established, as $kz \to \infty$:

$$E_x(r\mathbf{u}; \omega) = -i2\pi k u_z A_x(k u_x, k u_y; \omega) \frac{\exp(ikr)}{r},$$

$$E_y(r\mathbf{u}; \omega) = -i2\pi k u_z A_y(k u_x, k u_y; \omega) \frac{\exp(ikr)}{r},$$ (4.57)

$$E_z(r\mathbf{u}; \omega) = i2\pi k[u_x A_x(k u_x, k u_y; \omega) + u_y A_y(k u_x, k u_y; \omega)] \frac{\exp(ikr)}{r},$$

where, as before, $\mathbf{u} = (u_x, u_y, u_z)$ is the unit vector in the direction of the position vector $\mathbf{r}$, $r = |\mathbf{r}|$.

If the electric field is of stochastic nature then the angular spectrum representation may be established for the electric cross-spectral density matrix. In fact, on substituting from Eqs. (4.54) to the definition of the cross-spectral density matrix we find that

$$W_{\alpha\beta}(\mathbf{r}_1, \mathbf{r}_2; \omega) = \int_{-\infty}^{\infty} \int_{-\infty}^{\infty} \int_{-\infty}^{\infty} \int_{-\infty}^{\infty} A_{\alpha\beta}(k_{1x}, k_{1y}, k_{2x}, k_{2y}; \omega)$$

$$\times \exp[i(k_{2x}x_2 - k_{1x}x_1 + k_{2y}y_2 - k_{1y}y_1)] \qquad (4.58)$$

$$\times \exp[i(k_{2z}z_2 - k_{1z}^*z_1)] dk_{1x} dk_{1y} dk_{2x} dk_{2y},$$

$$(\alpha, \beta = x, y),$$

where

$$A_{\alpha\beta}(k_{1x}, k_{1y}, k_{2x}, k_{2y}; \omega) = \frac{1}{(2\pi)^4} \int_{-\infty}^{\infty} \int_{-\infty}^{\infty} \int_{-\infty}^{\infty} \int_{-\infty}^{\infty} W_{\alpha\beta}(x_1, y_1, 0, x_2, y_2, 0; \omega)$$

$$\times \exp[-i(k_{2x}x_2 - k_{1x}x_1 + k_{2y}y_2 - k_{1y}y_1)] dx_1 dy_1 dx_2 dy_2,$$

$$(\alpha, \beta = x, y),$$

(4.59)

are the elements of the 2×2 *angular-correlation matrix*. In the case when the field is scalar and is characterized by the cross-spectral density function, this matrix reduces to the angular correlation function (see Chapter 3). The representation (4.58)–(4.59) is sufficient for describing the stochastic electromagnetic fields in the paraxial limit. Further, it follows from Maxwell's equations that the angular correlation functions of the field involving its z-component can be related to $A_{\alpha\beta}$ ($\alpha, \beta = x, y$) as follows [25]:

$$A_{zx}(k_{1x}, k_{1y}, k_{2x}, k_{2y}; \omega) = -\frac{1}{k_{1z}^*}[k_{1x}A_{xx}(k_{1x}, k_{1y}, k_{2x}, k_{2y}; \omega)$$

$$+ k_{1y}A_{yx}(k_{1x}, k_{1y}, k_{2x}, k_{2y}; \omega)],$$

$$A_{xz}(k_{1x}, k_{1y}, k_{2x}, k_{2y}; \omega) = -\frac{1}{k_{2z}}\big[k_{2x}A_{xx}(k_{1x}, k_{1y}, k_{2x}, k_{2y}; \omega)$$
$$+k_{2y}A_{xy}(k_{1x}, k_{1y}, k_{2x}, k_{2y}; \omega)\big],$$

$$A_{zy}(k_{1x}, k_{1y}, k_{2x}, k_{2y}; \omega) = -\frac{1}{k_{1z}^*}\big[k_{1x}A_{xy}(k_{1x}, k_{1y}, k_{2x}, k_{2y}; \omega)$$
$$+k_{1y}A_{yy}(k_{1x}, k_{1y}, k_{2x}, k_{2y}; \omega)\big],$$

$$A_{yz}(k_{1x}, k_{1y}, k_{2x}, k_{2y}; \omega) = -\frac{1}{k_{2z}}\big[k_{2x}A_{yx}(k_{1x}, k_{1y}, k_{2x}, k_{2y}; \omega)$$
$$+k_{2y}A_{yy}(k_{1x}, k_{1y}, k_{2x}, k_{2y}; \omega)\big],$$

$$A_{zz}(k_{1x}, k_{1y}, k_{2x}, k_{2y}; \omega) = \frac{1}{k_{1z}^* k_{2z}}\big[k_{1x}k_{2x}A_{xx}(k_{1x}, k_{1y}, k_{2x}, k_{2y}; \omega)$$
$$+k_{1y}k_{2y}A_{yy}(k_{1x}, k_{1y}, k_{2x}, k_{2y}; \omega)$$
$$+k_{1x}k_{2y}A_{xy}(k_{1x}, k_{1y}, k_{2x}, k_{2y}; \omega)$$
$$+k_{1y}k_{2x}A_{yx}(k_{1x}, k_{1y}, k_{2x}, k_{2y}; \omega)\big].$$

It follows from Eqs. (4.57) that the asymptotic forms of the cross-spectral density functions in the far field are

$$W_{\alpha\beta}(r_1\mathbf{u}_1, r_2\mathbf{u}_2; \omega) = (2\pi k)^2 u_{z1}u_{z2}A_{\alpha\beta}(ku_{\perp 1}; ku_{\perp 2})$$
$$\times \frac{\exp[ik(r_2 - r_1)]}{r_1 r_2}, \quad (\alpha, \beta = x, y). \tag{4.60}$$

On the other hand, the elements of the cross-spectral density tensor in the far field involving the z-field component can be expressed in terms of the angular correlation functions of the x and y-components, namely

$$W_{xz}(r_1\mathbf{u}_1, r_2\mathbf{u}_2; \omega) = -(2\pi k)^2 u_{z1}[u_{x2}A_{xx}(ku_{\perp 1}, ku_{\perp 2}; \omega)$$
$$+ u_{y2}A_{xy}(ku_{\perp 1}, ku_{\perp 2}; \omega)]\frac{\exp[ik(r_2 - r_1)]}{r_1 r_2}, \tag{4.61}$$

$$W_{yz}(r_1\mathbf{u}_1, r_2\mathbf{u}_2; \omega) = -(2\pi k)^2 u_{z1}[u_{x2}A_{yx}(ku_{\perp 1}, ku_{\perp 2}; \omega)$$
$$+ u_{y2}A_{yy}(ku_{\perp 1}, ku_{\perp 2}; \omega)]\frac{\exp[ik(r_2 - r_1)]}{r_1 r_2}, \tag{4.62}$$

$$W_{zx}(r_1\mathbf{u}_1, r_2\mathbf{u}_2; \omega) = -(2\pi k)^2 u_{z2}[u_{x1}A_{xx}(ku_{\perp 1}, ku_{\perp 2}; \omega)$$
$$+ u_{y1}A_{yx}(ku_{\perp 1}, ku_{\perp 2}; \omega)]\frac{\exp[ik(r_2 - r_1)]}{r_1 r_2}, \tag{4.63}$$

$$W_{zy}(r_1\mathbf{u}_1, r_2\mathbf{u}_2; \omega) = -(2\pi k)^2 u_{z2}[u_{x1}A_{xy}(ku_{\perp 1}, ku_{\perp 2}; \omega)$$
$$+ u_{y1}A_{yy}(ku_{\perp 1}, ku_{\perp 2}; \omega)]\frac{\exp[ik(r_2 - r_1)]}{r_1 r_2}, \tag{4.64}$$

$$W_{zz}(r_1\mathbf{u}_1, r_2\mathbf{u}_2; \omega) = (2\pi k)^2 \{ u_{x1} [u_{x2} A_{xx}(k u_{\perp 1}, k u_{\perp 2}; \omega)$$
$$+ u_{y2} A_{xy}(k u_{\perp 1}, k u_{\perp 2}; \omega)]$$
$$+ u_{y1} [u_{x2} A_{yx}(k u_{\perp 1}, k u_{\perp 2}; \omega)$$
$$+ u_{y2} A_{yy}(k u_{\perp 1}, k u_{\perp 2}; \omega)] \}$$
$$\times \frac{\exp[ik(r_2 - r_1)]}{r_1 r_2}. \tag{4.65}$$

In the case when the field is electromagnetic, the classic measure that characterizes the radiated energy is the *Poynting vector* defined as ([8], Section 6.1)

$$S_p(\mathbf{r}; t) = \frac{c}{2\pi} Re \langle \mathbf{E}^*(\mathbf{r}; t) \circ \mathbf{H}(\mathbf{r}; t) \rangle_T, \tag{4.66}$$

where time average is used and $\mathbf{E}$ and $\mathbf{H}$ stand for the full electric and magnetic field vectors in a space-time domain, respectively. In terms of the cross-spectral density matrices the Poynting vector may be expressed as [25]

$$S_p(\mathbf{r}; \omega) = \frac{1}{2} Re \langle \mathbf{E}^*(\mathbf{r}; \omega) \circ \mathbf{H}(\mathbf{r}; \omega) \rangle. \tag{4.67}$$

Then the radiant intensity of the electromagnetic field can be defined as

$$J(\mathbf{u}; \omega) = \lim_{r \to \infty} [r^2 |S_p(\mathbf{r}; \omega)|]. \tag{4.68}$$

As in the scalar case, it provides the measure of the angular distribution of energy radiating into the far field. It has been proven in [25] that the radiant intensity may be expressed solely in terms of the angular correlation functions of the electric field components

$$J(\mathbf{u}; \omega) = \frac{1}{2Z_f} (2\pi k)^2 [u_z^2 A_{xx}(k u_\perp, k u_\perp; \omega) + u_z^2 A_{yy}(k u_\perp, k u_\perp; \omega)$$
$$+ u_x^2 A_{xx}(k u_\perp, k u_\perp; \omega) + u_y^2 A_{yy}(k u_\perp, k u_\perp; \omega)$$
$$+ 2 u_x u_y Re\{ A_{xy}(k u_\perp, k u_\perp; \omega) \}], \tag{4.69}$$

where $Z_f = \sqrt{\mu_0/\epsilon_0}$ is the impedance of free space. In the paraxial limit, $u_x \ll u_z$ and $u_y \ll u_z$ the last three terms of the previous expression become negligible and it reduces to the form

$$J(\mathbf{u}; \omega) = \frac{1}{2Z_f} (2\pi k)^2 [u_z^2 A_{xx}(k u_\perp, k u_\perp; \omega) + u_z^2 A_{yy}(k u_\perp, k u_\perp; \omega)]. \tag{4.70}$$

We note that if the field is monochromatic the radiant intensity reduces to the expression

$$J(\mathbf{u}; \omega) = \frac{1}{2Z_f} (2\pi k)^2 [u_z^2 |a_x(k u_\perp; \omega)|^2 + u_z^2 |a_y(k u_\perp; \omega)|^2$$
$$+ |u_x a_x(k u_\perp; \omega) + u_y a_y(k u_\perp; \omega)|^2]. \tag{4.71}$$

In the paraxial limit only the first two terms of the previous expression remain:

$$J(\mathbf{u};\omega) = \frac{1}{2Z_f}(2\pi k)^2 u_z^2[|a_x(k u_\perp;\omega)|^2 + |a_y(k u_\perp;\omega)|^2].\tag{4.72}$$

4.2 Electromagnetic quasi-homogeneous sources

The concept of quasi-homogeneity was developed a long time ago in connection with locally stationary random processes [26] and scattering media [27]. It was later adopted for scalar optical fields [8] and electromagnetic beam-like fields [28]. In this section we discuss the electromagnetic quasi-homogeneous sources and beam-like fields they generate. We also derive the electromagnetic version of the reciprocity relations which connects intensity and coherence properties of radiation in the source plane and in the far field, in the case when the sources are uniformly polarized, i.e., when they have the same polarization at each point. Conditions for spectral and polarization invariance conclude our analysis.

4.2.1 Far-field analysis and the reciprocity relations

Let us consider a fluctuating, planar, secondary electromagnetic source, located in the plane $z = 0$ and radiating into the half-space $z > 0$. We assume that the radiated field is beam-like, propagating close to the z axis, and that the source fluctuations are represented by a statistical ensemble that is stationary, at least in the wide sense. The second-order correlation properties of the source may be characterized by the electric cross-spectral density matrix (4.4) at the two-dimensional position vectors $\boldsymbol{\rho}_1' = (x_1',y_1')$ and $\boldsymbol{\rho}_2' = (x_2',y_2')$ of points in the source plane (see Fig. 3.2). Then the cross-spectral density matrix representing an electromagnetic quasi-homogeneous source may be obtained as a generalization of the cross-spectral density function of a scalar quasi-homogeneous source, as

$$W_{\alpha\beta}(\boldsymbol{\rho}_1',\boldsymbol{\rho}_2';\omega) = \sqrt{S_\alpha\left(\frac{\boldsymbol{\rho}_1'+\boldsymbol{\rho}_2'}{2};\omega\right)}\sqrt{S_\beta\left(\frac{\boldsymbol{\rho}_1'+\boldsymbol{\rho}_2'}{2};\omega\right)}\eta_{\alpha\beta}(\boldsymbol{\rho}_2'-\boldsymbol{\rho}_1';\omega),$$
$$(\alpha = x,y; \beta = x,y),$$
$$\tag{4.73}$$

where $S_\alpha(\boldsymbol{\rho}';\omega) = W_{\alpha\alpha}(\boldsymbol{\rho}',\boldsymbol{\rho}';\omega)$, $\alpha = (x,y)$ are the spectral densities of the electric field components E_α and $\eta_{\alpha\beta}(\boldsymbol{\rho}_2'-\boldsymbol{\rho}_1';\omega)$ are correlations between components E_α and E_β. Let $S_\alpha(\boldsymbol{\rho}';\omega)$, $\alpha = (x,y)$ vary much more slowly with position $\boldsymbol{\rho}'$ than $\eta_{\alpha\beta}(\boldsymbol{\rho}_d')$ vary with difference vector $\boldsymbol{\rho}_d'$, $\boldsymbol{\rho}_d' = \boldsymbol{\rho}_2'-\boldsymbol{\rho}_1'$, (see Fig. 4.2).

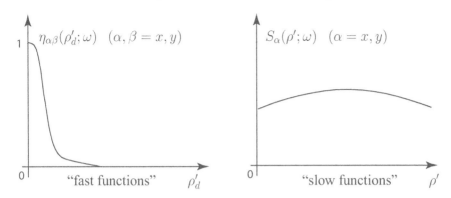

FIGURE 4.2
Illustration of the concept of an electromagnetic quasi-homogeneous source.

On substituting from Eq. (4.73) into definitions of the spectral density and the degree of coherence, Eqs. (4.12) and (4.13), respectively, we find that for the electromagnetic quasi-homogeneous source these quantities have the following forms:

$$S(\boldsymbol{\rho}';\omega) = S_x(\boldsymbol{\rho}';\omega) + S_y(\boldsymbol{\rho}';\omega) \tag{4.74}$$

and

$$
\eta(\boldsymbol{\rho}'_1,\boldsymbol{\rho}'_2;\omega)
$$
$$
= \frac{S_x\left(\frac{\boldsymbol{\rho}'_1+\boldsymbol{\rho}'_2}{2};\omega\right)\eta_{xx}(\boldsymbol{\rho}'_2-\boldsymbol{\rho}'_1;\omega) + S_y\left(\frac{\boldsymbol{\rho}'_1+\boldsymbol{\rho}'_2}{2};\omega\right)\eta_{yy}(\boldsymbol{\rho}'_2-\boldsymbol{\rho}'_1;\omega)}{S\left(\frac{\boldsymbol{\rho}'_1+\boldsymbol{\rho}'_2}{2};\omega\right)}. \tag{4.75}
$$

We will now restrict out analysis to sources for which the normalized spectral Stokes parameters are independent of $\boldsymbol{\rho}'$. The spectral degree of polarization and the spectral polarization ellipse associated with the polarized portion of the beam will then also have constant values across the source. Such sources may be referred to as *uniformly polarized*. It can be proven rigorously (see Appendix A of [28]) that the quasi-homogeneous source is uniformly polarized at frequency ω if and only if the spectral densities $S_x(\boldsymbol{\rho}';\omega)$ and $S_y(\boldsymbol{\rho}';\omega)$ are proportional to each other at given ω, i.e., if and only if

$$S_y(\boldsymbol{\rho}';\omega) = \Theta(\omega)S_x(\boldsymbol{\rho}';\omega), \tag{4.76}$$

and if, in addition, the correlation coefficient $\eta_{xy}(\boldsymbol{\rho}',\boldsymbol{\rho}';\omega)$ is independent of position, i.e.,

$$\eta_{xy}(\boldsymbol{\rho}',\boldsymbol{\rho}';\omega) \equiv \eta_{xy}(\omega). \tag{4.77}$$

A source that is linearly polarized along the y direction obviously corresponds to the limit $\Theta(\omega) \to \infty$. It follows from Eqs. (4.74) and (4.76) that

$$S_x(\boldsymbol{\rho}';\omega) = \frac{1}{1+\Theta(\omega)} S(\boldsymbol{\rho}';\omega), \qquad (4.78)$$

$$S_y(\boldsymbol{\rho}';\omega) = \frac{\Theta(\omega)}{1+\Theta(\omega)} S(\boldsymbol{\rho}';\omega). \qquad (4.79)$$

The last two equations if used in the Eq. (4.73) lead to the elements of the cross-spectral density matrix of the source that satisfy conditions (4.76) and (4.77) of the forms

$$W_{\alpha\beta}(\boldsymbol{\rho}'_1,\boldsymbol{\rho}'_2;\omega) = \Theta_{\alpha\beta}(\omega)S\left(\frac{\boldsymbol{\rho}'_1 + \boldsymbol{\rho}'_2}{2};\omega\right)\eta_{\alpha\beta}(\boldsymbol{\rho}'_2 - \boldsymbol{\rho}'_1;\omega), \qquad (4.80)$$

with the coefficients $\Theta_{\alpha\beta}(\omega)$

$$\Theta_{\alpha\beta}(\omega) = \begin{cases} \dfrac{1}{1+\Theta(\omega)}, & \alpha = \beta = x; \\[2ex] \dfrac{\sqrt{\Theta(\omega)}}{1+\Theta(\omega)}, & \alpha \neq \beta; \\[2ex] \dfrac{\Theta(\omega)}{1+\Theta(\omega)}, & \alpha = \beta = y. \end{cases} \qquad (4.81)$$

Then the elements of the cross-spectral density matrix generated by a uniformly polarized quasi-homogeneous source are determined in the far field by substituting from Eq. (4.80) into Eqs. (4.59) and (4.60) (see also [28]) and one finds that:

$$W_{\alpha\beta}^{(\infty)}(r_1\mathbf{u}_1, r_2\mathbf{u}_2;\omega) = (2\pi k)^2 \cos\theta_1 \cos\theta_2 \Theta_{\alpha\beta}(\omega)\tilde{S}(k(\mathbf{u}_{2\perp} - \mathbf{u}_{1\perp});\omega)$$
$$\times \tilde{\eta}_{\alpha\beta}\left(\frac{k(\mathbf{u}_{2\perp} + \mathbf{u}_{1\perp})}{2};\omega\right)\frac{\exp[ik(r_2 - r_1)]}{r_1 r_2}. \qquad (4.82)$$

Here θ_1 and θ_2 are the angles between the unit vectors $\mathbf{u}_1$ and $\mathbf{u}_2$ and the beam axis z (see Fig. 4.2). Hence, the spectral density of the far field reduces to the expression

$$S^{(\infty)}(r\mathbf{u};\omega) = Tr\mathbf{W}^{(\infty)}(r\mathbf{u}, r\mathbf{u};\omega)$$
$$= (2\pi k)^2 \frac{\cos^2\theta}{r^2}\tilde{S}(0;\omega) \qquad (4.83)$$
$$\times \left[\Theta_{xx}(\omega)\tilde{\eta}_{xx}(k\mathbf{u}_\perp;\omega) + \Theta_{yy}(\omega)\tilde{\eta}_{yy}(k\mathbf{u}_\perp;\omega)\right].$$

Further, on taking the Fourier transform of Eq. (4.75), we arrive at formula

$$\tilde{\eta}(k\mathbf{u}_\perp;\omega) = \Theta_{xx}(\omega)\tilde{\eta}_{xx}(k\mathbf{u}_\perp;\omega) + \Theta_{yy}(\omega)\tilde{\eta}_{yy}(k\mathbf{u}_\perp;\omega). \qquad (4.84)$$

On combining the last two formulas the following result is obtained:

$$S^{(\infty)}(r\mathbf{u};\omega) = (2\pi k)^2 \cos^2\theta \tilde{S}(0;\omega)\tilde{\eta}(k\mathbf{u}_\perp;\omega)\frac{1}{r^2}. \tag{4.85}$$

It is the mathematical form of the first reciprocity relation for beams generated by a planar, secondary, uniformly polarized electromagnetic quasi-homogeneous source: *the spectral density of the far field is proportional to the product of the Fourier transform of the spectral degree of coherence of the field across the source and of $\cos^2\theta/r^2$*. This relation is of utmost importance for developing sources with prescribed far fields.

Recall that the spectral degree of coherence of the far field at points $\mathbf{r}_1 = r_1\mathbf{u}_1$ and $\mathbf{r}_2 = r_2\mathbf{u}_2$, is given by expression

$$\eta^{(\infty)}(r_1\mathbf{u}_1, r_2\mathbf{u}_2;\omega) = \frac{Tr\mathbf{W}^{(\infty)}(r_1\mathbf{u}_1, r_2\mathbf{u}_2;\omega)}{\sqrt{S^{(\infty)}(r_1\mathbf{u}_1;\omega)}\sqrt{S^{(\infty)}(r_2\mathbf{u}_2;\omega)}}. \tag{4.86}$$

Due to Eq. (4.82) it reduces to

$$\begin{aligned}
\eta^{(\infty)} & (r_1\mathbf{u}_1, r_2\mathbf{u}_2;\omega) \\
&= [\Theta_{xx}(\omega)\tilde{\eta}_{xx}(k(\mathbf{u}_{1\perp}+\mathbf{u}_{2\perp})/2;\omega) + \Theta_{yy}(\omega)\tilde{\eta}_{yy}(k(\mathbf{u}_{1\perp}+\mathbf{u}_{2\perp})/2;\omega)] \\
&\times (\Theta_{xx}(\omega)\tilde{\eta}_{xx}(k\mathbf{u}_{1\perp};\omega) + \Theta_{yy}(\omega)\tilde{\eta}_{yy}(k\mathbf{u}_{1\perp};\omega))^{-1/2} \\
&\times (\Theta_{xx}(\omega)\tilde{\eta}_{xx}(k\mathbf{u}_{2\perp};\omega) + \Theta_{yy}(\omega)\tilde{\eta}_{yy}(k\mathbf{u}_{2\perp};\omega))^{-1/2} \\
&\times \frac{\tilde{S}(k(\mathbf{u}_{2\perp}-\mathbf{u}_{1\perp});\omega)}{\tilde{S}(0;\omega)} \exp[ik(r_2-r_1)].
\end{aligned} \tag{4.87}$$

Since the correlation coefficients $\eta_{\alpha\alpha}(\boldsymbol{\rho}'_d;\omega)$, $(\alpha = x, y)$ are "fast" functions of $\boldsymbol{\rho}'_d$ in the source plane their Fourier transforms $\tilde{\eta}_{\alpha\alpha}(\mathbf{f}';\omega)$ must be "slow" functions of $\mathbf{f}'$. Hence we can write:

$$\tilde{\eta}_{xx}(k\mathbf{u}_{1\perp};\omega) \approx \tilde{\eta}_{xx}(k\mathbf{u}_{2\perp};\omega) \approx \tilde{\eta}_{xx}(k(\mathbf{u}_{1\perp}+\mathbf{u}_{2\perp})/2;\omega), \tag{4.88}$$

$$\tilde{\eta}_{yy}(k\mathbf{u}_{1\perp};\omega) \approx \tilde{\eta}_{yy}(k\mathbf{u}_{2\perp};\omega) \approx \tilde{\eta}_{yy}(k(\mathbf{u}_{1\perp}+\mathbf{u}_{2\perp})/2;\omega). \tag{4.89}$$

These approximations imply that the first factor in (4.87) can be substituted by unity. Thus, the degree of coherence $\eta^{(\infty)}$ in the far field assunes the form

$$\eta^{(\infty)}(r_1\mathbf{u}_1, r_2\mathbf{u}_2;\omega) = \frac{\tilde{S}(k(\mathbf{u}_{2\perp}-\mathbf{su}_{1\perp});\omega)}{\tilde{S}(0;\omega)} \exp[ik(r_2-r_1)]. \tag{4.90}$$

This formula can be regarded as the second reciprocity relation for uniformly polarized planar quasi-homogeneous sources. The corresponding verbal statement can be put as *the spectral degree of coherence $\eta^{(\infty)}(r_1\mathbf{u}_1, r_2\mathbf{u}_2;\omega)$ of the far field generated by a uniformly polarized quasi-homogeneous source is proportional to the four-dimensional spatial Fourier transform of the spectral density of the electric field in the source plane, apart from the phase factor $k(r_2-r_1)$*.

For $\Theta \equiv 0$ the two reciprocity relations reduce to forms which can be also used for scalar quasi-homogeneous sources.

4.2.2   Conditions for spectral invariance

As we have seen from the discussion about scalar fields in Chapter 3, the spectrum of a stochastic light wave can suffer changes on propagation, even in free space, due to source correlations [29], [30]. This effect is referred to as the correlation-induced spectral shift, and should be distinguished from the Doppler-like spectral shifts. Only under special circumstances will the normalized spectrum of the far zone be equal to the normalized source spectrum: the source must be quasi-homogeneous and the degree of coherence of the source must satisfy the so-called scaling law. In this section we will formulate the conditions for spectral invariance in a more advanced setting of electromagnetic theory [31] and learn about the dependence of the spectral shifts on the polarization properties of the source [32].

Let us assume that the source occupies domain D with area A_D in the plane $z = 0$ and spectral densities of the two field components of the electromagnetic quasi-homogeneous source are the same at each source point, i.e., that $S_\alpha(\boldsymbol{\rho}'; \omega) \equiv S_\alpha(\omega)$, $(\alpha = x, y)$. Then the spectral density of the field in the source plane becomes

$$S(\omega) = S_x(\omega) + S_y(\omega). \qquad (4.91)$$

Its normalized version then takes form

$$S_N(\omega) = \frac{S_x(\omega) + S_y(\omega)}{\int [S_x(\omega) + S_y(\omega)] d\omega}. \qquad (4.92)$$

If the two spectral densities $S_\alpha(\boldsymbol{\rho}'; \omega)$, $(\alpha = x, y)$ are independent of position the far-field spectral density is

$$S^{(\infty)}(r\mathbf{u}; \omega) = (2\pi k)^2 A_D \cos^2 \theta \left[S_x(\omega) \tilde{\eta}_{xx}(k\mathbf{u}_\perp; \omega) + S_y(\omega) \tilde{\eta}_{yy}(k\mathbf{u}_\perp; \omega) \right] / r^2, \qquad (4.93)$$

while its normalized version is

$$
\begin{aligned}
S_N^{(\infty)}(r\mathbf{u}; \omega) &= \frac{S^{(\infty)}(r\mathbf{u}; \omega)}{\int S^{(\infty)}(r\mathbf{u}; \omega) d\omega} \\
&= \frac{k^2 [S_x(\omega) \tilde{\eta}_{xx}(k\mathbf{u}_\perp; \omega) + S_y(\omega) \tilde{\eta}_{yy}(k\mathbf{u}_\perp; \omega)]}{\int k^2 [S_x(\omega) \tilde{\eta}_{xx}(k\mathbf{u}_\perp; \omega) + S_y(\omega) \tilde{\eta}_{yy}(k\mathbf{u}_\perp; \omega)] d\omega},
\end{aligned}
\qquad (4.94)
$$

where in the last line we have substituted from Eq. (4.93) and used Eq. (4.91).

Equations (4.91) and (4.94) imply that generally the normalized far-zone spectrum differs from the normalized source spectrum. More specifically, the normalized spectrum of the far field generated by a quasi-homogeneous electromagnetic source depends on the spectral densities $S_x(\omega)$ and $S_y(\omega)$, of the field components in the source plane, on the direction $\mathbf{u}$ and on the correlation coefficients $\eta_{xx}(\boldsymbol{\rho}'_d; \omega)$ and $\eta_{yy}(\boldsymbol{\rho}'_d; \omega)$. Thus, the spectral, polarization and coherence properties of the source can influence the spectral composition of the beam in the far zone.

We will now discuss the situation when the normalized far-zone spectrum is the same throughout the far zone, i.e., independent of the direction $\mathbf{u}$. Equation (4.94) implies that a sufficient condition for such a case is that the Fourier transforms $\tilde{\eta}_{\alpha\alpha}(k\mathbf{u}_\perp;\omega)$, $(\alpha = x, y)$ of the correlation coefficients factorize, i.e.,

$$\tilde{\eta}_{\alpha\alpha}(k\mathbf{u}_\perp;\omega) = F_{\alpha\alpha}(\omega)\tilde{\Xi}(\mathbf{u}_\perp), \quad (\alpha = x, y), \tag{4.95}$$

Ξ being some function. Consequently,

$$S_N^{(\infty)}(r\mathbf{u};\omega) = \frac{k^2[S_x(\omega)F_{xx}(\omega) + S_y(\omega)F_{yy}(\omega)]}{\int k^2[S_x(\omega)F_{xx}(\omega) + S_y(\omega)F_{yy}(\omega)]d\omega}. \tag{4.96}$$

The right hand-side of this formula is not a function of $\mathbf{u}_\perp$. The inverse Fourier transform of Eq. (4.95) implies that the correlation coefficients then have the form

$$\eta_{\alpha\alpha}(\boldsymbol{\rho}_d';\omega) = F_{\alpha\alpha}(\omega)\int\int \tilde{\Xi}_{\alpha\alpha}(\mathbf{u}_\perp;\omega)\exp(-ik\mathbf{u}_\perp \cdot \boldsymbol{\rho}_d')d(k\mathbf{u}_\perp), \tag{4.97}$$

or,

$$\eta_{\alpha\alpha}(\boldsymbol{\rho}_d';\omega) = k^2 F_{\alpha\alpha}(\omega)\Xi(k\boldsymbol{\rho}_d'), \tag{4.98}$$

where

$$\Xi(k\boldsymbol{\rho}_d') = \frac{1}{4\pi^2}\int\int \tilde{\Xi}_{\alpha\alpha}(\mathbf{u}_\perp;\omega)\exp(-ik\mathbf{u}_\perp \cdot \boldsymbol{\rho}_d')d(k\mathbf{u}_\perp) \tag{4.99}$$

is the inverse Fourier transform of $\tilde{\Xi}_{\alpha\alpha}(\mathbf{u}_\perp;\omega)$.

Finally, we will derive an additional condition for the far field for having the same normalized spectral density as in the source plane. Since $|\eta_{\alpha\alpha}(k\mathbf{u}_\perp = 0;\omega)| \equiv 1$, Eq. (4.98) implies that

$$F_{\alpha\alpha}(\omega) = \frac{1}{k^2\Xi(0)}, \quad (\alpha = x, y). \tag{4.100}$$

On substituting from this equation into Eq. (4.98) we obtain the condition

$$\eta_{\alpha\alpha}(\boldsymbol{\rho}_d';\omega) = \frac{\Xi(k\boldsymbol{\rho}_d')}{\Xi(0)}, \tag{4.101}$$

meaning that the correlation coefficients must be functions of the variable $k\boldsymbol{\rho}_d'$ only. Formula (4.101), that can be termed the *electromagnetic scaling law*, expresses the sufficiency condition for a uniformly polarized quasi-homogeneous source for maintaining its spectral invariance everywhere in the far field.

4.2.3 Conditions for polarization invariance

In this section we will establish conditions for polarization invariance of electromagnetic quasi-homogeneous sources [33] (see also [34], [35]). The polarization properties, such as the degree of polarization $\wp$, the orientation angle of

144

the polarization ellipse ψ and the degree of ellipticity ϵ [see Eqs. (4.27), (4.28) and (4.30)] of such sources (4.73) can be readily found to be

$$\wp(\boldsymbol{\rho}';\omega) = \frac{\sqrt{[S_x(\boldsymbol{\rho}') - S_y(\boldsymbol{\rho}')]^2 + 4S_x(\boldsymbol{\rho}')S_y(\boldsymbol{\rho}')|\eta_{xy}(0)|^2}}{S_x(\boldsymbol{\rho}') + S_y(\boldsymbol{\rho}')}, \tag{4.102}$$

$$\psi(\boldsymbol{\rho}';\omega) = \frac{1}{2}\arctan\left[\frac{2\sqrt{S_x(\boldsymbol{\rho}')}\sqrt{S_y(\boldsymbol{\rho}')}Re[\eta_{xy}(0)]}{S_x(\boldsymbol{\rho}') - S_y(\boldsymbol{\rho}')}\right], \tag{4.103}$$

and

$$\epsilon(\boldsymbol{\rho}';\omega) = \sqrt{\frac{q_1 - q_2}{q_1 + q_2}}, \tag{4.104}$$

where

$$q_1 = \sqrt{[S_x(\boldsymbol{\rho}') - S_y(\boldsymbol{\rho}')]^2 + 4S_x(\boldsymbol{\rho}')S_y(\boldsymbol{\rho}')|\eta_{xy}(0)|^2}, \tag{4.105}$$

$$q_2 = \sqrt{[S_x(\boldsymbol{\rho}') - S_y(\boldsymbol{\rho}')]^2 + 4S_x(\boldsymbol{\rho}')S_y(\boldsymbol{\rho}')\Re[\eta_{xy}(0)]^2}, \tag{4.106}$$

and the dependence on ω in the right sides is ignored. We will only derive conditions for polarization invariance for sources with uniform polarization. Recall that to guarantee that a quasi-homogeneous source is uniformly polarized one must set [28]

$$S_y(\boldsymbol{\rho}';\omega) = \Theta(\omega)S_x(\boldsymbol{\rho}';\omega), \tag{4.107}$$

i.e., the spectral densities of the electric field components must be proportional to each other everywhere in the source plane. In addition, the correlation coefficient η_{xy} must take the same value across the source. If these conditions are met the polarization state in the source plane can be determined by the exoressions:

$$\wp(\boldsymbol{\rho}';\omega) = \frac{\sqrt{[1 - \Theta(\omega)]^2 + 4\Theta(\omega)|\eta_{xy}(0;\omega)|^2}}{1 + \Theta(\omega)}, \tag{4.108}$$

$$\psi(\boldsymbol{\rho}';\omega) = \frac{1}{2}\arctan\left[\frac{2\sqrt{\Theta(\omega)}Re[\eta_{xy}(0;\omega)]}{1 - \Theta(\omega)}\right], \tag{4.109}$$

and

$$\epsilon(\boldsymbol{\rho}';\omega) = \left[\sqrt{[1 - \Theta(\omega)]^2 + 4\Theta(\omega)|\eta_{xy}(0;\omega)|^2}\right.$$
$$\left. - \sqrt{[1 - \Theta(\omega)]^2 + 4\Theta(\omega)[Re\eta_{xy}(0;\omega)^2}\right]^{1/2}$$
$$\times \left[\sqrt{[1 - \Theta(\omega)]^2 + 4\Theta(\omega)|\eta_{xy}(0;\omega)|^2}\right.$$
$$\left. + \sqrt{[1 - \Theta(\omega)]^2 + 4\Theta(\omega)[Re\eta_{xy}(0;\omega)]^2}\right]^{-1/2}, \tag{4.110}$$

Let us now assume that the source (4.73) radiates a beam-like field into the far field, at a point with position vector $\mathbf{r}$. In this case the spectral density $S^{(\infty)}(r\mathbf{u};\omega) = S_x^{(\infty)}(r\mathbf{u};\omega) + S_y^{(\infty)}(r\mathbf{u};\omega)$, $r = |\mathbf{r}|$, $\mathbf{u} = \mathbf{r}/r$ in the far zone is appreciable only along directions $\mathbf{u}$ that are sufficiently close to the beam axis z (see Fig. 3.2). Then Eq. (4.82) implies that

$$W_{\alpha\beta}^{(\infty)}(r\mathbf{u}, r\mathbf{u};\omega) = \frac{(2\pi k)^2 \cos^2 \theta}{r^2} \Theta_{\alpha\beta}(\omega)\tilde{S}_x(0;\omega)\tilde{\eta}_{\alpha\beta}(k\mathbf{u}_\perp;\omega), \qquad (4.111)$$

where θ is the azymuthal angle, i.e. the angle between vector $\mathbf{u}$ and the positive z - axis, $\mathbf{u}_\perp$ is the transverse portion of vector $\mathbf{u}$ (projection onto the source plane $z = 0$), and $\Theta_{\alpha\beta}(\omega)$ is defined in Eq. (4.81). On substituting from Eq. (4.111) into Eqs. (4.27)–(4.30) one can see that the polarization properties of the far field generated by the quasi-homogeneous source are

$$\wp^{(\infty)}(r\mathbf{u};\omega)$$

$$= \frac{\sqrt{[\tilde{\eta}_{xx}(k\mathbf{u}_\perp) - \Theta(\omega)\tilde{\eta}_{yy}(k\mathbf{u}_\perp)]^2 + 4\Theta(\omega)|\tilde{\eta}_{xy}(k\mathbf{u}_\perp)|^2}}{\tilde{\eta}_{xx}(k\mathbf{u}_\perp) + \Theta(\omega)\tilde{\eta}_{yy}(k\mathbf{u}_\perp)}, \qquad (4.112)$$

$$\psi^{(\infty)}(r\mathbf{u};\omega) = \frac{1}{2}\arctan\left[\frac{2\sqrt{\Theta(\omega)}Re[\tilde{\eta}_{xy}(k\mathbf{u}_\perp)]}{\tilde{\eta}_{xx}(k\mathbf{u}_\perp) - \Theta(\omega)\tilde{\eta}_{yy}(k\mathbf{u}_\perp)}\right], \qquad (4.113)$$

and

$$\epsilon^{(\infty)}(r\mathbf{u};\omega)$$

$$= \left\{\sqrt{[\tilde{\eta}_{xx}(k\mathbf{u}_\perp) - \Theta(\omega)\tilde{\eta}_{yy}(k\mathbf{u}_\perp)]^2 + 4\Theta(\omega)[\tilde{\eta}_{xy}(k\mathbf{u}_\perp)]^2}\right.$$

$$\left. + \sqrt{[\tilde{\eta}_{xx}(k\mathbf{u}_\perp) - \Theta(\omega)\tilde{\eta}_{yy}(k\mathbf{u}_\perp)]^2 + 4\Theta(\omega)[\tilde{\eta}_{xy}(k\mathbf{u}_\perp)]^2}\right\}^{1/2} \qquad (4.114)$$

$$\times \left\{\sqrt{[\tilde{\eta}_{xx}(k\mathbf{u}_\perp) - \Theta(\omega)\tilde{\eta}_{yy}(k\mathbf{u}_\perp)]^2 + 4\Theta(\omega)\Re[\tilde{\eta}_{xy}(k\mathbf{u}_\perp)]^2}\right.$$

$$\left. - \sqrt{[\tilde{\eta}_{xx}(k\mathbf{u}_\perp) - \Theta(\omega)\tilde{\eta}_{yy}(k\mathbf{u}_\perp)]^2 + 4\Theta(\omega)Re[\tilde{\eta}_{xy}(k\mathbf{u}_\perp)]^2}\right\}^{-1/2}.$$

Thus in order for the polarization properties of a beam to be invariant throughout the far zone of the source it is necessary and sufficient that

$$\tilde{\eta}_{xx}(k\mathbf{u}_\perp;\omega) = q_3(\omega)\tilde{\eta}_{yy}(k\mathbf{u}_\perp;\omega) = q_4(\omega)\tilde{\eta}_{xy}(k\mathbf{u}_\perp;\omega), \qquad (4.115)$$

where $q_3(\omega)$ and $q_4(\omega)$ are complex numbers. Since the Fourier transforms of the correlation coefficients must be proportional for any $k\mathbf{u}_\perp$, the same must be true for the correlation coefficients themselves at vector $\boldsymbol{\rho}_1' - \boldsymbol{\rho}_2'$, i.e.,

$$\eta_{xx}(\boldsymbol{\rho}_1' - \boldsymbol{\rho}_2';\omega) = q_3(\omega)\eta_{yy}(\boldsymbol{\rho}_1' - \boldsymbol{\rho}_2';\omega)$$
$$= q_4(\omega)\eta_{xy}(\boldsymbol{\rho}_1' - \boldsymbol{\rho}_2';\omega). \qquad (4.116)$$

The constants $q_3(\omega)$ and $q_4(\omega)$ can be determined on setting $\boldsymbol{\rho}_1' = \boldsymbol{\rho}_2'$ in Eq.

(4.116). On the one hand, $q_3(\omega) \equiv 1$, because $\eta_{xx}(0;\omega) = \eta_{yy}(0;\omega) \equiv 1$, as self-correlations; on the other hand, $\eta_{xy}(0;\omega) \neq 0$, since $\eta_{xx}(0;\omega) = 1 = q_4(\omega)\eta_{xy}(0;\omega)$, hence $q_4(\omega) = 1/\eta_{xy}(0;\omega)$. Summarizing these results, in order for an electromagnetic quasi-homogeneous source to generate a far field with uniform polarization properties (and the same as across its source) it suffices to require that

$$\eta_{xy}(0;\omega)\tilde{\eta}_{xx}^{(0)}(k\mathbf{u}_\perp;\omega) = \eta_{xy}(0;\omega)\tilde{\eta}_{yy}(k\mathbf{u}_\perp;\omega)$$
$$= \tilde{\eta}_{xy}(k\mathbf{u}_\perp;\omega). \tag{4.117}$$

If $\eta_{xy}(0;\omega) = 0$, Eq. (4.116) implies that

$$\tilde{\eta}_{xx}(k\mathbf{u}_\perp;\omega) = \tilde{\eta}_{yy}(k\mathbf{u}_\perp;\omega). \tag{4.118}$$

As a final remark, we point out that in cases when the fluctuations in the electric field at any point in space are governed by Gaussian statistics, the conditions for polarization invariance also imply invariance in some higher-order single-point statistics, which we will consider below. For instance, since the probability density functions (pdf) of the normalized Stokes parameters can be expressed in terms of the average Stokes parameters and the degree of polarization the conditions that we established guarantee that the far-field pdf of such beams will remain the same as the source pdf.

4.3 Propagation in free space and linear media

As we have already pointed out in Chapter 3 the most intriguing phenomena occurring with the stochastic beams are associated with their free-space propagation. For scalar stochastic beams such effects include shifts of their spectrum and changes in their spectral degree of coherence. In addition to these phenomena the electromagnetic stochastic beams exhibit changes in polarization properties. In this section we will first discuss the general laws governing propagation of the cross-spectral density matrix in free space also discussing their implications, such as the conservation laws. Then propagation laws will be outlined for deterministic linear media with an arbitrary refractive index, including the cases of absorbing/gain media and negative phase materials.

4.3.1 Propagation in free space

It follows from Maxwell's equations that the space-dependent part $\mathbf{E}(\mathbf{r};\omega)$ of a monochromatic electric field vector $\mathbf{E}(\mathbf{r};\omega)\exp(-i\omega t)$ propagating in free space satisfies the Helmholtz equation ([13], Section 13.1.1)

$$\nabla^2 \mathbf{E}(\mathbf{r};\omega) + k^2 \mathbf{E}(\mathbf{r};\omega) = 0, \tag{4.119}$$

where $\mathbf{r} = (x, y, z)$. In free space and in linear isotropic media the Cartesian components E_x and E_y of the electric field $\mathbf{E}$ then propagate independently of each other in the sense that each satisfies the equation

$$\nabla^2 E_\alpha(\mathbf{r}; \omega) + k^2 E_\alpha(\mathbf{r}; \omega) = 0, \qquad (\alpha = x, y). \tag{4.120}$$

Suppose that the electromagnetic field is beam-like and propagates from the plane $z = 0$ into the half-space $z > 0$ close to the z-axis, in vacuum. Let $\mathbf{r} = (\boldsymbol{\rho}, z)$ be the position vector at a point in the half-space $z > 0$, $\boldsymbol{\rho}$ denoting a two-dimensional transverse vector perpendicular to the direction of propagation of the beam. If $E_\alpha(\boldsymbol{\rho}', 0; \omega)$ represents the electric field vector at the point $(\boldsymbol{\rho}', 0)$ in the plane $z = 0$, the components $E_\alpha(\boldsymbol{\rho}, z; \omega)$ can be expressed with the help of the Huygens-Fresnel principle [4]

$$E_\alpha(\boldsymbol{\rho}, z; \omega) = \iint E_\alpha(\boldsymbol{\rho}', 0; \omega) G(\boldsymbol{\rho}, \boldsymbol{\rho}', z; \omega) d^2 \rho', \qquad (\alpha = x, y), \tag{4.121}$$

the integration being performed over the source plane. Here, as before,

$$G(\boldsymbol{\rho}, \boldsymbol{\rho}', z; \omega) = -\frac{ik \exp(ikz)}{2\pi z} \exp\left[-ik\frac{(\boldsymbol{\rho} - \boldsymbol{\rho}')^2}{2z}\right] \tag{4.122}$$

is the free-space Green's function having the form of the spherical wave. In order to formulate the propagation laws for the cross-spectral density matrix components we substitute from Eq. (4.121) into Eq. (4.3) to obtain the formula

$$W_{\alpha\beta}(\mathbf{r}_1, \mathbf{r}_2; \omega) = \iint W_{\alpha\beta}(\boldsymbol{\rho}_1', \boldsymbol{\rho}_2', 0; \omega)$$
$$\times K(\boldsymbol{\rho}_1', \boldsymbol{\rho}_1, z_1, \boldsymbol{\rho}_2', \boldsymbol{\rho}_2, z_2; \omega) d^2 \rho_1' d^2 \rho_2', \tag{4.123}$$

where integration is performed twice over the source plane and K has the form

$$K(\boldsymbol{\rho}_1, \boldsymbol{\rho}_1', z_1, \boldsymbol{\rho}_2, \boldsymbol{\rho}_2', z_2; \omega) = G^*(\boldsymbol{\rho}_1', \boldsymbol{\rho}_1, z_1; \omega) G(\boldsymbol{\rho}_2', \boldsymbol{\rho}_2, z_2; \omega)$$
$$= \left(\frac{k}{2\pi}\right)^2 \frac{1}{z_1 z_2} \exp\left\{-\frac{ik}{2}\left[\frac{(\boldsymbol{\rho}_1 - \boldsymbol{\rho}_1')^2}{z_1} - \frac{(\boldsymbol{\rho}_2 - \boldsymbol{\rho}_2')^2}{z_2}\right]\right\}. \tag{4.124}$$

The formula (4.123) can also be obtained using the electromagnetic correlation matrix with the help of binomial approximation. A more general formulation, which includes the propagation of the z-component of the electric field, is discussed in Ref. [36].

4.3.2 Conservation laws for electromagnetic stochastic free fields

Generally, the conservation laws are of great importance in physics and in particular, in electromagnetic theory. For random statistically stationary electromagnetic fields propagating in free space, such laws were derived long time ago by Roman and Wolf [37] using the second-order coherence theory, in the space-time domain (see [8], Section 4.6). Such laws are valid only for quasi-monochromatic fields. In the framework of space-frequency coherence theory, the conservation laws for random stationary free fields with any spectral composition have been proven by Kowarz and Wolf [38]. By free fields we mean the fields that are considered far enough from their sources or scatterers [39]. The angular correlation function of a free field contains only homogeneous plane waves ([8], Section 4.7.2), while the evanescent plane waves are not present. In [38] it is shown that both the integrated spectrum (i.e., the spectral density distribution integrated in any transverse beam cross-section) and the integrated anti-spectrum (i.e., the integrated cross-spectral density function evaluated at points symmetric with respect to the origin) of the beam are invariant on free-space propagation.

The conservation laws for stochastic electromagnetic fields have been derived for the integrated spectrum/antispectrum and for the integrated Stokes parameters in Ref. [40] which we will now follow. The results of Ref. [38] follow directly from those in [40] as a special case of a linearly polarized field.

If in Eq. (4.54) $a_\alpha(u_\perp; \omega) \equiv 0$, $(\alpha = x, y)$ for $u_\perp^2 > 1$, i.e., evanescent waves are excluded from the angular spectra of both components of the electric vector then the region of integration reduces to the domain $u_\perp^2 \leq 1$, and the angular spectrum representation of each of the two components of the resulting *free electric field* $\mathbf{E}^{(f)}(\mathbf{r}; \omega)$ takes the form

$$E_\alpha^{(f)}(\mathbf{r}; \omega) = \int\limits_{u_\perp^2 \leq 1} a_\alpha(u_\perp; \omega) \exp(ik\mathbf{u} \cdot \mathbf{r}) d\mathbf{u}_\perp, \quad (\alpha = x, y). \qquad (4.125)$$

For a stochastic electromagnetic free beam-like field the cross-spectral density matrix can be formed as

$$\mathbf{W}^{(f)}(\mathbf{r}_1, \mathbf{r}_2; \omega) = \left[\langle E_\alpha^{(f)*}(\mathbf{r}_1; \omega) E_\beta^{(f)}(\mathbf{r}_2; \omega) \rangle \right], \quad (\alpha = x, y; \beta = x, y), \quad (4.126)$$

where $E_\alpha^{(f)}(\mathbf{r}; \omega)$ is defined in Eq. (4.125). On substituting from equation Eq. (4.125) into Eq. (4.126), we find that

$$\mathbf{W}^{(f)}(\mathbf{r}_1, \mathbf{r}_2; \omega) = \int\limits_{u_{1\perp}^2 \leq 1} \int\limits_{u_{2\perp}^2 \leq 1} \mathbf{A}(\mathbf{u}_{1\perp}, \mathbf{u}_{2\perp}; \omega) \exp[ik(\mathbf{u}_2 \cdot \mathbf{r}_2 - \mathbf{u}_1 \cdot \mathbf{r}_1)] d\mathbf{u}_{1\perp} d\mathbf{u}_{2\perp},$$

$$(4.127)$$

where $\mathbf{A}(\mathbf{u}_{1\perp}, \mathbf{u}_{2\perp}; \omega)$ is the angular correlation matrix [see Eq. (4.59)].

We will first derive the conservation laws for the cross-spectral density matrix elements in Eq. (4.126) of a stochastic electromagnetic free field propagating in free space. On making a transformation of two-dimensional transverse vectors $\boldsymbol{\rho}_1$ and $\boldsymbol{\rho}_2$,

$$\boldsymbol{\rho}_s = \frac{1}{2}(\boldsymbol{\rho}_1 + \boldsymbol{\rho}_2), \quad \boldsymbol{\rho}_d = \boldsymbol{\rho}_2 - \boldsymbol{\rho}_1, \qquad (4.128)$$

we obtain for the cross-spectral density matrix at points with position vectors $\boldsymbol{\rho}_s$ and $\boldsymbol{\rho}_d$ in a transverse plane $z = \text{const} > 0$ the expressions

$$\mathcal{W}^{(f)}(\boldsymbol{\rho}_s, \boldsymbol{\rho}_d, z; \omega) = \int\limits_{u_{1\perp}^2 \leq 1} \int\limits_{u_{2\perp}^2 \leq 1} \mathbf{A}(\mathbf{u}_{1\perp}, \mathbf{u}_{2\perp}; \omega) \exp[ik(\mathbf{u}_{2\perp} - \mathbf{u}_{1\perp}) \cdot \boldsymbol{\rho}_s]$$

$$\times \exp[ik(\mathbf{u}_{2\perp} + \mathbf{u}_{1\perp}) \cdot \boldsymbol{\rho}_d/2] \exp[ik(\mathbf{u}_{2z} - \mathbf{u}_{1z})z] d\mathbf{u}_{1\perp} d\mathbf{u}_{2\perp}. \tag{4.129}$$

The change from symbol $\mathbf{W}^{(f)}$ to symbol $\mathcal{W}^{(f)}$ is used for stressing that they are different (in form) functions of their arguments. The representation of a two-dimensional delta-function,

$$\frac{1}{(2\pi)^2} \int \exp[i\boldsymbol{\kappa} \cdot \boldsymbol{\nu}] d\boldsymbol{\nu} = \delta^2(\boldsymbol{\kappa}) \tag{4.130}$$

and integration of each element of matrix (4.129) over the planes $\boldsymbol{\rho}_s$ and $\boldsymbol{\rho}_d$, leads to expressions

$$\int \mathcal{W}^{(f)}(\boldsymbol{\rho}_s, \boldsymbol{\rho}_d, z; \omega) d\boldsymbol{\rho}_s = \frac{4\pi^2}{k^2} \int\limits_{u_\perp^2 \leq 1} \mathbf{A}(\mathbf{u}_\perp, \mathbf{u}_\perp; \omega) \exp(ik\mathbf{u}_\perp \cdot \boldsymbol{\rho}_d) d\mathbf{u}_\perp, \tag{4.131}$$

and

$$\int \mathcal{W}^{(f)}(\boldsymbol{\rho}_s, \boldsymbol{\rho}_d, z; \omega) d\boldsymbol{\rho}_d = \frac{4\pi^2}{k^2} \int\limits_{u_\perp^2 \leq 1} \mathbf{A}(-\mathbf{u}_\perp, \mathbf{u}_\perp; \omega) \exp(2ik\mathbf{u}_\perp \cdot \boldsymbol{\rho}_s) d\mathbf{u}_\perp.$$

$$\tag{4.132}$$

The right-hand sides of Eqs. (4.131) and (4.132) and, hence, the left-hand sides of these expressions are z-independent, i.e., conserved on propagation. Thus, we have proven that the integrated values of the elements of the cross-spectral density matrix remain invariant on free space propagation. For the analysis below it will be convenient to employ the following notations for the conserved quantities:

$$W_+^{(f)}(\boldsymbol{\rho}_d; \omega) = \int \mathcal{W}^{(f)}(\boldsymbol{\rho}_s, \boldsymbol{\rho}_d; \omega) d\boldsymbol{\rho}_s, \tag{4.133}$$

$$W_-^{(f)}(\boldsymbol{\rho}_s; \omega) = \int \mathcal{W}^{(f)}(\boldsymbol{\rho}_s, \boldsymbol{\rho}_d; \omega) d\boldsymbol{\rho}_d, \tag{4.134}$$

while omitting the z-dependence.

Since the spectral density of an electromagnetic stochastic free beam at a point $\mathbf{r} = (\boldsymbol{\rho}, z)$ is defined as

$$S^{(f)}(\boldsymbol{\rho}, z; \omega) = Tr\mathbf{W}^{(f)}(\boldsymbol{\rho}, \boldsymbol{\rho}, z; \omega) = Tr\mathcal{W}^{(f)}(\boldsymbol{\rho}, 0, z; \omega), \tag{4.135}$$

Eq. (4.133) with $\rho_1 = \rho_2 = \rho$, $(\rho_d \equiv 0, \rho_s \equiv \rho)$ imply that

$$
\int S^{(f)}(\rho, z; \omega) d\rho = \int Tr\mathcal{W}^{(f)}(\rho, 0, z; \omega) d\rho
$$

$$
= \int Tr\mathcal{W}^{(f)}(\rho, 0; \omega) d\rho = Tr\mathcal{W}_+^{(f)}(0; \omega). \tag{4.136}
$$

The right side of Eq. (4.136), and hence, its left-hand side is the linear combination of two z-independent quantities, which proves the fact that the spectral density integrated over any transverse plane, is invariant as the beam propagates.

The same result holds for a somewhat less physically important quantity, known in scalar theory as an "anti-spectrum", first introduced in [38] [see Eq. (3.16)]. For electromagnetic stochastic free fields the anti-spectrum $\overline{S}^{(f)}$ at a point $\mathbf{r} = (\rho, z)$ can be defined by the expression

$$
\overline{S}^{(f)}(\rho, z; \omega) = Tr\mathbf{W}^{(f)}(\rho, -\rho, z; \omega) = Tr\mathcal{W}^{(f)}(0, \rho, z; \omega). \tag{4.137}
$$

Integration of both sides of Eq. (4.137) and use of (4.134) with $\rho_2 = -\rho_1 \equiv \rho/2$, $(\rho_s \equiv 0, \rho_d \equiv \rho)$ leads to the expression

$$
\int \overline{S}^{(f)}(\rho, z; \omega) d\rho = \int Tr\mathcal{W}^{(f)}(0, \rho, z; \omega) d\rho
$$

$$
= \int Tr\mathcal{W}^{(f)}(0, \rho; \omega) d\rho = Tr\mathcal{W}_-^{(f)}(0; \omega), \tag{4.138}
$$

confirming that the anti-spectrum is also invariant on free-space propagation.

Using the preceding analysis we will now show that some polarization properties of the beam integrated over the transverse cross-sections are also conserved on free-space propagation. On recalling that the Stokes parameters of a stochastic beam can be expressed as linear combinations of the elements of the cross-spectral density matrix, we can write for the free fields the following relations

$$
\begin{aligned}
S_0^{(f)}(\rho, z; \omega) &= W_{xx}^{(f)}(\rho, \rho, z; \omega) + W_{yy}^{(f)}(\rho, \rho, z; \omega) \\
&= \mathcal{W}_{xx}^{(f)}(\rho, 0, z; \omega) + \mathcal{W}_{yy}^{(f)}(\rho, 0, z; \omega),
\end{aligned} \tag{4.139}
$$

$$
\begin{aligned}
S_1^{(f)}(\rho, z; \omega) &= W_{xx}^{(f)}(\rho, \rho, z; \omega) - W_{yy}^{(f)}(\rho, \rho, z; \omega) \\
&= \mathcal{W}_{xx}^{(f)}(\rho, 0, z; \omega) - \mathcal{W}_{yy}^{(f)}(\rho, 0, z; \omega),
\end{aligned} \tag{4.140}
$$

$$
\begin{aligned}
S_2^{(f)}(\rho, z; \omega) &= W_{xy}^{(f)}(\rho, \rho, z; \omega) + W_{yx}^{(f)}(\rho, \rho, z; \omega) \\
&= \mathcal{W}_{xy}^{(f)}(\rho, 0, z; \omega) + \mathcal{W}_{yx}^{(f)}(\rho, 0, z; \omega),
\end{aligned} \tag{4.141}
$$

$$
\begin{aligned}
S_3^{(f)}(\rho, z; \omega) &= i[W_{yx}^{(f)}(\rho, \rho, z; \omega) + W_{yx}^{(f)}(\rho, \rho, z; \omega)] \\
&= i[\mathcal{W}_{xy}^{(f)}(\rho, 0, z; \omega) - \mathcal{W}_{yx}^{(f)}(\rho, 0, z; \omega)],
\end{aligned} \tag{4.142}
$$

where the first Stokes parameter $S_0^{(f)}$ is defined by the same formula as the spectral density $S^{(f)}$ of the beam and, hence, is conserved. On substituting from Eq. (4.133) with $\boldsymbol{\rho}_1 = \boldsymbol{\rho}_2 \equiv \boldsymbol{\rho}$ ($\boldsymbol{\rho}_d \equiv 0$, $\boldsymbol{\rho}_s \equiv \boldsymbol{\rho}$) into Eqs. (4.139)–(4.142) we find that

$$
\int S_0^{(f)}(\boldsymbol{\rho}, z; \omega) d\boldsymbol{\rho} = \mathcal{W}_{xx+}^{(f)}(0; \omega) + \mathcal{W}_{yy+}^{(f)}(0; \omega),
$$

$$
\int S_1^{(f)}(\boldsymbol{\rho}, z; \omega) d\boldsymbol{\rho} = \mathcal{W}_{xx+}^{(f)}(0; \omega) - \mathcal{W}_{yy+}^{(f)}(0; \omega),
$$

$$
\int S_2^{(f)}(\boldsymbol{\rho}, z; \omega) d\boldsymbol{\rho} = \mathcal{W}_{xy+}^{(f)}(0; \omega) + \mathcal{W}_{yx+}^{(f)}(0; \omega),
$$

$$
\int S_3^{(f)}(\boldsymbol{\rho}, z; \omega) d\boldsymbol{\rho} = i[\mathcal{W}_{yx+}^{(f)}(0; \omega) - \mathcal{W}_{xy+}^{(f)}(0; \omega)].
$$

(4.143)

Since the right-hand sides of these expressions are z-independent, the Stokes parameters integrated over any transverse plane are also propagation-invariant:

$$
\int S_n^{(f)}(\boldsymbol{\rho}, z; \omega) d\boldsymbol{\rho} = \int S_n^{(f)}(\boldsymbol{\rho}; \omega) d\boldsymbol{\rho}, \quad (n = 0, 1, 2, 3). \tag{4.144}
$$

Thus, using the angular spectrum representation of stochastic electromagnetic free beam-like fields propagating in free space we have shown that their integrated spectral density, anti-spectral density and the Stokes parameters are conserved. Similar laws can also be obtained for quantities that could be called "anti-Stokes parameters," which are defined in analogy with the anti-spectrum of the beam. However, we do not include their conservation laws here because of their rare use.

4.3.3 Propagation in linear deterministic media with arbitrary index of refraction

The propagation of deterministic and random beams in gain or absorbing media has been of significant interest for several decades (cf. [41]–[46]). For random scalar beams the changes in the degree of coherence propagating in such media were discussed in depth in [41]–[42]. In the case of electromagnetic beams the analysis of changes in spectral, coherence and polarization properties has been carried out in [46].

Suppose that the medium in which the beam propagates has arbitrary refractive and absorptive properties which, in addition, may be anisotropic. This implies that the wave numbers in the medium, in $x-$ and $y-$directions, are different:

$$
k_x = k_{(r)x} + ik_{(i)x}, \qquad k_y = k_{(r)y} + ik_{(i)y}, \tag{4.145}
$$

where $k_{(r)x}$, $k_{(i)x}$ and $k_{(r)y}$, $k_{(i)y}$ are the real and the imaginary parts of k_x and k_y. The real parts are associated with the refractive properties of the medium. Namely, since $k = 2\pi/\lambda$ and $k_r = 2\pi/\lambda_r$, the relation between k_r and the refractive index n is $k_{(r)} = k\lambda/\lambda_{(r)} = kn$. Further, the imaginary parts k_{ix} and k_{iy} characterize the gain or absorption. If $k_{(i)x} > 0$ and $k_{(i)y} > 0$ the medium has gain, if $k_{(i)x} < 0$ and $k_{(i)y} < 0$ the medium is absorbing. The generalized Huygens-Fresnel integral for the elements of the cross-spectral density matrix of a beam propagating in such a medium at points with position vectors $\mathbf{r}_1 = (x_1, y_1, z)$ and $\mathbf{r}_2 = (x_2, y_2, z)$ can be obtained from the Collins formula [47] on taking field correlations in Eq. (1.73):

$$W_{\alpha\beta}(x_1, y_1, x_2, y_2, z; \omega) = \frac{|k_x||k_y|}{(2\pi z)^2} \exp[(k_{(i)x} + k_{(i)y})z]$$

$$\times \iiiint W_{\alpha\beta}(x_1', y_1', x_2', y_2'; \omega)$$

$$\times \exp\left[-\frac{i}{2z}(k_x x_1'^2 - k_x^* x_1'^2 + k_y y_1'^2 - k_y^* y_1'^2)\right]$$

$$\times \exp\left[-\frac{i}{z}(k_x x_1 x_1' - k_x^* x_2 x_2' + k_y y_1 y_1' - k_y^* y_2 y_2')\right]$$

$$\times \exp\left[-\frac{i}{2z}(k_x x_1^2 - k_x^* x_2^2 + k_y y_1^2 - k_y^* y_2^2)\right]$$

$$\times dx_1' dx_2' dy_1' dy_2', \quad (\alpha, \beta = x, y),$$

$$(4.146)$$

where $W_{\alpha\beta}(x_1', y_1', x_2', y_2'; \omega)$ are the components of the of the cross-spectral density matrix in the source plane $z = 0$.

4.4   Generalized Jones-Mueller calculus

4.4.1   Transmission through deterministic devices

The description of interaction of stochastic electromagnetic fields with thin, linear, non-image forming devices at a single point was developed by Jones [48] and Mueller [49] a long time ago (see also Chapter 1). Such devices can be classified into deterministic: polarizers, absorbers, rotators and retarders; and random: diffusers and spatial light modulators. While the Jones calculus deals with monochromatic fields and uses 2×2 matrices transforming two transverse components of the electric field, the Mueller calculus describes transmission of the Stokes parameters of monochromatic or stochastic field and uses 4×4 matrices. Both techniques may be employed for determining the transmission of spectral and polarization properties of fields. However, since both Jones calculus and Mueller calculus are limited to transmission of the field characteristics defined at a single point, the information about its correlation properties at two spatial positions is lost and transmission of such statistics as the spectral degree of coherence or the spectral degree of cross-polarization is not possible.

The Jones-Mueller calculus was recently generalized in [50] to two-point quantities which accounts for transmission of various correlation properties of fields. This technique describes how the cross-spectral density matrix or, alternatively, the generalized Stokes parameters are transmitted through the linear non-image-forming devices.

Since the generalized Jones-Mueller calculus can be developed in both space-time and space-frequency domains of coherence theory, throughout this section we will only retain the spatial counterparts of the arguments of all the functions and omit time/frequency counterparts. If the transformation is considered within the space-time coherence theory then the arguments involve time delay τ. In such a case the average is, of course, taken over a sufficiently long time interval. On the other hand, if the transformation is given in the space-frequency domain then the argument becomes ω and the average is assumed to be taken over an ensemble of monochromatic realizations.

A brief review of the classic Jones calculus, describing transmission of monochromatic beams through deterministic polarization devices was included in Chapter 1. As above suppose a thin polarization device is located between two parallel planes $\Pi^{(i)}$, $(z = z_i)$ and $\Pi^{(t)}$, $(z = z_i)$, perpendicular to the direction of propagation z, as shown in Fig. 4.3. Let $\boldsymbol{\rho} = (\rho_x, \rho_y)$ be a vector in either of two planes transverse to the direction of beam axis. Then the incident (monochromatic) electric field $\mathbf{E}^{(i)}$ and the transmitted electric field $\mathbf{E}^{(t)}$ may be written as 1×2 vectors, i.e.,

$$\mathbf{E}^{(i)}(\boldsymbol{\rho}) = \begin{bmatrix} E_x^{(i)}(\boldsymbol{\rho}) \\ E_y^{(i)}(\boldsymbol{\rho}) \end{bmatrix}, \qquad \mathbf{E}^{(t)}(\boldsymbol{\rho}) = \begin{bmatrix} E_x^{(t)}(\boldsymbol{\rho}) \\ E_y^{(t)}(\boldsymbol{\rho}) \end{bmatrix}, \qquad (4.147)$$

where for brevity we omitted the dependence on frequency.

According to the Jones calculus for each deterministic polarization device there is a 2×2 local, linear, frequency dependent transformation (matrix)

$$\mathbf{T}(\boldsymbol{\rho}) = \begin{bmatrix} T_{xx}(\boldsymbol{\rho}) & T_{xy}(\boldsymbol{\rho}) \\ T_{yx}(\boldsymbol{\rho}) & T_{yy}(\boldsymbol{\rho}) \end{bmatrix}, \qquad (4.148)$$

known as the Jones matrix, which relates the incident and the transmitted field vectors via the formula

$$\mathbf{E}^{(t)}(\boldsymbol{\rho}) = \mathbf{T}(\boldsymbol{\rho})\mathbf{E}^{(i)}(\boldsymbol{\rho}). \qquad (4.149)$$

The Jones matrix is Hermitian with generally complex off-diagonal elements. Jones matrices of typical devices of polarization optics, such as polarizers, rotators, absorbers and compensators were summarized in Chapter 1.

We will now consider the case when the incident field is stochastic and derive the formula describing the transformation of its cross-spectral density matrix through a deterministic polarization device (see Fig. 4.4). If the cross-spectral density functions of the initial and the transmitted fields are

$$\mathbf{W}^{(i)}(\boldsymbol{\rho}_1, \boldsymbol{\rho}_2) = \begin{bmatrix} \langle E_x^{*(i)}(\boldsymbol{\rho}_1) E_x^{(i)}(\boldsymbol{\rho}_2) \rangle & \langle E_x^{*(i)}(\boldsymbol{\rho}_1) E_y^{(i)}(\boldsymbol{\rho}_2) \rangle \\ \langle E_y^{*(i)}(\boldsymbol{\rho}_1) E_x^{(i)}(\boldsymbol{\rho}_2) \rangle & \langle E_y^{*(i)}(\boldsymbol{\rho}_1) E_y^{(i)}(\boldsymbol{\rho}_2) \rangle \end{bmatrix} \qquad (4.150)$$

154

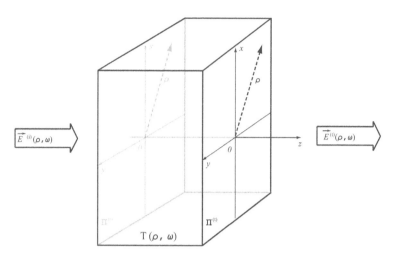

FIGURE 4.3
Transmission of a deterministic field through a thin device of polarization optics.

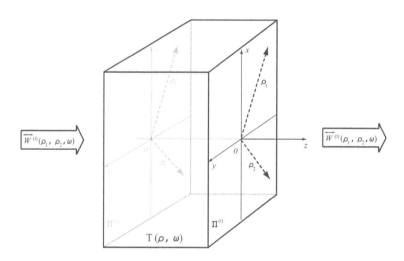

FIGURE 4.4
Transmission of a random beam through a thin device of polarization optics.

and

$$\mathbf{W}^{(t)}(\boldsymbol{\rho}_1,\boldsymbol{\rho}_2) = \begin{bmatrix} \langle E_x^{*(t)}(\boldsymbol{\rho}_1)E_x^{(i)}(\boldsymbol{\rho}_2)\rangle & \langle E_x^{*(t)}(\boldsymbol{\rho}_1)E_y^{(i)}(\boldsymbol{\rho}_2)\rangle \\ \langle E_y^{*(t)}(\boldsymbol{\rho}_1)E_x^{(i)}(\boldsymbol{\rho}_2)\rangle & \langle E_y^{*(t)}(\boldsymbol{\rho}_1)E_y^{(i)}(\boldsymbol{\rho}_2)\rangle \end{bmatrix}, \tag{4.151}$$

on substituting from Eq. (4.149) into Eq. (4.151) we can readily express matrix $\mathbf{W}^{(t)}$ in terms of matrices $\mathbf{W}^{(i)}$ and $\mathbf{T}$ as follows:

$$\begin{aligned} \mathbf{W}^{(t)}(\boldsymbol{\rho}_1,\boldsymbol{\rho}_2) &= \langle \mathbf{T}^*(\boldsymbol{\rho}_1)\mathbf{E}^{*(i)}(\boldsymbol{\rho}_1)\left[\mathbf{T}(\boldsymbol{\rho}_2)\mathbf{E}^{(i)}(\boldsymbol{\rho}_2)\right]^T \rangle \\ &= \mathbf{T}^*(\boldsymbol{\rho}_1)\mathbf{W}^{(i)}(\boldsymbol{\rho}_1,\boldsymbol{\rho}_2)\mathbf{T}^T(\boldsymbol{\rho}_2), \end{aligned} \tag{4.152}$$

the superscript T denoting the matrix transpose. It will be important for the subsequent analysis to write down the elements of the matrix $\mathbf{W}^{(t)}(\boldsymbol{\rho}_1,\boldsymbol{\rho}_2)$ explicitly:

$$\begin{aligned} W_{xx}^{(t)}(\boldsymbol{\rho}_1,\boldsymbol{\rho}_2) &= \mathfrak{T}_{xx}^{xx}(\boldsymbol{\rho}_1,\boldsymbol{\rho}_2)W_{xx}^{(i)}(\boldsymbol{\rho}_1,\boldsymbol{\rho}_2) + \mathfrak{T}_{xy}^{xx}(\boldsymbol{\rho}_1,\boldsymbol{\rho}_2)W_{xy}^{(i)}(\boldsymbol{\rho}_1,\boldsymbol{\rho}_2) \\ &\quad + \mathfrak{T}_{yx}^{xx}(\boldsymbol{\rho}_1,\boldsymbol{\rho}_2)W_{yx}^{(i)}(\boldsymbol{\rho}_1,\boldsymbol{\rho}_2) + \mathfrak{T}_{yy}^{xx}(\boldsymbol{\rho}_1,\boldsymbol{\rho}_2)W_{yy}^{(i)}(\boldsymbol{\rho}_1,\boldsymbol{\rho}_2) \\ W_{xy}^{(t)}(\boldsymbol{\rho}_1,\boldsymbol{\rho}_2) &= \mathfrak{T}_{xx}^{xy}(\boldsymbol{\rho}_1,\boldsymbol{\rho}_2)W_{xx}^{(i)}(\boldsymbol{\rho}_1,\boldsymbol{\rho}_2) + \mathfrak{T}_{xy}^{xy}(\boldsymbol{\rho}_1,\boldsymbol{\rho}_2)W_{xy}^{(i)}(\boldsymbol{\rho}_1,\boldsymbol{\rho}_2) \\ &\quad + \mathfrak{T}_{yx}^{xy}(\boldsymbol{\rho}_1,\boldsymbol{\rho}_2)W_{yx}^{(i)}(\boldsymbol{\rho}_1,\boldsymbol{\rho}_2) + \mathfrak{T}_{yy}^{xy}(\boldsymbol{\rho}_1,\boldsymbol{\rho}_2)W_{yy}^{(i)}(\boldsymbol{\rho}_1,\boldsymbol{\rho}_2) \\ W_{yx}^{(t)}(\boldsymbol{\rho}_1,\boldsymbol{\rho}_2) &= \mathfrak{T}_{xx}^{yx}(\boldsymbol{\rho}_1,\boldsymbol{\rho}_2)W_{xx}^{(i)}(\boldsymbol{\rho}_1,\boldsymbol{\rho}_2) + \mathfrak{T}_{xy}^{yx}(\boldsymbol{\rho}_1,\boldsymbol{\rho}_2)W_{xy}^{(i)}(\boldsymbol{\rho}_1,\boldsymbol{\rho}_2) \\ &\quad + \mathfrak{T}_{yx}^{yx}(\boldsymbol{\rho}_1,\boldsymbol{\rho}_2)W_{yx}^{(i)}(\boldsymbol{\rho}_1,\boldsymbol{\rho}_2) + \mathfrak{T}_{yy}^{yx}(\boldsymbol{\rho}_1,\boldsymbol{\rho}_2)W_{yy}^{(i)}(\boldsymbol{\rho}_1,\boldsymbol{\rho}_2) \\ W_{yy}^{(t)}(\boldsymbol{\rho}_1,\boldsymbol{\rho}_2) &= \mathfrak{T}_{xx}^{yy}(\boldsymbol{\rho}_1,\boldsymbol{\rho}_2)W_{xx}^{(i)}(\boldsymbol{\rho}_1,\boldsymbol{\rho}_2) + \mathfrak{T}_{xy}^{yy}(\boldsymbol{\rho}_1,\boldsymbol{\rho}_2)W_{xy}^{(i)}(\boldsymbol{\rho}_1,\boldsymbol{\rho}_2) \\ &\quad + \mathfrak{T}_{yx}^{yy}(\boldsymbol{\rho}_1,\boldsymbol{\rho}_2)W_{yx}^{(i)}(\boldsymbol{\rho}_1,\boldsymbol{\rho}_2) + \mathfrak{T}_{yy}^{yy}(\boldsymbol{\rho}_1,\boldsymbol{\rho}_2)W_{yy}^{(i)}(\boldsymbol{\rho}_1,\boldsymbol{\rho}_2), \end{aligned} \tag{4.153}$$

The sixteen transmission coefficients $\mathfrak{T}_{kl}^{pq}$, where subscript "kl" refers to an element of matrix $\mathbf{W}^{(i)}$ and superscript "pq" to an element of matrix $\mathbf{W}^{(t)}$, are then given by the expressions:

$$\begin{aligned} \mathfrak{T}_{xx}^{xx}(\boldsymbol{\rho}_1,\boldsymbol{\rho}_2) &= T_{xx}^*(\boldsymbol{\rho}_1)T_{xx}(\boldsymbol{\rho}_2), & \mathfrak{T}_{xy}^{xx}(\boldsymbol{\rho}_1,\boldsymbol{\rho}_2) &= T_{xx}^*(\boldsymbol{\rho}_1)T_{xy}(\boldsymbol{\rho}_2), \\ \mathfrak{T}_{yx}^{xx}(\boldsymbol{\rho}_1,\boldsymbol{\rho}_2) &= T_{xy}^*(\boldsymbol{\rho}_1)T_{xx}(\boldsymbol{\rho}_2), & \mathfrak{T}_{yy}^{xx}(\boldsymbol{\rho}_1,\boldsymbol{\rho}_2) &= T_{xy}^*(\boldsymbol{\rho}_1)T_{xy}(\boldsymbol{\rho}_2), \\ \mathfrak{T}_{xx}^{xy}(\boldsymbol{\rho}_1,\boldsymbol{\rho}_2) &= T_{xx}^*(\boldsymbol{\rho}_1)T_{yx}(\boldsymbol{\rho}_2), & \mathfrak{T}_{xy}^{xy}(\boldsymbol{\rho}_1,\boldsymbol{\rho}_2) &= T_{xx}^*(\boldsymbol{\rho}_1)T_{yy}(\boldsymbol{\rho}_2), \\ \mathfrak{T}_{yx}^{xy}(\boldsymbol{\rho}_1,\boldsymbol{\rho}_2) &= T_{xy}^*(\boldsymbol{\rho}_1)T_{yx}(\boldsymbol{\rho}_2), & \mathfrak{T}_{yy}^{xy}(\boldsymbol{\rho}_1,\boldsymbol{\rho}_2) &= T_{xy}^*(\boldsymbol{\rho}_1)T_{yy}(\boldsymbol{\rho}_2), \\ \mathfrak{T}_{xx}^{yx}(\boldsymbol{\rho}_1,\boldsymbol{\rho}_2) &= T_{yx}^*(\boldsymbol{\rho}_1)T_{xx}(\boldsymbol{\rho}_2), & \mathfrak{T}_{xy}^{yx}(\boldsymbol{\rho}_1,\boldsymbol{\rho}_2) &= T_{yx}^*(\boldsymbol{\rho}_1)T_{xy}(\boldsymbol{\rho}_2), \\ \mathfrak{T}_{yx}^{yx}(\boldsymbol{\rho}_1,\boldsymbol{\rho}_2) &= T_{yy}^*(\boldsymbol{\rho}_1)T_{xx}(\boldsymbol{\rho}_2), & \mathfrak{T}_{yy}^{yx}(\boldsymbol{\rho}_1,\boldsymbol{\rho}_2) &= T_{yy}^*(\boldsymbol{\rho}_1)T_{xy}(\boldsymbol{\rho}_2), \\ \mathfrak{T}_{xx}^{yy}(\boldsymbol{\rho}_1,\boldsymbol{\rho}_2) &= T_{yx}^*(\boldsymbol{\rho}_1)T_{yx}(\boldsymbol{\rho}_2), & \mathfrak{T}_{xy}^{yy}(\boldsymbol{\rho}_1,\boldsymbol{\rho}_2) &= T_{yx}^*(\boldsymbol{\rho}_1)T_{yy}(\boldsymbol{\rho}_2), \\ \mathfrak{T}_{yx}^{yy}(\boldsymbol{\rho}_1,\boldsymbol{\rho}_2) &= T_{yy}^*(\boldsymbol{\rho}_1)T_{yx}(\boldsymbol{\rho}_2), & \mathfrak{T}_{yy}^{yy}(\boldsymbol{\rho}_1,\boldsymbol{\rho}_2) &= T_{yy}^*(\boldsymbol{\rho}_1)T_{yy}(\boldsymbol{\rho}_2). \end{aligned} \tag{4.154}$$

The knowledge of these sixteen transmission coefficients guarantees the ability to predict modulation of the cross-spectral density matrix, and, consequently, all the second-order statistical properties of the beam on passage through deterministic linear devices of polarization optics.

As was mentioned previously, the classic Mueller calculus [49] deals with transformation of the classic Stokes parameters of a beam passing through a linear non-image forming device, which may affect a beam in a deterministic or random manner. In the case when a device is deterministic the transformation of the classic Stokes parameters assumes the following form

$$\mathbf{S}_l^{(t)}(\boldsymbol{\rho}) = \mathbf{M}(\boldsymbol{\rho})\mathbf{S}_l^{(i)}(\boldsymbol{\rho}), \qquad (l = 0, 1, 2, 3), \tag{4.155}$$

or, more explicitly,

$$\begin{bmatrix} S_0^{(t)}(\boldsymbol{\rho}) \\ S_1^{(t)}(\boldsymbol{\rho}) \\ S_2^{(t)}(\boldsymbol{\rho}) \\ S_3^{(t)}(\boldsymbol{\rho}) \end{bmatrix} = \begin{bmatrix} M_{11}(\boldsymbol{\rho}) & M_{12}(\boldsymbol{\rho}) & M_{13}(\boldsymbol{\rho}) & M_{14}(\boldsymbol{\rho}) \\ M_{21}(\boldsymbol{\rho}) & M_{22}(\boldsymbol{\rho}) & M_{23}(\boldsymbol{\rho}) & M_{24}(\boldsymbol{\rho}) \\ M_{31}(\boldsymbol{\rho}) & M_{32}(\boldsymbol{\rho}) & M_{33}(\boldsymbol{\rho}) & M_{34}(\boldsymbol{\rho}) \\ M_{41}(\boldsymbol{\rho}) & M_{42}(\boldsymbol{\rho}) & M_{43}(\boldsymbol{\rho}) & M_{44}(\boldsymbol{\rho}) \end{bmatrix} \begin{bmatrix} S_0^{(i)}(\boldsymbol{\rho}) \\ S_1^{(i)}(\boldsymbol{\rho}) \\ S_2^{(i)}(\boldsymbol{\rho}) \\ S_3^{(i)}(\boldsymbol{\rho}) \end{bmatrix}. \tag{4.156}$$

This transformation acting on the classic Stokes vector, being a single-position quantity, can be generalized to transformation at two spatial arguments. On recalling an alternative description of a stochastic electromagnetic beam via four generalized (two-point) Stokes parameters, we will write transformation of the incident generalized Stokes vector $S_l^{(i)}(\boldsymbol{\rho}_1, \boldsymbol{\rho}_2)$ into the transmitted generalized Stokes vector $S_l^{(t)}(\boldsymbol{\rho}_1, \boldsymbol{\rho}_2)$ as

$$\mathbf{S}_l^{(t)}(\boldsymbol{\rho}_1, \boldsymbol{\rho}_2) = \mathbf{M}(\boldsymbol{\rho}_1, \boldsymbol{\rho}_2)\mathbf{S}_l^{(i)}(\boldsymbol{\rho}_1, \boldsymbol{\rho}_2), \qquad (l = 0, 1, 2, 3), \tag{4.157}$$

or, more explicitly,

$$\begin{bmatrix} S_0^{(t)}(\boldsymbol{\rho}_1, \boldsymbol{\rho}_2) \\ S_1^{(t)}(\boldsymbol{\rho}_1, \boldsymbol{\rho}_2) \\ S_2^{(t)}(\boldsymbol{\rho}_1, \boldsymbol{\rho}_2) \\ S_3^{(t)}(\boldsymbol{\rho}_1, \boldsymbol{\rho}_2) \end{bmatrix} = \begin{bmatrix} M_{11}(\boldsymbol{\rho}_1, \boldsymbol{\rho}_2) & M_{12}(\boldsymbol{\rho}_1, \boldsymbol{\rho}_2) & M_{13}(\boldsymbol{\rho}_1, \boldsymbol{\rho}_2) & M_{14}(\boldsymbol{\rho}_1, \boldsymbol{\rho}_2) \\ M_{21}(\boldsymbol{\rho}_1, \boldsymbol{\rho}_2) & M_{22}(\boldsymbol{\rho}_1, \boldsymbol{\rho}_2) & M_{23}(\boldsymbol{\rho}_1, \boldsymbol{\rho}_2) & M_{24}(\boldsymbol{\rho}_1, \boldsymbol{\rho}_2) \\ M_{31}(\boldsymbol{\rho}_1, \boldsymbol{\rho}_2) & M_{32}(\boldsymbol{\rho}_1, \boldsymbol{\rho}_2) & M_{33}(\boldsymbol{\rho}_1, \boldsymbol{\rho}_2) & M_{34}(\boldsymbol{\rho}_1, \boldsymbol{\rho}_2) \\ M_{41}(\boldsymbol{\rho}_1, \boldsymbol{\rho}_2) & M_{42}(\boldsymbol{\rho}_1, \boldsymbol{\rho}_2) & M_{43}(\boldsymbol{\rho}_1, \boldsymbol{\rho}_2) & M_{44}(\boldsymbol{\rho}_1, \boldsymbol{\rho}_2) \end{bmatrix}$$

$$\times \begin{bmatrix} S_0^{(i)}(\boldsymbol{\rho}_1, \boldsymbol{\rho}_2) \\ S_1^{(i)}(\boldsymbol{\rho}_1, \boldsymbol{\rho}_2) \\ S_2^{(i)}(\boldsymbol{\rho}_1, \boldsymbol{\rho}_2) \\ S_3^{(i)}(\boldsymbol{\rho}_1, \boldsymbol{\rho}_2) \end{bmatrix}. \tag{4.158}$$

On performing straightforward calculations it is possible to show that the elements of the generalized Mueller matrix $\mathbf{M}(\boldsymbol{\rho}_1, \boldsymbol{\rho}_2)$ are linear combinations of the coefficients $\mathfrak{T}_{kl}^{pq}$. Such relations for all sixteen elements can be readily found in terms of coefficients $\mathfrak{T}$ [50].

4.4.2 Transmission through random devices

In some instances a linear, polarization-modulating device introduces random perturbations to the electric field components, capable of changing their amplitudes, phases or both. To account for such a random modulation we will consider, instead of a deterministic Jones matrix $\mathbf{T}(\boldsymbol{\rho})$, a random Jones matrix whose elements fluctuate in time or can be thought as realizations of a statistically stationary ensemble. Under the assumption that the fluctuations in the beam and in the device are statistically independent, the transformation (4.152) can be now generalized to

$$
\begin{aligned}
\mathbf{W}^{(t)}(\boldsymbol{\rho}_1,\boldsymbol{\rho}_2) &= \langle \mathbf{T}^*(\boldsymbol{\rho}_1)\mathbf{E}^{*(i)}(\boldsymbol{\rho}_1)\left[\mathbf{T}(\boldsymbol{\rho}_2)\mathbf{E}^{(i)}(\boldsymbol{\rho}_2)\right]^T \rangle \\
&= \langle \mathbf{T}^*(\boldsymbol{\rho}_1)\mathbf{W}^{(i)}(\boldsymbol{\rho}_1,\boldsymbol{\rho}_2)\mathbf{T}^T(\boldsymbol{\rho}_2,)\rangle,
\end{aligned}
\tag{4.159}
$$

where the averaging process may be considered either in time or in frequency domain. Written in terms of the components of the matrices this transformation takes the form

$$
\begin{aligned}
W_{xx}^{(t)}(\boldsymbol{\rho}_1,\boldsymbol{\rho}_2) &= \langle \mathfrak{T}_{xx}^{xx}(\boldsymbol{\rho}_1,\boldsymbol{\rho}_2)\rangle W_{xx}^{(i)}(\boldsymbol{\rho}_1,\boldsymbol{\rho}_2) + \langle \mathfrak{T}_{xy}^{xx}(\boldsymbol{\rho}_1,\boldsymbol{\rho}_2)\rangle W_{xy}^{(i)}(\boldsymbol{\rho}_1,\boldsymbol{\rho}_2) \\
&\quad + \langle \mathfrak{T}_{yx}^{xx}(\boldsymbol{\rho}_1,\boldsymbol{\rho}_2)\rangle W_{yx}^{(i)}(\boldsymbol{\rho}_1,\boldsymbol{\rho}_2) + \langle \mathfrak{T}_{yy}^{xx}(\boldsymbol{\rho}_1,\boldsymbol{\rho}_2)\rangle W_{yy}^{(i)}(\boldsymbol{\rho}_1,\boldsymbol{\rho}_2), \\
W_{xy}^{(t)}(\boldsymbol{\rho}_1,\boldsymbol{\rho}_2) &= \langle \mathfrak{T}_{xx}^{xy}(\boldsymbol{\rho}_1,\boldsymbol{\rho}_2)\rangle W_{xx}^{(i)}(\boldsymbol{\rho}_1,\boldsymbol{\rho}_2) + \langle \mathfrak{T}_{xy}^{xy}(\boldsymbol{\rho}_1,\boldsymbol{\rho}_2)\rangle W_{xy}^{(i)}(\boldsymbol{\rho}_1,\boldsymbol{\rho}_2) \\
&\quad + \langle \mathfrak{T}_{yx}^{xy}(\boldsymbol{\rho}_1,\boldsymbol{\rho}_2)\rangle W_{yx}^{(i)}(\boldsymbol{\rho}_1,\boldsymbol{\rho}_2) + \langle \mathfrak{T}_{yy}^{xy}(\boldsymbol{\rho}_1,\boldsymbol{\rho}_2)\rangle W_{yy}^{(i)}(\boldsymbol{\rho}_1,\boldsymbol{\rho}_2), \\
W_{yx}^{(t)}(\boldsymbol{\rho}_1,\boldsymbol{\rho}_2) &= \langle \mathfrak{T}_{xx}^{yx}(\boldsymbol{\rho}_1,\boldsymbol{\rho}_2)\rangle W_{xx}^{(i)}(\boldsymbol{\rho}_1,\boldsymbol{\rho}_2) + \langle \mathfrak{T}_{xy}^{yx}(\boldsymbol{\rho}_1,\boldsymbol{\rho}_2)\rangle W_{xy}^{(i)}(\boldsymbol{\rho}_1,\boldsymbol{\rho}_2) \\
&\quad + \langle \mathfrak{T}_{yx}^{yx}(\boldsymbol{\rho}_1,\boldsymbol{\rho}_2)\rangle W_{yx}^{(i)}(\boldsymbol{\rho}_1,\boldsymbol{\rho}_2) + \langle \mathfrak{T}_{yy}^{yx}(\boldsymbol{\rho}_1,\boldsymbol{\rho}_2)\rangle W_{yy}^{(i)}(\boldsymbol{\rho}_1,\boldsymbol{\rho}_2), \\
W_{yy}^{(t)}(\boldsymbol{\rho}_1,\boldsymbol{\rho}_2) &= \langle \mathfrak{T}_{xx}^{yy}(\boldsymbol{\rho}_1,\boldsymbol{\rho}_2)\rangle W_{xx}^{(i)}(\boldsymbol{\rho}_1,\boldsymbol{\rho}_2) + \langle \mathfrak{T}_{xy}^{yy}(\boldsymbol{\rho}_1,\boldsymbol{\rho}_2)\rangle W_{xy}^{(i)}(\boldsymbol{\rho}_1,\boldsymbol{\rho}_2) \\
&\quad + \langle \mathfrak{T}_{yx}^{yy}(\boldsymbol{\rho}_1,\boldsymbol{\rho}_2)\rangle W_{yx}^{(i)}(\boldsymbol{\rho}_1,\boldsymbol{\rho}_2) + \langle \mathfrak{T}_{yy}^{yy}(\boldsymbol{\rho}_1,\boldsymbol{\rho}_2)\rangle W_{yy}^{(i)}(\boldsymbol{\rho}_1,\boldsymbol{\rho}_2).
\end{aligned}
\tag{4.160}
$$

Here the sixteen correlations $\langle \mathfrak{T}_{kl}^{pq}\rangle$ between the elements of the random Jones matrix of a device at two points, $\boldsymbol{\rho}_1$ and $\boldsymbol{\rho}_2$, completely determine the transformations. As before the averaging can be performed over a long time interval or over the statistical ensemble of realization of a device, depending on the theoretical approach or measurement procedure used.

For transformation of generalized Stokes parameters by a random device, the two-point Mueller matrix $\langle \mathbf{M}\rangle$ should be used where an average is taken over the fluctuations of the device (temporal or via statistical ensemble) and we then obtain the relation

$$
\mathbf{S}_l^{(t)}(\boldsymbol{\rho}_1,\boldsymbol{\rho}_2) = \langle \mathbf{M}(\boldsymbol{\rho}_1,\boldsymbol{\rho}_2)\rangle \mathbf{S}_l^{(i)}(\boldsymbol{\rho}_1,\boldsymbol{\rho}_2), \qquad (l=0,1,2,3).
\tag{4.161}
$$

4.4.3 Combination of several devices

In the situations when the beam is incident on a system of thin, linear, non-image forming devices, located sufficiently close to each other and all aligned

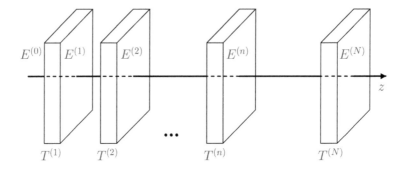

FIGURE 4.5
Cascaded system of linear non-image-forming devices.

in the planes transverse to direction of propagation of the beam (see Fig. 4.5) one can extend the calculus by applying the laws given by Eqs. (4.152), (4.169), (4.159) and (4.161) subsequently for each of the devices in the system [50].

Let us first consider the system of N deterministic devices with Jones matrices $T^{(n)}$, $n = 1, ..., N$. After passing through all N devices the incident electric field $\mathbf{E}^{(i)}(\boldsymbol{\rho})$ is modified by the following transformation

$$\mathbf{E}^{(N)}(\boldsymbol{\rho}) = \mathbb{T}(\boldsymbol{\rho})\mathbf{E}^{(i)}(\boldsymbol{\rho}), \tag{4.162}$$

where

$$\mathbb{T}(\boldsymbol{\rho}) = \prod_{n=1}^{N} T^{(n)}(\boldsymbol{\rho}) \tag{4.163}$$

is the Jones matrix of the system.

The relation between the cross-spectral densities of the incident field $\mathbf{W}^{(i)}$ and transmitted field $\mathbf{W}^{(N)}$ then has the form

$$\begin{aligned}
\mathbf{W}^{(N)}(\boldsymbol{\rho}_1, \boldsymbol{\rho}_2) &= \mathbb{T}^*(\boldsymbol{\rho}_1)\mathbf{E}^{*(i)}(\boldsymbol{\rho}_1) \left[\mathbb{T}(\boldsymbol{\rho}_2)\mathbf{E}^{(i)}(\boldsymbol{\rho}_2)\right]^T \\
&= \mathbb{T}^*(\boldsymbol{\rho}_1)\mathbf{W}^{(i)}(\boldsymbol{\rho}_1, \boldsymbol{\rho}_2)\mathbb{T}^T(\boldsymbol{\rho}_2).
\end{aligned} \tag{4.164}$$

Further, if the system consists of random devices then Eq. (4.163) becomes

$$\begin{aligned}
\langle \mathbb{T}(\boldsymbol{\rho})\rangle &= \langle \prod_{n=1}^{N} T^{(n)}(\boldsymbol{\rho})\rangle_d \\
&= \prod_{n=1}^{N} \langle T^{(n)}(\boldsymbol{\rho})\rangle_d,
\end{aligned} \tag{4.165}$$

where subscript d denotes the average over realizations of a random device, and the last equality was made under the assumption that the individual devices in the system are statistically independent. It then follows from Eqs. (4.159) and (4.165) that

$$
\begin{aligned}
\mathbf{W}^{(N)}(\boldsymbol{\rho}_1,\boldsymbol{\rho}_2) &= \langle \mathbb{T}^*(\boldsymbol{\rho}_1)\mathbf{E}^{*(i)}(\boldsymbol{\rho}_1)\left[\mathbb{T}(\boldsymbol{\rho}_2)\mathbf{E}^{(i)}(\boldsymbol{\rho}_2)\right]^T \rangle \\
&= \langle \mathbb{T}^*(\boldsymbol{\rho}_1)\mathbf{W}^{(i)}(\boldsymbol{\rho}_1,\boldsymbol{\rho}_2)\mathbb{T}^T(\boldsymbol{\rho}_2) \rangle.
\end{aligned} \tag{4.166}
$$

Transformations (4.164) and (4.166) can also be written more explicitly via the sixteen transmission coefficients that we have introduced earlier.

In a similar fashion the transmission through the system of devices may be described with the help of products of the generalized Mueller matrices of the individual devices. In particular, for a deterministic system of devices the relation between the generalized Stokes vectors $\mathbf{S}_l^{(i)}$ and $\mathbf{S}_l^{(N)}$ of the incident and transmitted fields, respectively takes the form

$$
\mathbf{S}_l^{(N)}(\boldsymbol{\rho}_1,\boldsymbol{\rho}_2) = \mathbb{M}(\boldsymbol{\rho}_1,\boldsymbol{\rho}_2)\mathbf{S}_l^{(i)}(\boldsymbol{\rho}_1,\boldsymbol{\rho}_2), \qquad (l=0,1,2,3), \tag{4.167}
$$

where $\mathbb{M}(\boldsymbol{\rho}_1,\boldsymbol{\rho}_2)$ is the generalized Mueller matrix of the system, i.e.,

$$
\mathbb{M}(\boldsymbol{\rho}_1,\boldsymbol{\rho}_2) = \prod_{n=1}^{N} \mathbf{M}_n(\boldsymbol{\rho}_1,\boldsymbol{\rho}_2) \tag{4.168}
$$

If the system of devices is random then the relation (4.169) becomes

$$
\mathbf{S}_l^{(N)}(\boldsymbol{\rho}_1,\boldsymbol{\rho}_2) = \langle \mathbb{M}(\boldsymbol{\rho}_1,\boldsymbol{\rho}_2) \rangle \mathbf{S}_l^{(i)}(\boldsymbol{\rho}_1,\boldsymbol{\rho}_2), \qquad (l=0,1,2,3), \tag{4.169}
$$

where

$$
\langle \mathbb{M}(\boldsymbol{\rho}_1,\boldsymbol{\rho}_2) \rangle_d = \prod_{n=1}^{N} \langle M_n(\boldsymbol{\rho}_1,\boldsymbol{\rho}_2) \rangle_d, \tag{4.170}
$$

where we again make use of the assumption that the devices are statistically independent.

4.5 Electromagnetic Gaussian Schell-model sources and beams

4.5.1 Realizability and beam conditions

An electromagnetic Schell-model beam is simply a vectorial generalization of the scalar Gaussian Schell-model beam. Its special cases were suggested by James [1], Seshadri [51], [52] and Gori et al. [53]. The general form of the elements of the cross-spectral density matrix of the electromagnetic Schell-model source is the following product:

$$W_{\alpha\beta}(\boldsymbol{\rho}_1', \boldsymbol{\rho}_2'; \omega) = \sqrt{S_\alpha(\boldsymbol{\rho}_1'; \omega)} \sqrt{S_\beta(\boldsymbol{\rho}_2'; \omega)} \eta_{\alpha\beta}(\boldsymbol{\rho}_2' - \boldsymbol{\rho}_1'; \omega),$$

$$(\alpha, \beta = x, y). \tag{4.171}$$

Here x and y are two mutually orthogonal directions in the Cartesian coordinate system, S_α is the spectral density of the α-th component of the electric vector-field and $\eta_{\alpha\beta}$ is the correlation coefficient between the α and β field components.

So far, the electromagnetic Gaussian Schell-model beam is the only well-explored class of electromagnetic stochastic beam-like fields. In this case both the spectral density and the spectral degree of coherence are Gaussian functions and hence the elements of the cross-spectral density matrix in the source plane at two spatial positions $\boldsymbol{\rho}_1'$ and $\boldsymbol{\rho}_2'$ have the form [53], [54] and [55]

$$W_{\alpha\beta}(\boldsymbol{\rho}_1', \boldsymbol{\rho}_2'; \omega) = A_\alpha A_\beta B_{\alpha\beta} \exp\left[-\left(\frac{\rho_1'^2}{2\sigma_\alpha^2} + \frac{\rho_2'^2}{2\sigma_\beta^2} \right) \right] \exp\left[-\frac{(\boldsymbol{\rho}_1' - \boldsymbol{\rho}_2')^2}{2\delta_{\alpha\beta}^2} \right],$$

$$(\alpha = x, y; \beta = x, y). \tag{4.172}$$

It is assumed that all the parameters of the model, i.e., spectral amplitudes A_α, $(\alpha = x, y)$, correlation coefficient $B_{\alpha\beta}$, $(\alpha = x, y; \beta = x, y)$ of x and y-components (in general complex-valued), root-mean-square (r.m.s.) widths σ_α, $(\alpha = x, y)$ of intensity along x and y-components and r.m.s. widths of correlation functions $\delta_{\alpha\beta}$ of x and y-components of the beam are independent of position but might depend on frequency.

From the intrinsic properties of the cross-spectral density matrix (see Section 4.1.2) we find that several relations should be satisfied by the parameters of the model source (4.172) in order to produce a physically realizable beam.

First of all it is a consequence of the scalar theory (see [8], Section 4.3.2) that the single-point correlation coefficients are equal for $\alpha = \beta$:

$$B_{xx} = B_{yy} \equiv 1, \tag{4.173}$$

since the diagonal components of the matrix represent the scalar cross-spectral density functions.

Further, from the quasi-Hermiticity of the cross-spectral density matrix [see Eq. (4.7)] it follows at once that [56]

$$\delta_{xy} = \delta_{yx}, \tag{4.174}$$

$$B_{xy} = B_{yx}^*. \tag{4.175}$$

We will now list several restrictions that are the implications of the non-negative definiteness condition on the cross-spectral density matrix. The first condition for the electromagnetic Gaussian Schell-model source to generate a physically realizable beam, which is both necessary and sufficient, follows at once from the non-negative definiteness condition at $\boldsymbol{\rho}_1' = \boldsymbol{\rho}_2'$ [56]:

$$|B_{xy}| \le 1, \quad |B_{yx}| \le 1. \tag{4.176}$$

The second (sufficient only) condition for radiation of a physically realizable field has the form of a double inequality

$$\max(\delta_{xx}, \delta_{yy}) \le \delta_{xy} \le \min\left(\frac{\delta_{xx}}{\sqrt{|B_{xy}|}}, \frac{\delta_{yy}}{\sqrt{|B_{xy}|}}\right). \tag{4.177}$$

In order to prove this inequality we recall that the condition for the correlation matrix to be non-negative definite can be expressed by the integral inequality (4.8). Let us choose function $g_\alpha(\boldsymbol{\rho}')$ to be of the form

$$g_\alpha(\boldsymbol{\rho}') = \frac{f_\alpha(\boldsymbol{\rho}')}{A_\alpha} \exp\left[-\frac{\rho'^2}{4\sigma_\alpha^2}\right], \quad (\alpha = x, y). \tag{4.178}$$

Each of the four terms in the integrand of Eq. (4.8) then can be expressed as

$$f_\alpha^*(\boldsymbol{\rho}_1')f_\beta(\boldsymbol{\rho}_2')W_{\alpha\beta}(\boldsymbol{\rho}_1', \boldsymbol{\rho}_2'; \omega) = |B_{\alpha\beta}|g_\alpha^*(\boldsymbol{\rho}_1')g_\alpha(\boldsymbol{\rho}_2') \exp\left[-\frac{(\boldsymbol{\rho}_2' - \boldsymbol{\rho}_1')^2}{2\delta_{\alpha\beta}^2}\right],$$

$$(\alpha = x, y; \beta = x, y). \tag{4.179}$$

On substituting from Eq. (4.179) into Eq. (4.8) and recalling relations (4.173) and (4.175), we obtain the inequality

$$\iint \left\{ g_x^*(\boldsymbol{\rho}_1')g_x(\boldsymbol{\rho}_2') \exp\left[-\frac{(\boldsymbol{\rho}_2' - \boldsymbol{\rho}_1')^2}{2\delta_{xx}^2}\right]\right.$$

$$+ |B_{xy}|g_x^*(\boldsymbol{\rho}_1')g_y(\boldsymbol{\rho}_2') \exp\left[-\frac{(\boldsymbol{\rho}_2' - \boldsymbol{\rho}_1')^2}{2\delta_{xy}^2}\right]$$

$$+ |B_{yx}|g_y^*(\boldsymbol{\rho}_1')g_x(\boldsymbol{\rho}_2') \exp\left[-\frac{(\boldsymbol{\rho}_2' - \boldsymbol{\rho}_1{}')^2}{2\delta_{yx}^2}\right]$$

$$\left.+ g_y^*(\boldsymbol{\rho}_1')g_y(\boldsymbol{\rho}_2') \exp\left[-\frac{(\boldsymbol{\rho}_2' - \boldsymbol{\rho}_1')^2}{2\delta_{yy}^2}\right]\right\} d^2\rho_1' d^2\rho_2' \ge 0. \tag{4.180}$$

We can represent the Gaussian functions $\exp\left[-\frac{(\boldsymbol{\rho}_2' - \boldsymbol{\rho}_1')^2}{2\delta_{\alpha\beta}^2}\right]$ $(\alpha = x, y; \beta = x, y)$, appearing in the integrand of Eq. (4.180) by means of their two-dimensional

Fourier transforms, viz.

$$\exp\left[-\frac{(\boldsymbol{\rho}_2' - \boldsymbol{\rho}_1')^2}{2\delta_{\alpha\beta}^2}\right]$$

$$= \int \delta_{\alpha\beta}^2 \exp[-2\pi^2 \delta_{\alpha\beta}^2 \kappa^2] \exp[2\pi i (\boldsymbol{\rho}_2' - \boldsymbol{\rho}_1') \cdot \boldsymbol{\kappa}] d^2 \kappa, \tag{4.181}$$

where $\boldsymbol{\kappa}$ is a two-dimensional vector in the Fourier domain, with magnitude $|\boldsymbol{\kappa}| = \kappa$. On substituting from the last equation into inequality (4.180) and on applying the restrictions (4.174) and (4.175) one obtains the inequality

$$\iiint \left\{ g_x^*(\boldsymbol{\rho}_1') g_x(\boldsymbol{\rho}_2') \delta_{xx}^2 \exp[-2\pi^2 \delta_{xx}^2 \kappa^2] \right.$$

$$+ [g_x^*(\boldsymbol{\rho}_1') g_y(\boldsymbol{\rho}_2') + g_y^*(\boldsymbol{\rho}_1') g_x(\boldsymbol{\rho}_2')] |B_{xy}| \delta_{xy}^2 \exp[-2\pi^2 \delta_{xy}^2 \kappa^2]$$

$$\left. g_y^*(\boldsymbol{\rho}_1') g_y(\boldsymbol{\rho}_2') \delta_{yy}^2 \exp[-2\pi^2 \delta_{yy}^2 \kappa^2] \right\} \tag{4.182}$$

$$\times \exp[2\pi i (\boldsymbol{\rho}_2' - \boldsymbol{\rho}_1') \cdot \boldsymbol{\kappa}] d^2 \rho_1' d^2 \rho_2' d^2 \kappa \geq 0.$$

We can now perform the two-dimensional Fourier transform of the expression above with respect to variables $\boldsymbol{\rho}_1'$ and $\boldsymbol{\rho}_2'$ which yields the following inequality:

$$\iiint \left\{ |G_x(\boldsymbol{\kappa})|^2 \delta_{xx}^2 \exp[-2\pi^2 \delta_{xx}^2 \kappa^2] \right.$$

$$+ 2Re[G_x^*(\boldsymbol{\kappa}) G_y(\boldsymbol{\kappa})] |B_{xy}| \delta_{xy}^2 \exp[-2\pi^2 \delta_{xy}^2 \kappa^2] \tag{4.183}$$

$$\left. + |G_y(\boldsymbol{\kappa})|^2 \delta_{yy}^2 \exp[-2\pi^2 \delta_{yy}^2 \kappa^2] \right\} d^2 \kappa \geq 0,$$

where functions $G_j(\boldsymbol{\kappa})$ are two-dimensional Fourier transforms of $g_j(\boldsymbol{\rho}')$, viz.,

$$G_j(\boldsymbol{\kappa}') = \int g_j(\boldsymbol{\rho}') \exp[-2\pi i \boldsymbol{\rho}' \cdot \boldsymbol{\kappa}] d^2 \rho'. \tag{4.184}$$

If we introduce the positive real functions q_1, q_2 and q_3 by the formulas

$$q_1(\boldsymbol{\kappa}) = \delta_{xx}^2 \exp(-2\pi^2 \delta_{xx}^2 \kappa^2),$$
$$q_2(\boldsymbol{\kappa}) = |B_{xy}| \delta_{xy}^2 \exp(-2\pi^2 \delta_{xy}^2 \kappa^2), \tag{4.185}$$
$$q_3(\boldsymbol{\kappa}) = \delta_{yy}^2 \exp(-2\pi^2 \delta_{yy}^2 \kappa^2),$$

then the integrand of Eq. (4.183) becomes

$$|G_x|^2 q_1 + 2Re[G_x^* G_y] q_2 + |G_y|^2 q_3$$

$$= |G_x + G_y|^2 q_2 + |G_x|^2 (q_1 - q_2) + |G_y|^2 (q_3 - q_2). \tag{4.186}$$

where the argument κ of all functions is omitted. One can see at once that in order for the right-hand side of the Eq. (4.186) to be non-negative for any values of G_x and G_y, it suffices to require that

$$q_1 \geq q_2, \quad \text{and} \quad q_3 \geq q_2, \tag{4.187}$$

for any values of the argument κ. Using the formulas (4.185) we obtain the inequalities

$$\begin{aligned}
\delta_{xx}^2 \exp(-2\pi^2 \delta_{xx}^2 \kappa^2) &\geq |B_{xy}| \delta_{xy}^2 \exp(-2\pi^2 \delta_{xy}^2 \kappa^2), \\
\delta_{yy}^2 \exp(-2\pi^2 \delta_{yy}^2 \kappa^2) &\geq |B_{xy}| \delta_{xy}^2 \exp(-2\pi^2 \delta_{xy}^2 \kappa^2).
\end{aligned} \tag{4.188}$$

Since, on the interval $[0; \infty)$ all functions in the inequalities (4.188) vary monotonically with κ, it suffices to consider their behavior only on the boundaries of that interval. Hence, letting $\kappa = 0$ in both inequalities we find that

$$\delta_{xx}^2 \geq |B_{xy}| \delta_{xy}^2, \quad \text{and} \quad \delta_{yy}^2 \geq |B_{xy}| \delta_{xy}^2. \tag{4.189}$$

On the other hand, letting $\kappa \to \infty$ leads to

$$\delta_{xx}^2 \leq \delta_{xy}^2, \quad \text{and} \quad \delta_{yy}^2 \leq \delta_{xy}^2. \tag{4.190}$$

Finally, combining the last two sets of inequalities, i.e., Eqs. (4.189) and (4.190), and using the fact that all parameters in these expressions are positive numbers we arrive at the fork inequality (4.177).

We note here that an alternative inequality representing a sufficient condition was derived in [10] and has the form

$$\sqrt{\frac{\delta_{xx}^2 + \delta_{yy}^2}{2}} \leq \delta_{xy} \leq \sqrt{\frac{\delta_{xx}\delta_{yy}}{|B_{xy}|}}. \tag{4.191}$$

The third condition, which is only a *necessary* realizability condition, is formed by a set of two inequalities [56]

$$\frac{A_x^2 \sigma_x^4 \delta_{xx}^2}{\delta_{xx}^2 + 4\sigma_x^2} - \frac{2 A_x A_y |B_{xy}| \sigma_x^2 \sigma_y^2 \delta_{xy}^2}{\delta_{xy}^2 + 2\sigma_x^2 + 2\sigma_y^2} + \frac{A_y^2 \sigma_y^4 \delta_{yy}^2}{\delta_{yy}^2 + 4\sigma_y^2} \geq 0, \tag{4.192}$$

and

$$\frac{2\sigma_x^2 \delta_{xx}^2}{\delta_{xx}^2 + 4\sigma_x^2} - \frac{(\sigma_x^2 + \sigma_y^2)\delta_{xy}^2}{\delta_{xy}^2 + 2\sigma_x^2 + 2\sigma_y^2} + \frac{2\sigma_y^2 \delta_{yy}^2}{\delta_{yy}^2 + 4\sigma_y^2} \leq 0. \tag{4.193}$$

As we see, the necessary condition involves the r.m.s. widths of the spectral densities of x- and y–components of the beam, in addition to the r.m.s. widths of correlation coefficients. In order to prove the inequalities (4.192) and (4.193) we can chose in the non-negative definiteness condition (4.8) the following form for the functions $f_x(\boldsymbol{\rho}')$ and $f_y(\boldsymbol{\rho}')$:

$$f_x(\boldsymbol{\rho}') = f_y(\boldsymbol{\rho}') = f(\boldsymbol{\rho}') = \exp(-2i\pi\boldsymbol{\kappa} \cdot \boldsymbol{\rho}'). \tag{4.194}$$

On substituting from Eq. (4.172) for the elements of the cross-spectral density matrix and Eq. (4.194) for functions $f_x(\boldsymbol{\rho}')$ and $f_y(\boldsymbol{\rho}')$ into each of the four terms of the Eq. (4.8) we obtain the formula

$$
\iint f_\alpha^*(\boldsymbol{\rho}_1') f_\beta(\boldsymbol{\rho}_2') W_{\alpha\beta}(\boldsymbol{\rho}_1', \boldsymbol{\rho}_2'; \omega) d^2\rho_1' d^2\rho_2'
$$

$$
= A_\alpha A_\beta |B_{\alpha\beta}| \iint \exp[-2i\pi\boldsymbol{\kappa}\cdot(\boldsymbol{\rho}_2' - \boldsymbol{\rho}_1')] \exp\left[-\left(\frac{\rho_1'^2}{2\sigma_\alpha^2} + \frac{\rho_2'^2}{2\sigma_\beta^2}\right)\right] \qquad (4.195)
$$

$$
\times \exp\left[-\frac{(\boldsymbol{\rho}_1' - \boldsymbol{\rho}_2')^2}{2\delta_{\alpha\beta}^2}\right] d^2\rho_1' d^2\rho_2'.
$$

After integrating twice over the source, the right-hand side of expression (4.195) becomes

$$
\text{sign}(\alpha, \beta) \frac{16 A_\alpha A_\beta |B_{\alpha\beta}| \sigma_\alpha^2 \sigma_\beta^2 \delta_{\alpha\beta}^2}{\delta_{\alpha\beta}^2 + 2\sigma_\alpha^2 + 2\sigma_\beta^2} exp\left[-\frac{4\pi^2\kappa^2\delta_{\alpha\beta}^2(\sigma_\alpha^2 + \sigma_\beta^2)}{\delta_{\alpha\beta}^2 + 2\sigma_\alpha^2 + 2\sigma_\beta^2}\right], \qquad (4.196)
$$

where $\text{sign}(\alpha, \beta) = 1$ when $\alpha = \beta$ and $\text{sign}(\alpha, \beta) = -1$ if $\alpha \neq \beta$. On substituting each of the four expressions (4.196) into Eq. (4.8) one then obtains the inequality

$$
\frac{A_x^2\sigma_x^4\delta_{xx}^2}{\delta_{xx}^2 + 4\sigma_x^2} \exp\left[-\frac{8\pi^2\kappa^2\delta_{xx}^2\sigma_x^2}{\delta_{xx}^2 + 4\sigma_x^2}\right] - \frac{2A_x A_y |B_{xy}|\sigma_x^2\sigma_y^2\delta_{xy}^2}{\delta_{xy}^2 + 2\sigma_x^2 + 2\sigma_y^2}
$$

$$
\times \exp\left[-\frac{4\pi^2\kappa^2\delta_{xy}^2(\sigma_x^2 + \sigma_y^2)}{\delta_{xy}^2 + 2\sigma_x^2 + 2\sigma_y^2}\right] \frac{A_y^2\sigma_y^4\delta_{yy}^2}{\delta_{yy}^2 + 4\sigma_y^2} \exp\left[-\frac{8\pi^2\kappa^2\delta_{yy}^2\sigma_y^2}{\delta_{yy}^2 + 4\sigma_y^2}\right] \geq 0. \qquad (4.197)
$$

Since the left-hand side of (4.197) is the sum of three monotonic functions of κ, to ensure that the inequality folds for all values of κ in the interval $[0; \infty)$, it suffices to check its validity on the boundaries of that interval. Letting $\kappa = 0$ and $\kappa \to \infty$ we arrive at inequalities (4.192) and (4.193), respectively.

In a particular case when

$$
\sigma_x = \sigma_y, \qquad (4.198)
$$

the polarization becomes uniform across the source (see also [53]), and hence the conditions (4.192) and (4.193) reduce to the following two inequalities

$$
\frac{A_x^2\delta_{xx}^2}{\delta_{xx}^2 + 4\sigma^2} - \frac{2A_x A_y |B_{xy}|\delta_{xy}^2}{\delta_{xy}^2 + 4\sigma^2} + \frac{A_y^2\delta_{yy}^2}{\delta_{yy}^2 + 4\sigma^2} \geq 0, \qquad (4.199)
$$

and

$$
\frac{2\delta_{xx}^2}{\delta_{xx}^2 + 4\sigma^2} - \frac{2\delta_{xy}^2}{\delta_{xy}^2 + 4\sigma^2} + \frac{\delta_{yy}^2}{\delta_{yy}^2 + 4\sigma_y} \leq 0. \qquad (4.200)
$$

We note here that the use of sufficient and necessary conditions should not be made at the same time. While the sufficient condition may be used to verify that the set of parameters does represent the physically realizable source, the necessary condition may be used in situations when a certain set of model parameters needs to be proven to represent the source that cannot be physically realized.

The elements of the angular correlation matrix of the electromagnetic Gaussian Schell-model source with uniform polarization can be obtained in a manner similar to that for the scalar source [54]. Further, since the spectral density of the beam in the far zone is just the trace of the correlation matrix, in order to derive the beam conditions it is necessary and sufficient to require that both on-diagonal elements of the cross-spectral density matrix satisfy the scalar beam conditions, expressed by Eq. (3.67). Then one obtains the inequalities:

$$\frac{1}{4\sigma^2} + \frac{1}{\delta_{xx}^2} \ll \frac{2\pi^2}{\lambda^2}, \quad \frac{1}{4\sigma^2} + \frac{1}{\delta_{yy}^2} \ll \frac{2\pi^2}{\lambda^2}. \qquad (4.201)$$

4.5.2   Methods of generation

Two fundamentally different methods were proposed so far for synthesis of the electromagnetic Gaussian Schell-model sources, one being based on the increase in coherence of light on propagation via the van Cittert-Zernike theorem [57] and the other employing phase diffusers or spatial light modulators in order to change statistical properties of laser radiation [58]. Both methods use the Mach-Zender interferometer for creating vectorial beams from scalar. We will now briefly discuss both procedures.

The experimental setting shown in Fig. 4.6 (a) uses the Argon laser source producing a linearly polarized beam which impinges on the beam splitter B of the Mach-Zender interferometer. In one of the arms of the interferometer a beam is rotated by a half-wave plate and acquires a polarization that is orthogonal to that of the beam in the other arm. After being reflected by mirrors M the beams pass through two microscope objectives and two rotating ground glass diffusers denoted by MO and RGG, respectively, in order to make the beams spatially incoherent. After being randomized, the two beams interact with intensity masks F, and then are sent to the beam combiner and imaged by another microscope MO'. At this stage an electromagnetic spatially incoherent beam with uncorrelated electric field components (described by a diagonal correlation matrix) is synthesized. With the help of a lens L placed exactly at its focal distance f from the ready image and the intensity filter GF, located right after the lens, a partially coherent beam is produced, still with uncorrelated field components. Finally, a quarter-plate rotator can be used for their mutual correlation.

This technique was recently extended for synthesis of the most general Schell-model beams, i.e., the beams not necessarily obeying Gaussian distribution of intensities and Gaussian correlation functions [59].

Alternatively, the electromagnetic Gaussian Schell-model beam may be produced as shown in Fig. 4.6 (b). Here a linearly polarized plane wave is generated by a single-mode laser and is sent to the Mach-Zender interferometer. After being split and sent to two arms by a beam-splitter $BS1$ the polarization of a plane wave in one of the arms is being changed to the orthogonal by a half-wave plate HWP. After being reflected from two mirrors $M1$ and $M2$ the coherent mutually orthogonal waves are being transmitted through the two liquid crystal [LC] spatial light modulators [SLM] $SLM1$ and $SLM2$. Such devices act as random phase screens, whose individual and joint statistical properties can be controlled by a computer (see Section 3.3 for more details on interaction of the scalar beam with an SLM). It is therefore sufficient to generate, for each point on the SLMs with position vector $\boldsymbol{\rho}$ real random sequences $\varpi_x(\boldsymbol{\rho};\omega)$ and $\varpi_y(\boldsymbol{\rho};\omega)$, assuming that LC directors for each of the SLMs are arranged along x and y directions. We now assume that the random phases obey Gaussian statistics with zero mean, and that their second-order autocorrelations at two positions $\boldsymbol{\rho}_1$ and $\boldsymbol{\rho}_2$ are given by the expressions

$$\langle \varpi_x(\boldsymbol{\rho}_1;\omega)\varpi_x(\boldsymbol{\rho}_2;\omega)\rangle = \varpi_x^{(0)2} \exp\left[-\frac{(\boldsymbol{\rho}_1 - \boldsymbol{\rho}_2)^2}{2\delta_{\varpi x}^2} \right], \qquad (4.202)$$

$$\langle \varpi_y(\boldsymbol{\rho}_1;\omega)\varpi_y(\boldsymbol{\rho}_2;\omega)\rangle = \varpi_y^{(0)2} \exp\left[-\frac{(\boldsymbol{\rho}_1 - \boldsymbol{\rho}_2)^2}{2\delta_{\varpi y}^2} \right], \qquad (4.203)$$

with $\varpi_x^{(0)} = \langle|\varpi_x(\boldsymbol{\rho};\omega)|^2\rangle$, $\varpi_y^{(0)} = \langle|\varpi_y(\boldsymbol{\rho};\omega)|^2\rangle$; and $\delta_{\varpi x}^2$ and $\delta_{\varpi y}^2$ being the r.m.s. correlation widths of the random phases. We further assume that the mutual correlation of the phase on the two SLMs is

$$\langle \varpi_x(\boldsymbol{\rho}_1;\omega)\varpi_y(\boldsymbol{\rho}_2;\omega)\rangle = \varpi_{xy}^{(0)2} \exp\left[-\frac{(\boldsymbol{\rho}_1 - \boldsymbol{\rho}_2)^2}{2\delta_{\varpi xy}^2} \right], \qquad (4.204)$$

with $\varpi_{xy}^{(0)} = \langle \varpi_x(\boldsymbol{\rho};\omega)\varpi_x(\boldsymbol{\rho};\omega)\rangle$, and $\delta_{\varpi xy}^2$ being the r.m.s. width of the cross-correlation between ϖ_x and ϖ_y. The phases of the SLMs, are controlled by electric signals: the constant phase being associated with each pixel of the SLMs which is proportional to the two-dimensional array of the control electric signal. After passing through the SLMs followed by the Gaussian amplitude filters $F1$ and $F2$ the produced beam-like fields are being superimposed on the combiner $BS2$ and sent through an optional compensator C, which produces a constant phase delay between the two field components.

A version of the method discussed in Ref. [58] was introduced and experimentally implemented in Ref. [60], where instead of the SLM's two diffusers with known properties were employed. While the procedure involving the SLM is more expensive it provides the opportunity of modulating the statistical properties of the produced random source virtually in real time, without changing any of the optical elements and of prescribing the phase patterns that are hard to obtain with a phase diffuser.

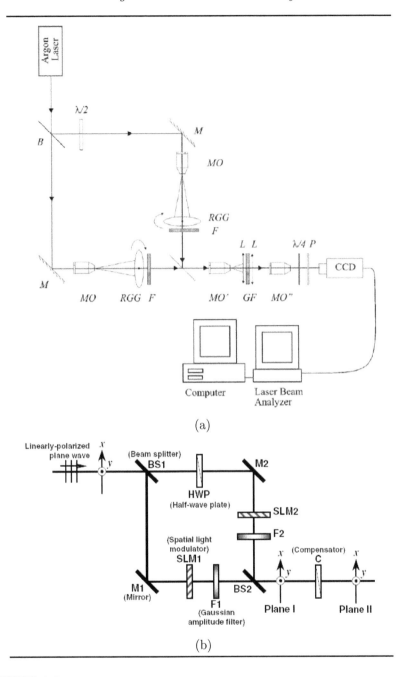

FIGURE 4.6
Optical arrangements for producing electromagnetic Gaussian Schell-model beams: (a) from Ref. [57]; (b) from Ref. [58].

4.5.3 Propagation in free space

The general expressions for free-space propagation of the cross-spectral density matrix elements of the beam were given in Section 4.3.1. We will now apply these laws for determination of the cross-spectral density matrix of the propagating electromagnetic Gaussian Schell-model beam. We will restrict ourselves here to consideration of the uniformly polarized sources only, i.e., sources with $\sigma_\alpha = \sigma_\beta = \sigma$. For such a source the cross-spectral density matrix takes the form:

$$W_{\alpha\beta}(\boldsymbol{\rho}_1', \boldsymbol{\rho}_2'; \omega) = A_\alpha A_\beta B_{\alpha\beta} exp\left[-\frac{\boldsymbol{\rho}_1'^2 + \boldsymbol{\rho}_2'^2}{4\sigma^2}\right] \exp\left[-\frac{(\boldsymbol{\rho}_1' - \boldsymbol{\rho}_2')^2}{2\delta_{\alpha\beta}^2}\right], \quad (4.205)$$

$$(\alpha = x, y; \beta = x, y),$$

On substituting from Eq. (4.205) into Eqs. (4.123) and (4.124), we find that [61]

$$W_{\alpha\beta}(\boldsymbol{\rho}_1, z_1, \boldsymbol{\rho}_2, z_2; \omega) = \frac{k^2}{(2\pi)^2} \frac{A_\alpha A_\beta B_{\alpha\beta}}{z_1 z_2} \exp\left[-\frac{ik}{2}\left(\frac{\boldsymbol{\rho}_1^2}{z_1} - \frac{\boldsymbol{\rho}_2^2}{z_2}\right)\right]$$

$$\times \iiiint \exp\left[-\frac{\boldsymbol{\rho}_1'^2 + \boldsymbol{\rho}_2'^2}{4\sigma^2}\right] \exp\left[-\frac{(\boldsymbol{\rho}_2' - \boldsymbol{\rho}_1')^2}{2\delta_{\alpha\beta}^2}\right]$$

$$\times \exp\left[-ik\left(\frac{\boldsymbol{\rho}_1'^2}{2z_1} - \frac{\boldsymbol{\rho}_2'^2}{2z_2} - \frac{\boldsymbol{\rho}_1\boldsymbol{\rho}_1'}{z_1} + \frac{\boldsymbol{\rho}_2\boldsymbol{\rho}_2'}{z_2}\right)\right] d^2\rho_1' d^2\rho_2'.$$

$$(4.206)$$

In an alternative form this formula can be written as

$$W_{\alpha\beta}(\boldsymbol{\rho}_1, z_1, \boldsymbol{\rho}_2, z_2; \omega) = \frac{k^2}{(2\pi)^2} \frac{A_\alpha A_\beta B_{\alpha\beta}}{z_1 z_2} \exp\left[-\frac{ik}{2}\left(\frac{\boldsymbol{\rho}_1^2}{z_1} - \frac{\boldsymbol{\rho}_2^2}{z_2}\right)\right]$$

$$\times \iint \exp[-q_{\alpha\beta1}\boldsymbol{\rho}_1'^2] \exp\left[-2\pi i\left(\frac{i\boldsymbol{\rho}_2'}{2\pi\delta_{\alpha\beta}^2} - \frac{\boldsymbol{\rho}_1}{\lambda z_1}\right)\right] d^2\rho_1'$$

$$\times \iint \exp\left[-q_{\alpha\beta2}\boldsymbol{\rho}_2'^2 - ik\frac{\boldsymbol{\rho}_2\boldsymbol{\rho}_2'}{z_2}\right] d^2\rho_2',$$

$$(4.207)$$

where

$$q_{\alpha\beta1} = \frac{1}{4\sigma^2} + \frac{1}{2\delta_{\alpha\beta}^2} + \frac{ik}{2z_1}, \quad q_{\alpha\beta2} = \frac{1}{4\sigma^2} + \frac{1}{2\delta_{\alpha\beta}^2} - \frac{ik}{2z_2}. \quad (4.208)$$

The integrals with respect to $\boldsymbol{\rho}_1'$ and $\boldsymbol{\rho}_2'$ can be evaluated with the help of the well-known expression for the two-dimensional Fourier transform of a Gaussian function, $g(\rho) = \exp(-q\rho^2)$, namely

$$
\mathcal{F}\{g(\rho)\} = \int_{-\infty}^{\infty} \exp(-qx^2) \exp(-2\pi i w_x x) dx \int_{-\infty}^{\infty} \exp(-qy^2) \exp(-2\pi i w_y y) dy
$$

$$
= \frac{\pi}{q} \exp\left(-\frac{\pi^2 w^2}{q}\right).
$$

(4.209)

The integration over vector ρ'_1 in Eq. (4.207) then leads to the expression

$$
\iint \exp[-q_{\alpha\beta 1}\rho'^2_1] \exp\left[-2\pi i \left(\frac{i\rho'_2}{2\pi\delta^2_{\alpha\beta}} - \frac{\rho_1}{\lambda z_1}\right)\right] d^2\rho'_1
$$

$$
= \frac{\pi}{q_{\alpha\beta 1}} \exp\left[\frac{\pi^2\rho^2_1}{q_{\alpha\beta 1}\lambda^2 z^2_1}\right] \exp\left[\frac{\rho^2_2}{q_{\alpha\beta 1}4\delta^4_{\alpha\beta}} + \frac{\pi i \rho_1 \rho'_2}{q_{\alpha\beta 1}\lambda z_1\delta^2_{\alpha\beta}}\right].
$$

(4.210)

Hence, Eq. (4.207) becomes

$$
W_{\alpha\beta}(\boldsymbol{\rho}_1, z_1, \boldsymbol{\rho}_2, z_2; \omega) = \frac{k^2}{4\pi^2} \frac{A_\alpha A_\beta B_{\alpha\beta}}{z_1 z_2} \frac{\pi}{q_{\alpha\beta 1}} \exp\left[-\frac{ik}{2}\left(\frac{\rho^2_1}{z_1} - \frac{\rho^2_2}{z_2}\right)\right]
$$

$$
\times \exp\left[\frac{\pi^2\rho^2_1}{q_{\alpha\beta 1}\lambda^2 z^2_1}\right] \iint \exp\left[-\left(q_{\alpha\beta 2} - \frac{1}{q_{\alpha\beta 1}4\delta^4_{\alpha\beta}}\right)\rho'^2_2\right] \quad (4.211)
$$

$$
\times \exp\left[-2\pi i\left(\frac{\boldsymbol{\rho}_2}{\lambda z_2} - \frac{\boldsymbol{\rho}_1}{2q_{\alpha\beta 1}\lambda z_1\delta^2_{\alpha\beta}}\right)\rho'^2_2\right] d^2\rho'_2.
$$

Finally, after integrating with respect to vector ρ'_2 and performing the simplifications of the result we arrive at the formula:

$$
W_{\alpha\beta}(\boldsymbol{\rho}_1, z_1, \boldsymbol{\rho}_2, z_2; \omega) = \frac{A_\alpha A_\beta B_{\alpha\beta}}{\Delta^{(G)^2}_{\alpha\beta}(z_1, z_2; \omega)}
$$

$$
\times \exp\left\{-\left[\frac{1}{8\sigma^2} + i\frac{z_2 - z_1}{8k\sigma^2}\left(\frac{1}{4\sigma^2} + \frac{1}{\delta^2_{\alpha\beta}}\right)\right]\frac{(\boldsymbol{\rho}_1 + \boldsymbol{\rho}_2)^2}{\Delta^{(G)^2}_{\alpha\beta}(z_1, z_2; \omega)}\right\}
$$

$$
\times \exp\left[-\frac{(\boldsymbol{\rho}_1 - \boldsymbol{\rho}_2)^2}{2\Delta^{(G)^2}_{\alpha\beta}(z_1, z_2; \omega)}\left(\frac{1}{4\sigma^2} + \frac{1}{\delta^2_{\alpha\beta}}\right)\right] \exp\left[-i\frac{k(\rho^2_1 - \rho^2_2)}{2R^{(G)}_{\alpha\beta}(z_1, z_2; \omega)}\right],
$$

(4.212)

where

$$
\Delta^{(G)^2}_{\alpha\beta}(z_1, z_2; \omega) = 1 + \frac{z_1 z_2}{k^2\sigma^2}\left(\frac{1}{4\sigma^2} + \frac{1}{\delta^2_{\alpha\beta}}\right) + i\frac{z_2 - z_1}{k}\left(\frac{1}{2\sigma^2} + \frac{1}{\delta^2_{\alpha\beta}}\right) \quad (4.213)
$$

and

$$R_{\alpha\beta}^{(G)}(z_1, z_2; \omega) = \sqrt{z_1 z_2} \left[1 + \frac{k^2 \sigma^2}{z_1 z_2} \left(\frac{1}{4\sigma^2} + \frac{1}{\delta_{\alpha\beta}^2} \right)^{-1} \right]. \tag{4.214}$$

The terms $\Delta_{\alpha\beta}^{(G)2}(z_1, z_2; \omega)$ and $R_{\alpha\beta}^{(G)}(z_1, z_2; \omega)$ entering expression (4.212) can be regarded as the generalized spreading coefficient of the beam and the generalized front radius of curvature, the two parameters that can completely characterize propagation characteristics of the lowest order Gaussian beam. Equation (4.212) can be readily reduced to account for transverse and longitudinal counterparts for $z_1 = z_2 = z$ and $\boldsymbol{\rho}_1 = \boldsymbol{\rho}_2 = \boldsymbol{\rho}$, respectively.

The derived formulas can be used for illustration of polarization changes in electromagnetic random beams. For instance, it can be shown (see Figs. 4.7) that the degree of polarization, and the parameters of polarization ellipse of the beam generated by source (4.205) can evolve on propagation in free space sufficiently close to the source but tend to constant values at large distances. Alternatively, the evolution of the Stokes parameters can also be illustrated (see Fig. 4.8). The saturation value are closely related to source correlation properties.

Spectral changes in electromagnetic Gaussian Schell-model beams occurring on free-space propagation are shown to depend on both coherence and polarization of the source [32] (see Fig. 4.9). Here the normalized spectra $S_N(r\mathbf{u}; \omega)$ are given as functions of $\varrho_\omega + 1$ for two different directions of observation ($\theta = 0^o$ and $\theta = 0.3^o$) and compared with the normalized spectral density of the source. (A)–(B) unpolarized source, (C)–(D) $\wp = 1/3$ and (E)–(F) polarized source; $\delta_{yy} = 0.05$ mm; $\delta_{xx} = 0.5$ mm (left column), $\delta_{xx} = 0.1$ mm (right column); $\Omega = 0.2\bar{\omega}/\sqrt{2}$.

4.6 Electromagnetic beams with Gaussian statistics

4.6.1 Higher-order statistical moments of fields

Although in the majority of situations the second-order statistical properties provide with a fairly complete picture about the general behavior of an electromagnetic stochastic beam, in certain applications the statistical moments of orders higher than the second should be considered, in addition. The most general form of the Nth moment of the electromagnetic field in the space-time domain is given by the formula [19]

$$\Gamma_{\alpha_1 \dots \alpha_N}^{(N)}(\mathbf{r}_1, \dots, \mathbf{r}_N; t_1, \dots, t_N) = \langle E_{\alpha_1}^*(\mathbf{r}_1; t_1) \dots E_{\alpha_N}(\mathbf{r}_N; t_N) \rangle_T, \tag{4.215}$$

where $(\alpha_1, \dots, \alpha_N = x, y)$, and the conjugate sign is applied to the odd-numbered entries. In most practical situations the analysis is restricted to

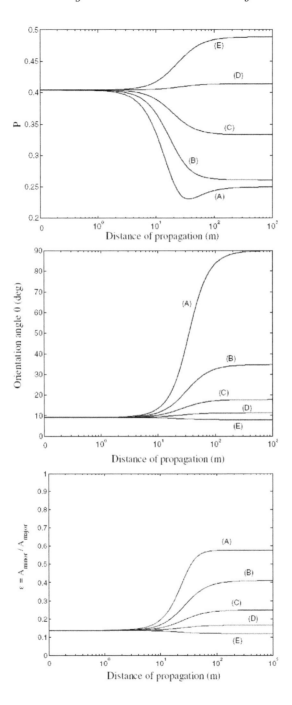

FIGURE 4.7
Evolution of the polarization properties of a typical electromagnetic Gaussian
Schell-model beam propagating in free space. From Ref. [12]. Different curves
corresponds to different sets of the r.m.s. widths of correlation functions δ_{xx},
δ_{xy} and δ_{yy}.

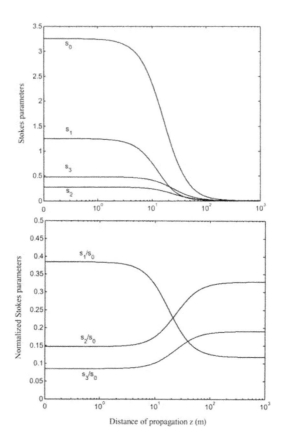

FIGURE 4.8

Evolution of the Stokes parameters of the polarization ellipse of a typical electromagnetic Gaussian Schell-model beam propagating in free space. From Ref. [20].

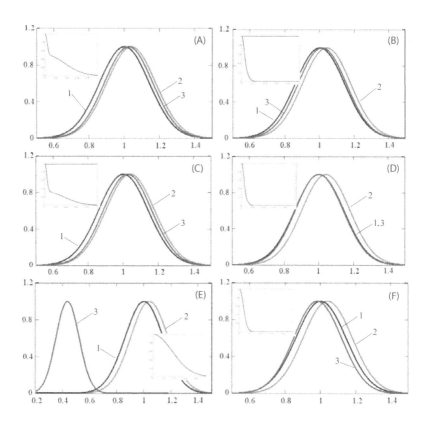

FIGURE 4.9

Spectral changes of electromagnetic Gaussian Schell-model beams generated by sources with several different coherence and polarization states, and propagating in free space. Subfigures show the source degree of coherence. Curves: 1 — source; 2 — $\theta = 0^{o}$; 3 — $\theta = 0.3^{o}$. From Ref. [32].

the fourth-order moment at two spatial positions and at two time moments, with delay τ. Then the following quantities become important

$$\Gamma^{(4)}_{\alpha_1\alpha_2\alpha_3\alpha_4}(\mathbf{r}_1,\mathbf{r}_2;\tau) = \langle E^*_{\alpha_1}(\mathbf{r}_1;t)E^*_{\alpha_2}(\mathbf{r}_1;t)E_{\alpha_3}(\mathbf{r}_2;t+\tau)E_{\alpha_4}(\mathbf{r}_2;t+\tau)\rangle_T,$$
(4.216)

where $(\alpha_1,\alpha_2,\alpha_3,\alpha_4) = (x,y)$. In what follows we will be primarily interested in the correlations of the beam's intensity. Consider the increment in the intensity

$$\Delta I(\mathbf{r};t) = I(\mathbf{r};t) - \langle I(\mathbf{r})\rangle_T,$$
(4.217)

where the instantaneous intensity $I(\mathbf{r};t)$ and the average intensity $\langle I(\mathbf{r})\rangle_T$ are defined by the expressions

$$I(\mathbf{r};t) = Tr\left[\mathbf{E}^\dagger(\mathbf{r};t)\mathbf{E}(\mathbf{r};t)\right],$$
(4.218)

the dagger denoting a Hermitian adjoint, and

$$\langle I(\mathbf{r})\rangle_T = \langle Tr\left[\mathbf{E}^\dagger(\mathbf{r};t)\mathbf{E}(\mathbf{r};t+\tau)\right]\rangle_T.$$
(4.219)

The covariance function of the fluctuating intensity of a random electromagnetic beam at a pair of points, $\mathbf{r}_1$ and $\mathbf{r}_2$, and time lag τ may be defined by the expression [62]

$$\begin{aligned}B_I(\mathbf{r}_1,\mathbf{r}_2;\tau) &= \langle\Delta I(\mathbf{r}_1;t)\Delta I(\mathbf{r}_2;t+\tau)\rangle_T\\&= \langle I(\mathbf{r}_1;t)I(\mathbf{r}_2;t+\tau)\rangle_T - \langle I(\mathbf{r}_1;t)\rangle\langle I(\mathbf{r}_2;t+\tau)\rangle_T.\end{aligned}$$
(4.220)

It can also be expressed in terms of the traces of the matrices $\mathbf{\Gamma}^{(4)}(\mathbf{r}_1,\mathbf{r}_2;\tau)$ and $\mathbf{\Gamma}(\mathbf{r}_1,\mathbf{r}_2;\tau)$ as

$$B_I(\mathbf{r}_1,\mathbf{r}_2;\tau) = Tr[\mathbf{\Gamma}^{(4)}(\mathbf{r}_1,\mathbf{r}_2;\tau)] - Tr[\mathbf{\Gamma}(\mathbf{r}_1,\mathbf{r}_1;\tau)]Tr[\mathbf{\Gamma}(\mathbf{r}_2,\mathbf{r}_2;\tau)].$$
(4.221)

From the covariance function of the intensity two useful quantities can be deduced. The correlation coefficient $b_I(\mathbf{r}_1,\mathbf{r}_2;\tau)$, defined by the normalized expression

$$b_I(\mathbf{r}_1,\mathbf{r}_2;\tau) = \frac{B_I(\mathbf{r}_1,\mathbf{r}_2;\tau)}{Tr[\mathbf{\Gamma}(\mathbf{r}_1,\mathbf{r}_1;\tau)]Tr[\mathbf{\Gamma}(\mathbf{r}_2,\mathbf{r}_2;\tau)]},$$
(4.222)

shows how the beam's intensity is correlated at a pair of points if compared with full correlation ($b_I = 1$) at $\mathbf{r}_1 \equiv \mathbf{r}_2$. The contrast of intensity fluctuations $c_I(\mathbf{r})$, also known as the *scintillation index* [66], for the electromagnetic beam can be defined by the expression

$$\begin{aligned}c_I(\mathbf{r}) &= b_I(\mathbf{r},\mathbf{r};0) - 1\\&= \frac{Tr[\mathbf{\Gamma}^{(4)}(\mathbf{r},\mathbf{r};\tau)]}{(Tr[\mathbf{\Gamma}(\mathbf{r},\mathbf{r};\tau)])^2} - 1.\end{aligned}$$
(4.223)

It indicates the typical amplitude of the intensity fluctuations of the beam at a given point $\mathbf{r}$. It can also be shown that the scintillation index of the electromagnetic beam can be expressed via the scintillation indexes of the individual components of the field, i.e., [64]

$$c_I(\mathbf{r}) = \frac{c_{xx}(\mathbf{r})\langle I_x(\mathbf{r})\rangle_T^2 + 2c_{xy}\langle I_x(\mathbf{r})\rangle_T\langle I_y(\mathbf{r})\rangle_T + c_{yy}(\mathbf{r})\langle I_y(\mathbf{r})\rangle_T^2}{[\langle I_x(\mathbf{r})\rangle_T + \langle I_y(\mathbf{r})\rangle_T]^2}, \quad (4.224)$$

where

$$c_{\alpha\beta}(\mathbf{r}) = \frac{\Gamma^{(4)}_{\alpha\alpha\beta\beta}(\mathbf{r},\mathbf{r};0) - \Gamma_{\alpha\alpha}(\mathbf{r},\mathbf{r};0)\Gamma_{\beta\beta}(\mathbf{r},\mathbf{r})}{\Gamma_{\alpha\alpha}(\mathbf{r},\mathbf{r};0)\Gamma_{\beta\beta}(\mathbf{r},\mathbf{r};0)} \quad (4.225)$$

and $\langle I_x(\mathbf{r})\rangle_T$, $\langle I_y(\mathbf{r})\rangle_T$ are the average intensities of the x- and y components of the electric field.

4.6.2 Higher-order moments of beams with Gaussian statistics

It is often the case that the electric field fluctuations are governed by Gaussian statistics. Under such circumstances the moments of any order of the field at any number of points in space can be expressed via the second–order statistical moment at two points, on the basis of the moments theorem (see Section 1.2.4). More precisely, when the electric field can be described as a statistically stationary, complex Gaussian random process, it is possible, by the use of the moment theorem [63], to express the elements of the matrix $\mathbf{\Gamma}^{(4)}$ in terms of the elements of matrix $\mathbf{\Gamma}$ as

$$\begin{aligned} \Gamma^{(4)}_{\alpha_1\alpha_2\alpha_3\alpha_4}(\mathbf{r}_1,\mathbf{r}_2;\tau) &= \langle E^*_{\alpha_1}(\mathbf{r}_1,t)E_{\alpha_2}(\mathbf{r}_2,t+\tau)\rangle_T\langle E^*_{\alpha_3}(\mathbf{r}_1,t)E_{\alpha_4}(\mathbf{r}_2,t+\tau)\rangle_T \\ &+ \langle E^*_{\alpha_1}(\mathbf{r}_1,t)E_{\alpha_4}(\mathbf{r}_2,t+\tau)\rangle_T\langle E^*_{\alpha_3}(\mathbf{r}_1,t)E_{\alpha_2}(\mathbf{r}_2,t+\tau)\rangle_T, \\ &\quad (\alpha_1,\alpha_2,\alpha_3,\alpha_4 = x,y). \end{aligned}$$
$$(4.226)$$

Hence, in this case, the formula for the covariance function [see Eq. (4.221)] can be expressed as

$$B_I(\mathbf{r}_1,\mathbf{r}_2;\tau) = Tr[\mathbf{\Gamma}(\mathbf{r}_1,\mathbf{r}_2;\tau)^2], \quad (4.227)$$

where the power sign implies matrix multiplication. Further, the correlation coefficient at a pair of points $\mathbf{r}_1$ and $\mathbf{r}_2$ takes the form

$$b_I(\mathbf{r}_1,\mathbf{r}_2) = \frac{Tr[\mathbf{\Gamma}(\mathbf{r}_1,\mathbf{r}_2;\tau)^2]}{(Tr[\mathbf{\Gamma}(0,0;\tau)])^2}, \quad (4.228)$$

and the contrast $c_I(\mathbf{r})$ of intensity fluctuations becomes

$$c_I(\mathbf{r}) = \frac{Tr[\mathbf{\Gamma}(\mathbf{r},\mathbf{r};0)^2]}{Tr[\mathbf{\Gamma}(\mathbf{r},\mathbf{r};0)]^2}. \quad (4.229)$$

In the framework of the scalar theory these two quantities, b_I and c_I are of particular interest only in studies of partially coherent beams whose fluctuations obey statistics other than Gaussian, i.e., beams propagating through random media and scattered from rough surfaces, for example. In such situations these quantities cannot be expressed in terms of the degree of coherence of the beam and, therefore, provide additional information about the fluctuating fields. In electromagnetic theory these quantities are of interest even though the field is Gaussian, for in this case, their values cannot be determined from the state of coherence of the beam alone, they also depend on its polarization properties.

In Figs. 4.10–4.11 three electromagnetic Gaussian Schell-model sources are chosen: $\lambda = 0.628$ μm, $\sigma = 1$ cm, (A) $A_x^2 = 0.5$, $A_y^2 = 0.5$, $B_{xy} = 0$, $\delta_{xx} = 0.1$ mm, $\delta_{yy} = 1$ mm (unpolarized beam with uncorrelated field components); (B) $A_x^2 = 0.95$, $A_y^2 = 0.05$, $B_{xy} = 0.3e^{\pi/6}$, $\delta_{xx} = 0.175$ mm, $\delta_{yy} = 0.225$ mm, $\delta_{xy} = 0.25$ mm (partially polarized beam with correlated field components); (C) $A_x^2 = 0.9$, $A_y^2 = 0.1$, $B_{xy} = 0$, $\delta_{xx} = 0.1$ mm, $\delta_{yy} = 0.5$ mm (partially polarized beam with uncorrelated field components). Figure 4.10 indicates the complete ρ–z cross-sections of both intensity (left column) and contrasts (right column) of an electromagnetic Gaussian Schell-model beam governed by Gaussian statistics. Figure 4.11 shows the possible changes of the contrasts of the same three beams on the optical axis.

Thus, the assumption of the Gaussian-distributed field allows for significant simplification of all the higher-order statistical properties of the electromagnetic beams. In finishing, we would like to include one more interesting relation: it can be shown that under such an assumption the intensity-intensity correlations (4.218) reduce to the form

$$\langle I(\mathbf{r}_1;t)I(\mathbf{r}_2;t+\tau)\rangle_T = \langle I(\mathbf{r}_1;t)\rangle_T\langle I(\mathbf{r}_2;t+\tau)\rangle_T[1 + \eta(\mathbf{r}_1,\mathbf{r}_2;\tau)]\mathcal{P}(\mathbf{r}_1,\mathbf{r}_2;\tau),$$
(4.230)

where $\eta(\mathbf{r}_1,\mathbf{r}_2;\tau)$ is the degree of coherence in the space-time domain and factor $\mathcal{P}(\mathbf{r}_1,\mathbf{r}_2;\tau)$ is the degree of cross-polarization [16], see also (4.32), which can be regarded as a generalization of the ordinary degree of polarization to two positions in space.

4.6.3 Fluctuations in power

We will now discuss how to calculate quantities relating to the fluctuations in power of the beam which is collected by an aperture of arbitrary size. The knowledge of the contrast $c_I(\mathbf{r})$ [see Eq. (4.223)] is sufficient only for applications where the beam is collected by a point aperture, i.e., by an aperture for which $\bar{\rho} \ll \rho_I$, $\bar{\rho}$ being its radius and ρ_I being the typical transverse correlation width of the beam intensity [65]. By analogy with scalar theory, ρ_I may be defined as an e^{-2} point of the correlation coefficient $b_I(\rho_d)$, i.e., the value of ρ_d at which $b_I(\rho_d) = e^{-2}$. However, when $\rho_I \ll \bar{\rho}$, spatial averaging of the fluctuating intensity must be taken into account [66]. In this latter case, instead of $c_I(\mathbf{r})$, the knowledge of the normalized variance $c_p(\mathbf{r})$ of the collected

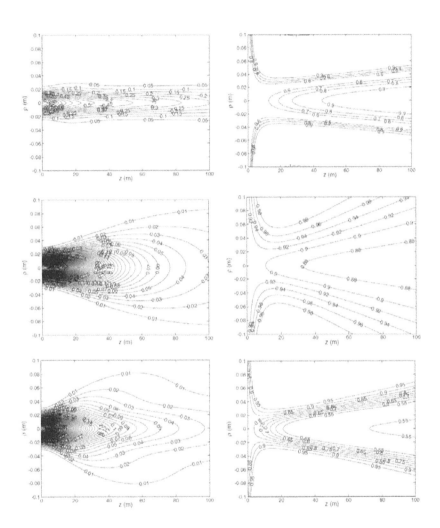

FIGURE 4.10
Contour plots of intensity (left) and contrasts (right) of the Gaussian Schell-model beam with Gaussian statistics. From Ref. [62].

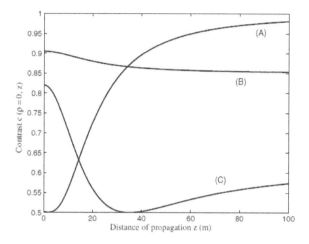

FIGURE 4.11
Changes in the contrast of the intensity of the Gaussian Schell-model beam with Gaussian statistics. From Ref. [62].

power is needed. This quantity can be expressed in terms of the covariance function $B_I(\mathbf{r}_1, \mathbf{r}_2)$ of the intensity fluctuations by a formula analogous to the one used in scalar theory, namely

$$c_p(\boldsymbol{\rho}, z) = \frac{\langle \Delta p_E^2(\boldsymbol{\rho}, z) \rangle_T}{\langle p_E(\boldsymbol{\rho}, z) \rangle_T^2} \qquad (4.231)$$

where the variance of power fluctuations and the average power are given by the expressions

$$
\begin{aligned}
\langle \Delta p_E^2(\boldsymbol{\rho}, z) \rangle_T &= \langle p_E^2(\boldsymbol{\rho}, z) \rangle_T - \langle p_E(\boldsymbol{\rho}, z) \rangle_T^2 \\
&= \int_{-\infty}^{\infty} \int_{-\infty}^{\infty} B_I(\boldsymbol{\rho}_1, \boldsymbol{\rho}_2, z)\xi_c(\boldsymbol{\rho}_1)\xi_c(\boldsymbol{\rho}_2)d^2\rho_1 d^2\rho_2,
\end{aligned}
\qquad (4.232)
$$

and

$$\langle p_E(\boldsymbol{\rho}, z) \rangle_T = \int_{-\infty}^{\infty} \langle I(\boldsymbol{\rho}, z) \rangle \xi_c(\boldsymbol{\rho})d^2\rho, \qquad (4.233)$$

$\xi_c(\boldsymbol{\rho})$ being the amplitude transmission function of a circular aperture of radius R. In cases when approximate analytic formulae are preferred to numerical integrations the amplitude transmission function may be approximated by a Gaussian function with soft radius ρ_g, i.e.,

$$\xi_c(\boldsymbol{\rho}) \approx \exp\left(-\frac{\rho^2}{\rho_g^2}\right). \qquad (4.234)$$

Such an approximation leads to results that are in good agreement with those obtained by numerical integration over an actual, hard aperture with diameter $2\sqrt{2}\rho_g$ [67].

The *aperture averaging factor* [68], which indicates how much the intensity fluctuations are reduced when the beam is collected by an aperture of radius R, compared with the intensity fluctuations when the beam is measured by a point aperture (located on its axis), can be calculated by use of the expression

$$A_\xi(\boldsymbol{\rho}, z) = \frac{c_p(\boldsymbol{\rho}, z)}{c_I(0, z)}. \tag{4.235}$$

4.6.4 Higher-order moments of Stokes parameters

In general it is very difficult to obtain the analytic form of the probability density function of any statistical property of the beam even in the case when the electric field obeys Gaussian statistics. Such a calculation was performed, however, for the classic Stokes parameters. Let us introduce the *instantaneous Stokes parameters* by the expressions [19], [69]

$$\begin{aligned}
s_0(\mathbf{r}, t) &= E_x^*(\mathbf{r}, t)E_x(\mathbf{r}, t) + E_y^*(\mathbf{r}, t)E_y(\mathbf{r}, t), \\
s_1(\mathbf{r}, t) &= E_x^*(\mathbf{r}, t)E_x(\mathbf{r}, t) - E_y^*(\mathbf{r}, t)E_y(\mathbf{r}, t), \\
s_2(\mathbf{r}, t) &= E_x^*(\mathbf{r}, t)E_y(\mathbf{r}, t) + E_y^*(\mathbf{r}, t)E_x(\mathbf{r}, t), \\
s_3(\mathbf{r}, t) &= i[E_y^*(\mathbf{r}, t)E_x(\mathbf{r}, t) - E_x^*(\mathbf{r}, t)E_y(\mathbf{r}, t)],
\end{aligned} \tag{4.236}$$

from which the classic Stokes parameters may be determined by taking the long time average

$$\langle s_m(\mathbf{r})\rangle_T = \lim_{T\to\infty} \frac{1}{2T} \int_{-T}^{T} s_m(\mathbf{r}, t)dt, \quad (m = 0, 1, 2, 3). \tag{4.237}$$

The probability density functions of the instantaneous Stokes parameters (4.236) have the forms [70]–[72]:

$$\begin{aligned}
\mathrm{pdf}[s_0(\mathbf{r})] = \frac{1}{\wp(\mathbf{r})\langle s_0(\mathbf{r})\rangle} &\left\{ \exp\left[-\frac{2s_0(\mathbf{r})}{(1 + \wp(\mathbf{r}))\langle s_0(\mathbf{r})\rangle} \right] \right. \\
&\left. - \exp\left[-\frac{2s_0(\mathbf{r})}{(1 - \wp(\mathbf{r}))\langle s_0(\mathbf{r})\rangle} \right] \right\},
\end{aligned} \tag{4.238}$$

and

$$\begin{aligned}
\mathrm{pdf}[s_m(\mathbf{r})] = &\frac{1}{\sqrt{\langle s_0(\mathbf{r})\rangle^2(1 - \wp^2(\mathbf{r})) + \langle s_m(\mathbf{r})\rangle^2}} \\
&\times \exp\left[2\frac{\langle s_m(\mathbf{r})\rangle s_m(\mathbf{r}) - |s_m(\mathbf{r})|\sqrt{\langle s_0(\mathbf{r})\rangle^2(1 - \wp^2(\mathbf{r})) + \langle s_m(\mathbf{r})\rangle^2}}{\langle s_0(\mathbf{r})\rangle^2(1 - \wp^2(\mathbf{r}))} \right], \\
&\hspace{9cm} (m = 1, 2, 3),
\end{aligned} \tag{4.239}$$

where $\wp$ is the ordinary degree of polarization of the beam.

Statistical moments of the instantaneous Stokes parameters of order $n \geq 1$ can be calculated from Eqs. (4.238)–(4.239). In particular, for the moments of the instantaneous Stokes parameter $S_0(\mathbf{r})$ one obtains the formula

$$\langle S_0^n(\mathbf{r}) \rangle = \int_0^\infty \mathrm{pdf}[S_0(\mathbf{r})] S_0^n(\mathbf{r}) dS_0(\mathbf{r})$$

$$= \frac{\langle I(\mathbf{r}) \Gamma(n+1) \rangle^n}{\wp(\mathbf{r}) 2^{n+1}} [(1 + \wp(\mathbf{r}))^{n+1} - (1 - \wp(\mathbf{r}))^{n+1}], \qquad (4.240)$$

where $\Gamma(x)$ is the Gamma function. The statistical moments for the other components of the Stokes vector were shown to be given by the formulas [72]

$$\langle S_m^n(\mathbf{r}) \rangle = \int_0^\infty \mathrm{pdf}[S_m(\mathbf{r})] S_m^n(\mathbf{r}) dS_m(\mathbf{r})$$

$$= \frac{n!}{2^{n+1} \sqrt{\langle S_0(\mathbf{r}) \rangle^2 (1 - \wp^2(\mathbf{r})) + \langle S_m(\mathbf{r}) \rangle^2}}$$

$$\times \left\{ \left[\sqrt{\langle S_0(\mathbf{r}) \rangle^2 (1 - \wp^2(\mathbf{r})) + \langle S_m(\mathbf{r}) \rangle^2} + \langle S_m(\mathbf{r}) \rangle^2 \right]^{n+1} \right.$$

$$\left. + (-1)^n \left[\sqrt{\langle S_0(\mathbf{r}) \rangle^2 (1 - \wp^2(\mathbf{r})) + \langle S_m(\mathbf{r}) \rangle^2} - \langle S_m(\mathbf{r}) \rangle \right]^{n+1} \right\}, \qquad (4.241)$$

where $m = 1, 2, 3$. As a matter of fact, the contrasts of the Stokes parameters can be defined by the formulas analogous to that for the intensity contrast (scintillation index), i.e., as

$$c_m(\mathbf{r}) = \frac{\langle S_m^2(\mathbf{r}) \rangle - \langle S_m(\mathbf{r}) \rangle^2}{\langle S_m(\mathbf{r}) \rangle^2}, \quad (m = 0, 1, 2, 3). \qquad (4.242)$$

Then it can readily be shown that they can be expressed in terms of the average Stokes parameters and the degree of polarization by the formulas [19]

$$c_0(\mathbf{r}) = \frac{1}{2}[1 + \wp^2(\mathbf{r})], \qquad (4.243)$$

and

$$c_m(\mathbf{r}) = 1 + \frac{\langle S_0(\mathbf{r}) \rangle^2}{2 \langle S_m(\mathbf{r}) \rangle^2}[1 - \wp^2(\mathbf{r})], \quad (m = 1, 2, 3). \qquad (4.244)$$

In cases when field fluctuations are governed by Gaussian statistics it is also possible to determine the probability density functions of the instantaneous Stokes parameters normalized by the instantaneous intensity, i.e.,

$$s_m^{(norm)}(\mathbf{r}) = \frac{S_m(\mathbf{r})}{S_0(\mathbf{r})}, \quad (m = 1, 2, 3). \qquad (4.245)$$

These quantities become useful when the fluctuations in the state of polarization of the beam must be considered independently of its intensity level. Following Ref. [73] we find that

$$\text{pdf}[s_m^{(norm)}(\mathbf{r})] = \frac{1}{2}$$

$$\times \frac{[1 - \wp^2(\mathbf{r})][1 - s_m^{(norm)}(\mathbf{r})\langle s_m^{(norm)}(\mathbf{r})\rangle]}{\left[1 + \langle s_m^{(norm)}(\mathbf{r})\rangle^2 - \wp^2(\mathbf{r}) - 2s_m^{(norm)}(\mathbf{r})\langle s_m^{(norm)}(\mathbf{r})\rangle + s_m^{(norm)2}(\mathbf{r})\wp^2(\mathbf{r})\right]},$$

$$(m = 1, 2, 3),$$

$$(4.246)$$

for $|s_m(\mathbf{r})| \le 1$ and zero otherwise. The statistical moments of orders $n \ge 1$ of the normalized instantaneous Stokes parameters can then be shown to be given by the expressions

$$\langle s_m^{(norm)n}(\mathbf{r})\rangle = \int_0^\infty \text{pdf}[s_m(\mathbf{r})]s_m^{(norm)}(\mathbf{r})ds_m(\mathbf{r}) \qquad (4.247)$$

$$= \frac{1 - \wp^2(\mathbf{r})}{2}[I_m^n - q_m(\mathbf{r})I_m^{(n+1)}],$$

where

$$q_m(\mathbf{r}) = \frac{\langle S_m(\mathbf{r})\rangle}{\langle S_0(\mathbf{r})\rangle} \qquad (4.248)$$

and

$$I_m^n = \int_{-1}^1 \frac{x^n dx}{[1 + q_m(\mathbf{r}) - \wp^2(\mathbf{r}) - 2xq_m(\mathbf{r}) + x^2\wp(\mathbf{r})]^{3/2}}. \qquad (4.249)$$

The contrasts of fluctuations in the normalized instantaneous Stokes parameters

$$c_m^{(norm)}(\mathbf{r}) = \frac{\langle s_m^{(norm)2}(\mathbf{r})\rangle - \langle s_m^{(norm)}(\mathbf{r})\rangle^2}{\langle s_m^{(norm)}(\mathbf{r})\rangle^2} \qquad (4.250)$$

can also be expressed entirely in terms of the average Stokes parameters and the degree of polarization as

$$c_m^{(norm)}(\mathbf{r}) = \frac{[1 - \wp(\mathbf{r})]^2}{q_m(\mathbf{r})}$$

$$\times \frac{2\Upsilon_m(\mathbf{r})[\wp^2(\mathbf{r}) - \wp(\mathbf{r})q_j^2(\mathbf{r})] - q_j^2(\mathbf{r})[1 - \wp^2(\mathbf{r})]\Upsilon_m(\mathbf{r}) - 4\wp^4(\mathbf{r}) + 8q_j^2(\mathbf{r})\wp^2(\mathbf{r})}{4\wp^2(\mathbf{r}) - 4[\wp(\mathbf{r}) - \wp^3(\mathbf{r})]\Upsilon_m(\mathbf{r}) + [1 - \wp^2(\mathbf{r})]^2\Upsilon_m^2(\mathbf{r})},$$

$$(m = 1, 2, 3),$$

$$(4.251)$$

where

$$\Upsilon_m(\mathbf{r}) = \sinh^{-1}\left[\frac{\wp^2(\mathbf{r}) + q_m(\mathbf{r})}{\sqrt{[1 - \wp^2(\mathbf{r})][\wp^2(\mathbf{r}) - q_m(\mathbf{r})]}}\right]$$
$$+ \sinh^{-1}\left[\frac{\wp^2(\mathbf{r}) - q_m(\mathbf{r})}{\sqrt{[1 - \wp^2(\mathbf{r})][\wp^2(\mathbf{r}) - q_m(\mathbf{r})]}}\right].$$

(4.252)

Thus, in the case when the fluctuations in the beam are governed by Gaussian statistics all the higher-order moments and the probability density functions of the Stokes parameters and of the normalized Stokes parameters can be expressed in terms of the average Stokes parameters and the degree of polarization, which, as a matter of fact, is also their function. Since the average Stokes parameters and the degree of polarization of stochastic beams generally change on propagation in free space it then follows that the higher-order moments and the probability density functions do not remain constants.

To illustrate the changes in these high-order statistical properties we will consider a set of figures relating to various statistical properties of the electromagnetic Gaussian Schell-model beams [see Eq. (4.172)] generated by uniformly polarized sources, i.e., in the absence of anisotropic features in the spectral density: $\sigma_\alpha = \sigma_\beta = \sigma$. On substituting from formula (4.212) into the expressions (4.34)–(4.35) one can then determine the average Stokes parameters and the degree of polarization. Finally the higher-order statistics can then be obtained with the help of the formulas introduced in the previous three sections.

Typical changes in the probability density functions of the Stokes parameters and in the normalized Stokes parameters of an electromagnetic Gaussian Schell-model beam with propagation distance from the source are illustrated in Figs. 4.12 and 4.13. The choice of the parameters for these situations corresponds to the set of parameters (B) in Figs. 4.10 and 4.11. While the modifications in all the probability density functions may be drastic quantitatively, their shapes remain the same. This is certainly not true if a beam propagates in a random medium whose fluctuations are governed by statistics other than Gaussian. One can see from Fig. 4.12 that the tails of distributions for Stokes parameters become narrower, with the increasing distance, implying that the fluctuations of the instantaneous Stokes parameters reduce. The larger the distance the more the pdfs resemble the delta-function. In contrast, the probability density functions of the normalized Stokes parameters (see Fig. 4.13) do not change indefinitely with the increasing distance but rather approach certain profiles depending on the correlation properties of the source.

The effects of the devices of polarization optics on the higher-order statistics of the random beams can be found in Ref. [74].

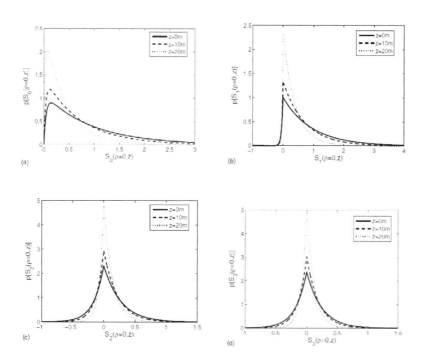

FIGURE 4.12

The changes in the probability density functions of the instantaneous Stokes parameters of a typical random beam: (a) pdf$[S_0(\rho = 0, z)]$, (b) pdf$[S_1(\rho = 0, z)]$, (c) pdf$[S_2(\rho = 0, z)]$ and (d) pdf$[S_3(\rho = 0, z)]$ at several distances z from the source. From Ref. [69].

184

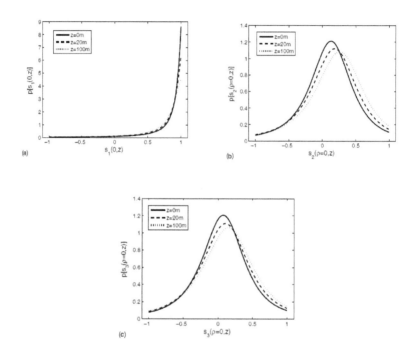

(a)

(b)

(c)

FIGURE 4.13
The changes in the probability density functions of the normalized instantaneous Stokes parameters of a typical random beam: (a) $\text{pdf}[s_1(\rho = 0, z)]$, (b)
$\text{pdf}[s_2(\rho = 0, z)]$, (c) $\text{pdf}[s_3(\rho = 0, z)]$ at several distances z from the source.
From Ref. [69].

4.7   Other stochastic electromagnetic beams

The extension of the scalar random source models to the electromagnetic domain is not an easy task because of the non-negative definiteness requirement on the cross-spectral density matrix. As we have shown so far only the class of electromagnetic Gaussian Schell-model sources has been analyzed in detail. Yet, two other models, the multi-Gaussian Schell-model beams and the non-uniformly correlated beams which we have considered in Chapter 3 have been recently extended to the electromagnetic domain. In the following two sections we will briefly introduce such novel beam classes.

4.7.1   Electromagnetic multi-Gaussian Schell-model beams

Recall that for the scalar multi-Gaussian Schell-model sources the intensity distribution might be quite arbitrary but the degree of coherence is always the sum of suitably weighted Gaussian functions with different variances and sign-alternating heights (see Chapter 3). The beams generated by the multi-Gaussian Schell-model sources and propagating in free space to the far field form plateaus in their central parts of the intensity distributions and steeply decay to zero at larger transverse positions. If such beams propagate in a random medium, the intensity plateau might still be formed but are destroyed at large distances or even might not be able to appear at all, if the medium fluctuations are too strong. Such optical fields are potentially useful for material processing at a distance, free-space communications and imaging.

In this section we extend the scalar multi-Gaussian Schell-model sources to electromagnetic domain and investigate the evolution of the polarization states of the produced beams on propagation in free space. The main objective of this discussion is to describe electromagnetic sources and beams which, remaining random, might form constant polarization areas in their central part on propagation to the far zone [75].

The electromagnetic counterpart of the scalar multi-Gaussian Schell-model source should be characterized by the 2×2 cross-spectral density matrix at two position vectors $\boldsymbol{\rho}_1$ and $\boldsymbol{\rho}_2$ of the source plane, $z = 0$, has the elements in the form (4.171). We then assume that the spectral densities have Gaussian profiles:

$$S_\alpha^{(G)}(\boldsymbol{\rho}'; \omega) = A_\alpha^2 \exp\left[-\frac{\rho'^2}{4\sigma_\alpha^2}\right], \quad (\alpha = x, y), \tag{4.253}$$

and the correlation coefficients have multi-Gaussian form:

$$\eta_{\alpha\beta}^{(G)}(\boldsymbol{\rho}_2' - \boldsymbol{\rho}_1'; \omega)$$
$$= \frac{B_{\alpha\beta}}{C_0} \sum_{m=1}^{M} \binom{M}{m} \frac{(-1)^{m-1}}{m} \exp\left[-\frac{|\boldsymbol{\rho}_2' - \boldsymbol{\rho}_1'|^2}{2m\delta_{\alpha\beta}^2}\right], \quad (\alpha, \beta = x, y). \tag{4.254}$$

Here C_0 is the same normalization factor as in the scalar case, A_α and σ_α are the amplitude and the r.m.s. width of the αth electric field component; $\delta_{\alpha\beta}$ and $B_{\alpha\beta}$ are the r.m.s. width of the correlation and the single-point correlation coefficient between αth and βth electric field components, respectively. On substituting from Eqs. (4.253) and (4.254) into Eq. (4.171) we find that the elements of the cross-spectral density matrix of the electromagnetic multi-Gaussian Schell-model source have the form:

$$
W_{\alpha\beta}^{(MG)}(\boldsymbol{\rho}_1',\boldsymbol{\rho}_2';\omega) = \frac{A_\alpha A_\beta B_{\alpha\beta}}{C_0} \exp\left[-\frac{\boldsymbol{\rho}_1'^2 + \boldsymbol{\rho}_2'^2}{4\sigma^2}\right]
$$
$$
\times \sum_{m=1}^{M} \binom{M}{m} \frac{(-1)^{m-1}}{m} \exp\left[-\frac{|\boldsymbol{\rho}_2' - \boldsymbol{\rho}_1'|^2}{2m\delta_{\alpha\beta}^2}\right],
$$
(4.255)

where it was assumed for simplicity that $\sigma_x = \sigma_y = \sigma$. Just like for the electromagnetic Gaussian-Schell model sources this condition implies that the polarization properties of the multi-Gaussian-Schell model beams are uniform in the source plane. We also assume here that summation limit M is the same for all four components of the correlation matrix.

We will now establish conditions for the source parameters guaranteeing that the mathematical model (4.255) describes a physically realizable field. From the quasi-Hermitian property of the correlation matrix it follows that

$$
B_{xx} = B_{yy} = 1, \quad |B_{xy}| = |B_{yx}|, \quad \delta_{xy} = \delta_{yx}.
$$
(4.256)

Next, by will use the integral representation (4.9) for the elements of the cross-spectral density matrix derived in Ref. [9], which is equivalent to the requirement for non-negative definiteness. Indeed on setting

$$
H_\alpha(\boldsymbol{\rho}', \mathbf{s}) = A_\alpha \exp\left(-\frac{\boldsymbol{\rho}'^2}{4\sigma^2}\right) \exp[-i\mathbf{s}\cdot\boldsymbol{\rho}'],
$$
(4.257)

and

$$
p_{\alpha\beta}(\mathbf{s}) = \frac{B_{\alpha\beta}\delta_{\alpha\beta}^2}{C_0} \sum_{m=1}^{M} (-1)^{m-1}\binom{M}{m} \exp\left[-\frac{m\delta_{\alpha\beta}^2|\mathbf{s}|^2}{2}\right],
$$
(4.258)

we find from inequality (4.11) that

$$
\delta_{xx}^2\delta_{xy}^2 \sum_{m=1}^{M} (-1)^{m-1}\binom{M}{m} \exp\left[-\frac{m\delta_{xx}^2|\mathbf{s}|^2}{2}\right]
$$
$$
\times \sum_{m=1}^{M} (-1)^{m-1}\binom{M}{m} \exp\left[-\frac{m\delta_{yy}^2|\mathbf{s}|^2}{2}\right]
$$
$$
\geq |B_{xy}|^2\delta_{xy}^4 \left[\sum_{m=1}^{M} (-1)^{m-1}\binom{M}{m} \exp\left[-\frac{m\delta_{xy}^2|\mathbf{s}|^2}{2}\right]\right]^2,
$$
(4.259)

or, equivalently

$$\frac{\delta_{xx}^2 \delta_{yy}^2}{|B_{xy}|^2 \delta_{xy}^4} \frac{\left\{1 - \left[1 - \exp\left(-\frac{\delta_{xx}^2 |\mathbf{s}|^2}{2}\right)\right]^M\right\} \left\{1 - \left[1 - \exp\left(-\frac{\delta_{yy}^2 |\mathbf{s}|^2}{2}\right)\right]^M\right\}}{\left\{1 - \left[1 - \exp\left(-\frac{\delta_{xy}^2 |\mathbf{s}|^2}{2}\right)\right]^M\right\}^2} \geq 1.$$

(4.260)

We will now derive the sufficiency conditions for parameters δ_{xx}, δ_{yy}, δ_{xy} and B_{xy} such that the inequality (4.260) holds. Since its left-hand side is a product of two ratios, it is sufficient to find δ_{xx}, δ_{yy}, δ_{xy} and B_{xy} such that both factors exceed unity simultaneously. If we set $\mathbf{s} = 0$, we immediately arrive at the inequality

$$\frac{\delta_{xx}^2 \delta_{yy}^2}{|B_{xy}|^2 \delta_{xy}^4} \geq 1.$$

(4.261)

The condition

$$\min\left\{1 - \left[1 - \exp\left(-\frac{\delta_{xx}^2 |\mathbf{s}|^2}{2}\right)\right]^M ; 1 - \left[1 - \exp\left(-\frac{\delta_{yy}^2 |\mathbf{s}|^2}{2}\right)\right]^M\right\}$$

(4.262)

$$\geq 1 - \left[1 - \exp\left(-\frac{\delta_{xx}^2 |\mathbf{s}|^2}{2}\right)\right]^M,$$

which leads to the inequality

$$\max\{\delta_{xx}, \delta_{yy}\} \leq \delta_{xy},$$

(4.263)

guarantees that the second factor of the left side of inequality (4.260) exceeds one. At last, after combining inequalities (4.261) and (4.263), we arrive at the double inequality

$$\max\{\delta_{xx}, \delta_{yy}\} \leq \delta_{xy} \leq \sqrt{\frac{\delta_{xx} \delta_{yy}}{|B_{xy}|}}.$$

(4.264)

We note that the realizability condition (4.264) does not involve summation index M. Also, that inequality (4.10) has been proven to hold in the discussion on scalar multi-Gaussian Schell-model beams in Chapter 3.

By a straightforward generalization of the beam conditions derived for the scalar multi-Gaussian Schell-model beams (see Chapter 3) we find that for the electromagnetic beams they become

$$\frac{1}{4\sigma^2} + \frac{1}{\delta_{xx}^2} \ll \frac{2\pi^2}{\lambda^2}, \quad \frac{1}{4\sigma^2} + \frac{1}{\delta_{yy}^2} \ll \frac{2\pi^2}{\lambda^2}.$$

(4.265)

Note that, just like for the scalar sources these conditions do not include summation index M.

On assigning subindexes α and β for the results (3.124)–(3.126) pertinent to the scalar multi-Gaussian Schell-model beams we find that

$$W_{\alpha\beta}^{(MG)}(\boldsymbol{\rho}_1, z_1, \boldsymbol{\rho}_2, z_2; \omega) = \frac{I_{\alpha\beta}}{C_0} \sum_{m=1}^{M} \binom{M}{m} \frac{(-1)^{m-1}}{m} \frac{1}{\Delta_{m\alpha\beta}^{(MG)2}(z_1, z_2; \omega)}$$

$$\exp\left\{-\frac{(\boldsymbol{\rho}_1 - \boldsymbol{\rho}_2)^2}{2\Delta_{m\alpha\beta}^{(MG)2}(z_1, z_2; \omega)}\left[\frac{1}{4\sigma^2} + \frac{1}{\delta_{m\alpha\beta}^2}\right]\right\}$$

$$\times \exp\left\{-\left[\frac{1}{8\sigma^2} + i\frac{z_2 - z_1}{8k\sigma^2}\left(\frac{1}{4\sigma^2} + \frac{1}{\delta_{m\alpha\beta}^2}\right)\right]\frac{(\boldsymbol{\rho}_1 + \boldsymbol{\rho}_2)^2}{\Delta_{m\alpha\beta}^{(MG)2}(z_1, z_2; \omega)}\right\}$$

$$\times \exp\left[-i\frac{k(\boldsymbol{\rho}_1^2 - \boldsymbol{\rho}_2^2)}{2R_{m\alpha\beta}^{(MG)}(z_1, z_2; \omega)}\right],$$

$$\text{(4.266)}$$

where

$$\Delta_{m\alpha\beta}^{(MG)2}(z_1, z_2; \omega) = 1 + \frac{z_1 z_2}{k^2\sigma^2}\left[\frac{1}{4\sigma^2} + \frac{1}{\delta_{m\alpha\beta}^2}\right] + i\frac{z_2 - z_1}{k}\left[\frac{1}{2\sigma^2} + \frac{1}{\delta_{m\alpha\beta}^2}\right],$$

$$\text{(4.267)}$$

and

$$R_{m\alpha\beta}^{(MG)}(z_1, z_2; \omega) = \sqrt{z_1 z_2}\left[1 + \frac{k^2\sigma^2}{z_1 z_2}\left(\frac{1}{4\sigma^2} + \frac{1}{\delta_{m\alpha\beta}^2}\right)\right]. \qquad \text{(4.268)}$$

Here we have set $\delta_{m\alpha\beta} = \sqrt{m}\delta_{\alpha\beta}$.

In Fig. 4.14 the comparison of polarization ellipse distributions of the electromagnetic multi-Gaussian Schell-model beams is made for indexes $M = 1$ and $M = 30$ at a sufficiently large distance from the source [75]. For $M = 30$ there exists a well-defined region into the central part of the cross-section where the ellipses have the same orientation and shape. The background is the transverse distribution of the degree of polarization of the beam that also forms a plateau for $M = 30$.

4.7.2 Electromagnetic non-uniformly correlated beams

In this section we use a scalar model for sources with non-uniform correlations discussed in Chapter 3 as a basis for developing electromagnetic sources with non-uniform correlations among the electric field components [77]. Recall that for scalar sources with spatially variable correlations their typical correlation width can attain the maximum value at an arbitrary location, not necessarily on the optical axis. Hence, on free-space propagation a field generated by such a source exhibit self-interference that resembles self-focusing exactly to

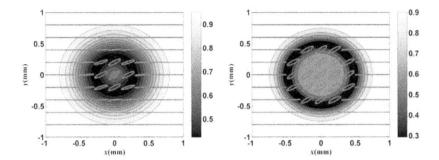

FIGURE 4.14
Changes in the polarization ellipse associated with the polarized part of the electromagnetic multi-Gaussian Schell-model beams for different values of index M (left $M = 1$), (right $M = 30$). From Ref. [75].

this lateral location. After extending the scalar model to the full electromagnetic domain and deriving the realizability conditions we will also explore the behavior of the spectral degree of polarization of such a beam on free-space propagation.

We begin by choosing functions $p_{\alpha\beta}(s)$ and $H_\alpha(\rho', s)$ needed for the integral representation (4.9) of the cross-spectral density matrix. As a straightforward generalization of the scalar non-uniformly correlated sources (see Section 3.2.6) we set:

$$p_{\alpha\beta}(s;\omega) = \frac{B_{\alpha\beta} k \delta_{\alpha\beta}^2}{2\sqrt{\pi}} \exp\left[-\frac{1}{4}k^2 \delta_{\alpha\beta}^4 s^2\right], \qquad (4.269)$$

and

$$H_\alpha(\rho', s;\omega) = A_\alpha \exp\left[-\frac{\rho'^2}{2\sigma_0^2}\right] \exp[-ik(\rho' - \gamma_\alpha)^2 s],$$

$$H_\beta(\rho', s;\omega) = A_\beta \exp\left[-\frac{\rho'^2}{2\sigma_0^2}\right] \exp[-ik(\rho' - \gamma_\beta)^2 s]. \qquad (4.270)$$

In Eqs. (4.269) and (4.270) A_α and A_β are the amplitudes of the field components, σ_0 is the r.m.s. source width, $\delta_{\alpha\beta}$ are the characteristic source correlations. We have assumed here that $\sigma_0 = \sigma_x = \sigma_y$ for simplicity. Also γ_α and γ_β are real-valued two-dimensional vectors that introduce projected shifts in the spectral densities of the electric field along the x and y axes, and $B_{\alpha\beta} = |B_{\alpha\beta}| \exp^{i\phi_{\alpha\beta}}$ is the single-point correlation coefficient, being a complex number, in general. Then explicitly the cross-spectral density matrix elements can be found after substituting from Eqs. (4.269) and (4.270) into

Eq. (4.171) to be:

$$W_{\alpha\beta}^{(NUC)}(\boldsymbol{\rho}_1', \boldsymbol{\rho}_2'; \omega) = A_\alpha A_\beta B_{\alpha\beta} \exp\left[-\frac{\rho_1'^2 + \rho_2'^2}{2\sigma_0^2}\right]$$

$$\times \exp\left\{-\frac{[(\boldsymbol{\rho}_1' - \boldsymbol{\gamma}_\alpha)^2 - (\boldsymbol{\rho}_2' - \boldsymbol{\gamma}_\beta)^2]^2}{\delta_{\alpha\beta}^4}\right\}. \tag{4.271}$$

In order to generate a physical electromagnetic non-uniformly correlated beam the cross-spectral density matrix must be quasi-Hermitian and non-negative definite [8]. The former condition is met when

$$B_{xx} = B_{yy} = 1, \quad |B_{xy}| = |B_{yx}|, \quad \phi_{xy} = \phi_{yx}, \quad \delta_{xy} = \delta_{yx}. \tag{4.272}$$

Substitution from Eq. (4.269) into Eq. (4.9) implies that the first two inequalities hold and that the last inequality reduces to

$$\delta_{xx}^2 \delta_{yy}^2 \exp\left[-\frac{k^2 s^2}{4}(\delta_{xx}^4 + \delta_{yy}^4)\right] \geq |B_{xy}|^2 \delta_{xy}^4 \exp\left[-\frac{k^2 \delta_{xy}^4 s^2}{2}\right]. \tag{4.273}$$

Function $\exp(-x^2)$ in both parts of this equation is even and monotonic on the real axis. Hence it is sufficient to find the restrictions only on the ends of the interval, $x = 0$ and $x \to \infty$. They are:

$$\delta_{xx}^2 \delta_{yy}^2 \geq |B_{xy}|^2 \delta_{xy}^4, \tag{4.274}$$

and

$$\frac{\delta_{xx}^4 + \delta_{yy}^4}{2} \leq \delta_{xy}^4, \tag{4.275}$$

and their combination results in the double inequality:

$$\sqrt[4]{\frac{\delta_{xx}^4 + \delta_{yy}^4}{2}} \leq \delta_{xy} \leq \sqrt{\frac{\delta_{xx}\delta_{yy}}{|B_{xy}|}}. \tag{4.276}$$

Note that in form these realizability conditions are somewhat similar to those for the classic electromagnetic Gaussian Schell-model sources.

It is important to point out that for the electromagnetic non-uniformly correlated sources the polarization properties are not uniform across the source, even if $\sigma_x = \sigma_y = \sigma_0$. For instance, the degree of polarization reduces to the expression

$$\wp(\boldsymbol{\rho}'; \omega) = \sqrt{1 - \frac{4 Det \mathbf{W}(\boldsymbol{\rho}', \boldsymbol{\rho}'; \omega)}{[Tr\mathbf{W}(\boldsymbol{\rho}', \boldsymbol{\rho}'; \omega)]^2}}$$

$$= \frac{\sqrt{(A_x^2 - A_y^2)^2 + 4 A_x^2 A_y^2 |B_{xy}|^2 \eta_{xy}^2 (\boldsymbol{\rho}'; \omega)}}{A_x^2 + A_y^2}, \tag{4.277}$$

with

$$\eta_{xy}(\boldsymbol{\rho}';\omega) = \exp\left\{-\frac{[(\boldsymbol{\rho}'-\boldsymbol{\gamma}_x)^2 - (\boldsymbol{\rho}'-\boldsymbol{\gamma}_y)^2]^2}{\delta_{xy}^4}\right\}. \qquad (4.278)$$

Hence, the degree of polarization depends on $\boldsymbol{\gamma}_x$ and $\boldsymbol{\gamma}_y$. Only provided $\boldsymbol{\gamma}_x = \boldsymbol{\gamma}_y$, $\eta_{xy}(\boldsymbol{\rho}';\omega) = 1$ for any $\boldsymbol{\rho}'$ and the distribution of the degree of polarization in the source plane becomes uniform.

We will now derive the propagation laws for the beams of our interest. Recall that according to the Huygens-Fresnel principle the elements of the cross-spectral density matrix at positions $\mathbf{r}_1 = (\boldsymbol{\rho}_1, z)$ and $\mathbf{r}_2 = (\boldsymbol{\rho}_2, z)$ are related to those in the source plane as (see Eqs. (4.123))

$$\begin{aligned}
W_{\alpha\beta}^{(NUC)}(\boldsymbol{\rho}_1,\boldsymbol{\rho}_2,z;\omega) &= \frac{k^2}{4\pi^2 z^2}\iint W_{\alpha\beta}^{(NUC)}(\boldsymbol{\rho}_1',\boldsymbol{\rho}_2';\omega) \\
&\times \exp\left[-ik\frac{(\boldsymbol{\rho}_1-\boldsymbol{\rho}_1')^2 - (\boldsymbol{\rho}_2-\boldsymbol{\rho}_2')^2}{2z}\right] d^2\boldsymbol{\rho}_1' d^2\boldsymbol{\rho}_2'.
\end{aligned} \qquad (4.279)$$

On substituting from Eq. (4.271) into Eq. (4.279) we obtain, after interchanging the orders of integrals, the expression

$$W_{\alpha\beta}^{(NUC)}(\boldsymbol{\rho}_1,\boldsymbol{\rho}_2,z;\omega) = \frac{k^2}{4\pi^2 z^2}\int p_{\alpha\beta}(s)H_\alpha^*(\boldsymbol{\rho}_1,s,z)H_\beta(\boldsymbol{\rho}_2,s,z)ds, \qquad (4.280)$$

where

$$H_\alpha(\boldsymbol{\rho}_1,s,z) = \int H_\alpha(\boldsymbol{\rho}_1',s)\exp\left[\frac{ik}{2z}(\boldsymbol{\rho}-\boldsymbol{\rho}')^2\right]d^2\boldsymbol{\rho}'. \qquad (4.281)$$

The structure of the elements of the cross-spectral density matrix constructed through the superposition (4.280) is invariant on propagation in free space. While the modes H_α evolve on propagation, the weighting coefficients $p_{\alpha\beta}$ remain constant. On substituting from Eq. (4.270) into Eq. (4.281) and considering the modes at a single point $\boldsymbol{\rho}_1 = \boldsymbol{\rho}_2 = \boldsymbol{\rho}$, we obtain, the formula

$$\begin{aligned}
H_\alpha^*(\boldsymbol{\rho},s,z)H_\beta(\boldsymbol{\rho},s,z) &= A_\alpha A_\beta \frac{4\pi^2 z^2}{k^2}\frac{\sigma_0^2}{h^2(z,s)} \\
&\times \exp[ik(\gamma_\alpha^2 - \gamma_\beta^2)s]\exp[-k^2\sigma_0^2(\gamma_\alpha - \gamma_\beta)^2 s^2] \\
&\times \exp\left\{-\frac{[\boldsymbol{\rho}-sz(\gamma_\alpha+\gamma_\beta)+ik\sigma_0^2 s(1-2sz)(\gamma_\alpha-\gamma_\beta)]^2}{h^2(z,s)}\right\},
\end{aligned} \qquad (4.282)$$

where

$$h^2(z,s) = \frac{z^2}{k^2\sigma_0^2} + \sigma_0^2(1-2sz)^2, \qquad (4.283)$$

is the expansion coefficient of each mode. One can find that the maximum intensity of each mode is located at point $\boldsymbol{\rho} = sz(\gamma_\alpha + \gamma_\beta)$ by neglecting the phase term. On substituting from Eq. (4.282) into Eq. (4.280) the cross-spectral density matrix, and hence the degree of polarization, can be evaluated numerically.

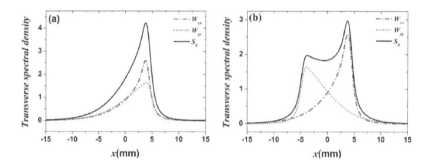

FIGURE 4.15
Spectral density of the propagating electromagnetic non-uniformly correlated beam in a transverse plane $z = 30$ m. (a) $\gamma_x = (0.8\sigma_0, 0)$, $\gamma_y = (0.9\sigma_0, 0)$ and (b) $\gamma_x = (0.8\sigma_0, 0)$, $\gamma_y = (-0.9\sigma_0, 0)$. From Ref. [76].

Figure 4.15 shows several distributions of the spectral density of the beam at the plane $z = 30$ m in case of two lateral shifts. Thus, unlike in the scalar case, due to the presence of two components of the electric field it is possible to generate two different intensity maxima at two different locations.

Figure 4.16 shows the contours of the degree of polarization of the beam propagating at several selected distances in the transverse planes. This perspective reveals that the degree of polarization in the source plane [Fig. 4.16 (a)] is not uniform but is symmetric with respect to the line defined by γ_x and γ_y. However, with growing propagation distance the spike of the degree of polarization forms in the center of symmetry, gradually evolving [Figs. 4.16 (b)–(d)]. At sufficiently large distances from the source the pattern changes only quantitatively, preserving its far-zone distribution and increasing in scale.

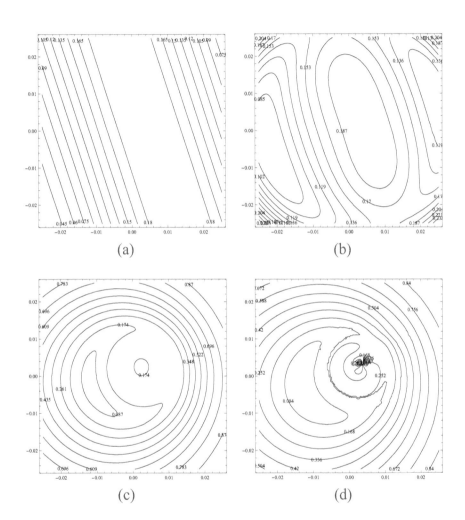

FIGURE 4.16

Distribution of the degree of polarization of the beam in transverse planes: (a) $z = 0$ m, (b) $z = 1$ m; (c) $z = 15$ m; (d) $z = 30$ m. From Ref. [77].

Bibliography

[1] D. F. V. James, "Change in polarization of light beams on propagation in free space," *J. Opt. Soc. Am. A* **11**, 1641–1643 (1994).

[2] F. Gori, M. Santarsiero, S. Vicalvi, R. Borghi, and G. Guattari, "Beam cohrence-polarization matrix," *Pure Appl. Opt.* **7**, 941–951 (1998).

[3] E. Wolf, "Unified theory of coherence and polarization of statistical electromagnetic beams," *Phys. Lett. A* **312**, 263–267 (2003).

[4] E. Wolf, *Introduction to Theories of Coherence and Polarization of Light*, Cambridge University Press, 2007.

[5] P. Meemon, M. Salem, K. S. Lee, M. Chopra, and J. P. Rolland, "Determination of the coherency matrix of a broadband stochastic electromagnetic light beam," *J. Mod. Opt.* **55**, 2765–2776 (2008).

[6] J. Tervo, T. Setälä, and A. T. Friberg, "Theory of partially coherent electromagnetic fields in the space-frequency domain," *J. Opt. Soc. Am. A* **21**, 2205–2215 (2004).

[7] M. A. Alonso and E. Wolf, "The cross-spectral density matrix of a planar, electromagnetic stochastic source as a correlation matrix," *Opt. Commun.* **281**, 2393–2396 (2008).

[8] L. Mandel and E. Wolf, *Optical Coherence and Quantum Optics*, Cambridge University Press, 1995.

[9] F. Gori, V. Ramírez-Sánchez, M. Santarsiero, and T. Shirai, "On genuine cross-spectral density matrices," *J. Opt. A: Pure Appl. Opt.* **11**, 085706 (2009).

[10] F. Gori, M. Santarsiero, R. Borghi, and V. Ramírez-Sánchez, "Realizability conditions for electromagnetic Schell-model sources," *J. Opt. Soc. Am. A* **25**, 1016–1021 (2008).

[11] H. Roychowdhury and E. Wolf, "Young's interference experiment with light of any state of coherence and of polarization," *Opt. Comm.* **252**, 268–274 (2005).

[12] O. Korotkova and E. Wolf, "Changes in the state of polarization of a random electro-magnetic beam on propagation," *Opt. Comm.* **246**, 35–43 (2005).

[13] M. Born and E. Wolf, *Principles of Optics*, Cambridge University Press, 7th Edition, 1999.

[14] D. Goldstein, *Polarized Light*, 2nd Ed., Marcel Dekker, 2003.

[15] O. Korotkova, T.D. Visser, and E. Wolf "Polarization properties of stochastic electromagnetic beams," *Opt. Comm.* **281**, 515–520 (2008).

[16] T. Shirai and E. Wolf, "Correlations between intensity fluctuations in stochastic electromagnetic beams of any state of coherence and polarization," *Opt. Commun.* **272**, 289–292 (2007).

[17] S. N. Volkov, D. F. V. James, T. Shirai, and E. Wolf, "Intensity fluctuations and the degree of cross-polarization of stochastic electromagnetic beams," *J. Opt. A: Pure Appl. Opt.* **10**, 05001 (2008).

[18] G. G. Stokes, "On the composition and resolution of streams of polarized light from different sources," *Trans. Cambridge Phil. Soc.* **9**, 399–416 (1852).

[19] C. Brosseau, *Fundamentals of Polarized Light*, Wiley, 1998.

[20] O. Korotkova and E. Wolf, "Generalized Stokes parameters of random electromagnetic beams," *Opt. Lett.* **30** 198–200 (2005).

[21] J. Tervo, T. Setälä, A. Roueff, P. Réfrégier, and A. T. Friberg, "Two-point Stokes parameters: interpretation and properties," *Opt. Lett.* **34**, 3074–3076 (2009).

[22] F. Gori, M. Santarsiero, R. Simon, G. Piquero, R. Borghi, and G. Guattari, "Coherent-mode decomposition of partially polarized, partially coherent sources," *J. Opt. Soc. Am. A* **20**, 78–83 (2003).

[23] J. Tervo, T. Setälä, and A. T. Friberg, "Theory of partially coherent electromagnetic fields in the spacefrequency domain," *J. Opt. Soc. A* **21**, 2205–2215 (2004).

[24] K. Kim and E. Wolf, "A scalar-mode representation of stochastic, planar, electromagnetic sources," *Opt. Commun.* **261**, 19–22 (2006).

[25] J. Tervo and J. Turunen, "Angular spectrum representation of partially coherent electromagnetic fields," *Opt. Commun.* **209**, 7-16 (2002).

[26] R. A. Silverman, "Locally stationary random process," *IRE Transactions Info. Theory* **3**, 182–187 (1957).

[27] R. A. Silverman, "Scattering of plane waves by locally homogeneous dielectric noise," *Proc. Cambridge Philos. Soc.* **54**, 530–537 (1958).

[28] O. Korotkova, B. G. Hoover, V. L. Gamiz, and E. Wolf, "Coherence and polarization properties of far fields generated by quasi-homogeneous planar electromagnetic sources," *J. Opt. Soc. Am. A* **22**, 2547–2556 (2005).

[29] E. Wolf, "Invariance of the spectrum of light on propagation," *Phys. Rev. Lett.* **56**, 1370-1372 (1986).

[30] E. Wolf and D. F. V. James, "Correlation-induced spectral changes," *Rep. Prog. Phys.* **59**, 771–818 (1996).

[31] J. Pu, O. Korotkova, and E. Wolf, "Invariance and non-invariance of the spectra of stochastic electromagnetic beams on propagation," *Opt. Lett.* **31**, 2097–2099 (2006).

[32] J. Pu, O. Korotkova, and E. Wolf, "Polarization-induced spectral changes on propagation of stochastic electromagnetic beams," *Phys. Rev. E* **75**, 056610 (2007).

[33] O. Korotkova, "Sufficient condition for polarization invariance of beams generated by quasi-homogeneous sources," *Opt. Lett.* **36**, 3768–3770 (2011).

[34] E. Wolf, "Polarization invariance in beam propagation," *Opt. Lett.* **32**, 3400–3401 (2007).

[35] X. Zhao, Y. Yao, Y. Sun, and C. Liu, "Condition for Gaussian Schell-model beam to maintain the state of polarization on the propagation in free space," *Opt. Express* **17**, 17888–17894 (2009).

[36] M. A. Alonso, O. Korotkova, and E. Wolf, "Propagation of the electric correlation matrix and the van Cittert–Zernike theorem for random electromagnetic fields," *J. Mod. Opt.* **53**, 969-978 (2006).

[37] P. Roman and E. Wolf, "Correlation theory of stationary electromagnetic fields, Part II: Conservation laws," *Nuovo Cimento* **17**, 477–490 (1960).

[38] M. W. Kowarz and E. Wolf, "Conservation laws for partially coherent free fields," *J. Opt. Soc. Am. A* **10**, 88–94 (1993).

[39] G. C. Sherman, "Diffracted wave fields expressible by planewave expansions containing only homogeneous waves," *Phys. Rev. Lett.* **9**, 761–764 (1968).

[40] O. Korotkova, "Conservation laws for stochastic electromagnetic free fields," *J. Opt. A: Pure Appl. Opt.* **10**, 025003 (2008).

[41] C. Palma and P. De Santis, "Propagation of partially coherent beams in absorbing media," *J. Mod. Opt.* **42**, 1123–1135 (1995).

[42] C. Palma and P. De Santis, "Propagation and coherence evolution of optical beams in gain media," *J. Mod. Opt.* **43** 139–153 (1995).

[43] B. Zhang, Q. Wen, and X. Guo, "Beam propagation factor of partially coherent beams in gain or absorbing media," *Optik* **117**, 123–127 (2006).

[44] M.G. Sicairos and J.G. Vega, "Propagation of Helmholtz-Gauss beams in absorbing and gain media," *J. Opt. Soc. Am. A* **23**, 1994–2001 (2006).

[45] Y.J. Cai, Q. Lin, and D. Ge, "Propagation of partially coherent twisted anisotropic Gaussian Schell-model beams in dispersive and absorbing media," *J. Opt. Soc. Am. A* **19** 2036–2042 (2002).

[46] J. Pu and O. Korotkova, "Spectral and polarization properties of stochastic electromagnetic beams propagating in gain or absorbing media," *Opt. Commun.* **283**, 1693–1706 (2010).

[47] S.A. Collins, "Lens-system diffraction integral written in terms of matrix optics," *J. Opt. Soc. Am.* **60**, 1168–1177 (1970).

[48] The papers by R. C. Jones are reprinted in W. Swindel, *Polarized Light*, Stroudsburg, Pennsylvania, Dowden, Hutchinson and Ross, 1975.

[49] H. Mueller, "Theory of polarimetric investigations of light scattering," Parts I, II. Contract W-18-035-CWS-1304. D.I.C. 2-6467. MIT (1946–1947); H. Mueller, "The Foundations of Optics," *J. Opt. Soc. Am.* **38**, 661–661 (1948).

[50] O. Korotkova and E. Wolf, "Effects of linear non-image-forming devices on sprecta and on coherence and polarization properties of stochastic electromagnetic beams. Part 1 General theory," *J . Mod. Opt.* **52**, 2659–2671 (2005).

[51] S. R. Seshadri, "Partially coherent Gaussian Schell-model electromagnetic beams," *J. Opt. Soc. Am. A* **16**, 1373–1380 (1999).

[52] S. R. Seshadri, "Polarization properties of partially coherent Gaussian Schell-model electromagnetic beams," *J. Appl. Phys.* **87**, 4084–4093 (2000).

[53] F. Gori, M. Santarsiero, G. Piquero, R. Borghi, A. Mondello, and R. Simon, "Partially polarized Gaussian Schell-model beams," *J. Opt. A: Pure Appl. Opt.* **3**, 1–9 (2001).

[54] O. Korotkova, M. Salem, and E. Wolf, "Beam conditions for radiation generated by an electromagnetic Gaussian Schell-model source," *Opt. Lett.* **29**, 1173–1175 (2004).

[55] O. Korotkova, M. Salem, and E. Wolf, "The far-zone behavior of the degree of polarization of partially coherent beams propagating through atmospheric turbulence," *Opt. Comm.* **233**, 225–230 (2004).

[56] H. Roychowdhury and O. Korotkova, "Realizability conditions for electromagnetic Gaussian Schell-model sources," *Opt. Comm.* **249**, 379–385 (2005).

[57] G. Piquero, F. Gori, P. Romanini, M. Santarsiero, R. Borghi, and A. Mondello, "Synthsis of partially polarized Gaussian Schell-model sources," *Opt. Comm.* **208**, 9–16 (2002).

[58] T. Shirai, O. Korotkova, and E. Wolf, "A method of generating electromagnetic Gaussian Schell-model beams," *J. Opt. A: Pure Appl. Opt.* **7** 232–237 (2005).

[59] M. Santarsiero, R. Borghi, and V. Ramírez-Sánchez, "Synthesis of electromagnetic Schell-model sources," *J. Opt. Soc. Am. A* **26**, 1437–1443 (2009).

[60] A. S. Ostrovsky, G. Rodríguez-Zurita, C. Meneses-Fabián, M. Á. Olvera-Santamaría, and C. Rickenstorff-Parrao, "Experimental generating the partially coherent and partially polarized electromagnetic source," *Opt. Express* **18**, 12864–12871 (2010).

[61] S. Sahin, O. Korotkova, G. Zhang, and J. Pu, "Free-space propagation of the spectral degree of cross-polarization of stochastic electromagnetic beams," *J. Opt. A: Pure Appl. Opt.* **11**, 085703 (2009).

[62] O. Korotkova, "Changes in the intensity fluctuations of a class of random electromagnetic beams on propagation," *J. Opt. A: Pure Appl. Opt.* **8**, 30–37 (2006).

[63] J. Goodman, *Statistical Optics*, New York, Wiley, 1985.

[64] O. Korotkova, "Scintillation index of a stochastic electromagnetic beam propagating in random media," *Opt. Commun.* **281**, 2342–2348 (2008).

[65] L. C. Andrews, R. L. Phillips, and C. Y. Hopen, *Laser Beam Scintillations with Applications*, SPIE Press, 2001.

[66] L.C. Andrews and R. L. Phillips, *Laser Beam Propagation through Random Media*, 2nd Ed., SPIE Press, 2005.

[67] S. J. Wang, Y. Baykal, and M. A. Plonus, "Receiver aperture averaging effects for the intensity fluctuation of the beam wave in the turbulent atmosphere," *J. Opt. Soc. Am.* **73**, 831–837 (1983).

[68] D. L. Fried, "Aperture averaging of scintillation," *J. Opt. Soc. Am.* **57**, 169–175 (1967).

[69] O. Korotkova, "Changes in statistics of the instantaneous Stokes parameters of a quasi-monochromatic electromagnetic beam on propagation," *Opt. Commun.* **261**, 218–224 (2006).

[70] R. Barakat, "Statistics of the Stokes parameters," *J. Opt. Soc. A* **4**, 1256–1263 (1987).

[71] C. Brosseau, R. Barakat, and E. Rockower, "Statistics of the Stokes parameters for Gaussian distributed fields," *Opt. Commun.* **82**, 204–208 (1991).

[72] D. Eliyahu, "Vector statistics of correlated Gaussian fields," *Phys. Rev. E* **47**, 2881–2892 (1993).

[73] C. Brosseau, "Statistics of the normalized Stokes parameters for a Gaussian stochastic plane wave field," *Appl. Opt.* **34**, 4788–4793 (1995).

[74] H. C. Jacks and O. Korotkova, "Polarization and intensity correlations in stochastic electromagnetic beams upon interaction with devices of polarization optics," *Appl. Phys. B* **103**, 413–419 (2011).

[75] Z. Mei, O. Korotkova, and E. Shchepakina, "Electromagnetic multi-Gaussian Schell-Model Beams," *J. Opt.* **15**, 025705 (2013).

[76] Z. Mei, Z. Tong, and O. Korotkova, "Electromagnetic non-uniformly correlated beams in turbulent atmosphere," *Opt. Express* **20**, 26458–26463 (2012).

[77] Z. Tong and O. Korotkova, "Electromagnetic non-uniformly correlated beams," *J. Opt. Soc. Am. A* **29**, 2154–2158 (2012).

5

Interaction of random electromagnetic beams with optical systems

CONTENTS

Regardless of a relatively short history of controllable stochastic beams the number and importance of their applications are remarkable, ranging from imaging to communications, from remote sensing to microscopy, from tomography to holography, just to list a few. In this chapter we will explore some of the applications that are based on passage of a beam through the image-forming optical systems, assuming that there is vacuum or a linear deterministic medium between the system elements. In particular, we will explore the evolution of random beams in the human eye, negative phase materials, telescopic systems and laser resonators. Then, after the detailed analysis of the behavior of beams propagating in random media is presented in Chapter 6, we will discuss, in Chapter 7, beam passage through optical systems in the presence of random medium between the optical elements.

5.1 $ABCD$ matrix method for beam interaction with image-forming optical systems

An optical element is called *aligned* if its axis of symmetry coincides with the direction of propagation, say z, of a light beam. Consider such an optical element, situated between planes $\mathcal{A}$ and $\mathcal{B}$ (see Fig. 5.1). A ray impinging onto the optical element at plane $\mathcal{A}$ can be completely characterized by position $\rho' = (x', y')$, and orientation angle $\theta' = (\theta'_x, \theta'_y)$ both measured from the z-

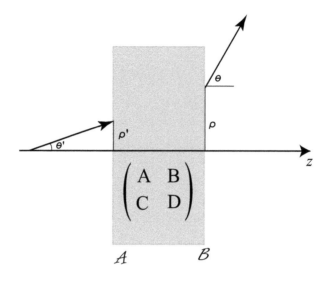

FIGURE 5.1
Aligned optical elements.

axis. Another ray, exiting the element at plane $\mathcal{B}$ can be described by position $\boldsymbol{\rho} = (x, y)$, and orientation angle $\boldsymbol{\theta} = (\theta_x, \theta_y)$.

In cases when the passage of rays through the element is treated under the paraxial approximation the initial and transmitted rays are related by the linear transformation [1, 2]

$$\begin{cases} \boldsymbol{\rho} = A\boldsymbol{\rho}' + B\boldsymbol{\theta}', \\ \boldsymbol{\theta} = C\boldsymbol{\rho}' + D\boldsymbol{\theta}', \end{cases} \tag{5.1}$$

which can equivalently be written as matrix equation

$$\mathbf{r} = \boldsymbol{\Omega}\mathbf{r}', \tag{5.2}$$

where

$$\mathbf{r} = \begin{bmatrix} \boldsymbol{\rho} \\ \boldsymbol{\theta} \end{bmatrix}, \qquad \mathbf{r}' = \begin{bmatrix} \boldsymbol{\rho}' \\ \boldsymbol{\theta}' \end{bmatrix}, \quad \text{and} \quad \boldsymbol{\Omega} = \begin{bmatrix} A & B \\ C & D \end{bmatrix}. \tag{5.3}$$

Matrix $\boldsymbol{\Omega}$ is also known as the *ray transfer matrix* or *ABCD matrix* [3, 4].

The most common procedure for derivation of the *ABCD* matrices stems from Eq. (5.1): for finding the four elements of the matrix it is sufficient to determine the corresponding derivatives:

$$A = \frac{\rho}{\rho'}|_{\theta'=0}, \quad B = \frac{\rho}{\theta'}|_{\rho'=0}, \quad C = \frac{\theta}{\rho'}|_{\theta'=0}, \quad D = \frac{\theta}{\theta'}|_{\rho'=0}. \tag{5.4}$$

For instance, the following matrices can be derived by such means:

- free-space propagation path of length L in a homogeneous medium with refractive index n which coincides with that of the surrounding medium:

$$\begin{bmatrix} 1 & L \\ 0 & 1 \end{bmatrix};$$ (5.5)

- spherical mirror with radius of curvature R:

$$\begin{bmatrix} 1 & 0 \\ 2/R & 1 \end{bmatrix};$$ (5.6)

- spherical thin lens with radius of curvature R inserted between media with refractive indexes n_1 and n_2:

$$\begin{bmatrix} 1 & 0 \\ \frac{n_1 - n_2}{n_2 R} & \frac{n_1}{n_2} \end{bmatrix};$$ (5.7)

- corner-cube retro-reflector:

$$\begin{bmatrix} -1 & 0 \\ 0 & -1 \end{bmatrix};$$ (5.8)

- Gaussian lens with lens parameter $\alpha_G = 2/(kW_G^2) + i/F_G$, W_G being the radius of the lens, F_G being its focal length:

$$\begin{bmatrix} 1 & 0 \\ i\alpha_G & 1 \end{bmatrix}.$$ (5.9)

On letting $R \to \infty$ in Eqs. (5.6) and (5.7) the corresponding $ABCD$ matrices can be obtained for planar mirror:

$$\begin{bmatrix} 1 & 0 \\ 0 & 1 \end{bmatrix},$$ (5.10)

and for planar interface between two media:

$$\begin{bmatrix} 1 & 0 \\ 0 & n_1/n_2 \end{bmatrix}.$$ (5.11)

In cases when several, say, N, optical elements are aligned in series along the propagation axis, the $ABCD$ matrix accounting for the whole system is expressed as a product of the matrices of individual elements, $\mathbf{\Omega}_n$, $n = 1, 2, ..., N$, as

$$\mathbf{\Omega} = \mathbf{\Omega}_N \mathbf{\Omega}_{N-1} ... \mathbf{\Omega}_2 \mathbf{\Omega}_1,$$ (5.12)

where left matrix multiplication is implied.

For example, in order to derive the $ABCD$ matrix of a thick lens, consisting

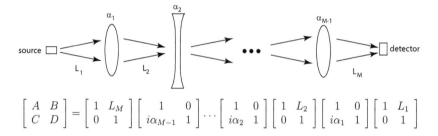

FIGURE 5.2
Diagram representing $ABCD$ matrix approach for optical systems with Gaussian lenses.

of two lenses with radii of curvature R_1 and R_2 and the medium of length L with refractive index n in between, it suffices to calculate the product

$$\Omega = \Omega_{R_2}\Omega_n\Omega_{R_1}, \tag{5.13}$$

where Ω_{R_1} and Ω_{R_2} are the matrices of the lenses and Ω_n is that of the medium. Then one obtains:

$$\Omega = \begin{bmatrix} 1 - \frac{(n-1)L}{nR_1} & \frac{L}{n} \\ (n-1)\left[\frac{1}{R_2} - \frac{1}{R_1} - \frac{(n-1)L}{nR_1R_2}\right] & 1 + \frac{(n-1)L}{nR_2} \end{bmatrix}. \tag{5.14}$$

Another example of an $ABCD$ matrix relates to a system of M Gaussian lenses with different values of parameter α_G separated by different distances is illustrated in Fig. 5.2.

So far we have described calculation of the $ABCD$ matrices and their systems in the case when the matrix elements depended only on propagation distance and on parameters responsible for variation in the plane transverse to the direction of propagation. However, in some situations the dependence of the optical properties of the media, and, hence, of the $ABCD$ matrices on coordinates parallel and perpendicular to the direction of propagation is general, i.e., each element may depend on vector $\mathbf{r} = (x, y, z)$. The typical examples of such media are Gradient-Index (GRIN) optical fibers and thick lenses having complex $\mathbf{r}$ profiles, such as the crystalline lens of the human eye, which we will treat in the following section.

Passage of light through optical fibers is based on the multiple total internal reflections of light. For the simple step-index cylindrically symmetric fibers [see Fig. 5.3 (a)] consisting of a core with an index of refraction n_1 and a cladding with index of refraction n_2, such that $n_1 > n_2$, light always remains in the core. Ideally, a beam entering the core at a sufficiently high angle of incidence (almost parallel to the fiber) remains in it indefinitely, provided that the relative index of refraction, $(n_1 - n_2)/n_1$ is small, several percent at most.

In the GRIN fibers, instead of a sharp transition of the refractive index

(a)

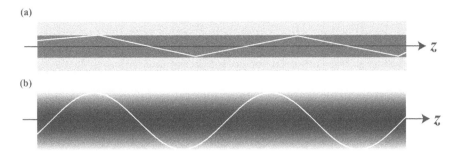

(b)

FIGURE 5.3
Optical fibers: (a) step-index fiber; (b) GRIN fiber.

between the core and the cladding, its smooth variation $n(x, y; \omega)$ is modeled
by some function, for instance a parabola [see Fig. 5.3 (b)] as [5], [6]

$$
n(x, y; \omega) = \begin{cases} n_0(\omega)\sqrt{1 - \alpha_g^2(\omega)(x^2 + y^2)}, & x^2 + y^2 \leq R_g^2, \\ n_0(\omega)\sqrt{1 - \alpha_g^2(\omega)R_g^2}, & x^2 + y^2 \geq R_g^2. \end{cases}
\tag{5.15}
$$

Here $n_0(\omega)$ is the refractive index in the center of the fiber, R_g is the radius
of the core, $\alpha_g(\omega)$ is the radial gradient of the refractive index:

$$
\alpha_g(\omega) = \frac{1}{R_g}\sqrt{1 - \frac{n_2^2(\omega)}{n_1^2(\omega)}}.
\tag{5.16}
$$

The $ABCD$ matrix for paraxial ray propagation through a GRIN fiber at
distance $z > 0$ then has the form

$$
\begin{bmatrix} A(z) & B(z) \\ C(z) & D(z) \end{bmatrix} =
$$

$$
\begin{bmatrix} \cos[\alpha_g(\omega)z] & \sin[\alpha_g(\omega)z]/n(x, y; \omega)\alpha_g(\omega) \\ -n(x, y; \omega)\alpha_g(\omega)\sin[\alpha_g(\omega)z] & \cos[\alpha_g(\omega)z]. \end{bmatrix}
\tag{5.17}
$$

The connection between ray optics and diffraction theory was first established by Collins [7]. According to this theory the propagation of two mutually-
orthogonal components of a monochromatic field $E_\alpha(\rho'; \omega)$, $(\alpha = x, y)$ in the
source plane $z = 0$ takes the following form after propagation through the
$ABCD$ system of total axial optical distance z:

$$
E_\alpha(\rho, z; \omega) = \frac{-ik\exp(ikz)}{2\pi B(z)}\iint E_\alpha(\rho'; \omega)
$$

$$
\times \exp\left[\frac{ik}{2B(z)}\{A(z)\rho'^2 - 2\rho' \cdot \rho + D(z)\rho^2\}\right]d^2\rho'.
\tag{5.18}
$$

$$
(\alpha = x, y).
$$

This formula is known as the *generalized Huygens-Fresnel integral* and reduces to Eq. (1.73) if the $ABCD$ matrix is chosen to be the one for free-space propagation at distance z. It is important to note that the same elements of the $ABCD$ matrix are used for both field components and, hence, they are not coupled on interaction with image-forming systems.

The generalized Huygens-Fresnel integral (5.18) can be readily generalized to random electromagnetic beams. If monochromatic realizations of the electric field components $E_x(\boldsymbol{\rho}, z; \omega)$ and $E_y(\boldsymbol{\rho}, z; \omega)$ are known then on correlating them at two points, $(\boldsymbol{\rho}_1, z)$ and $(\boldsymbol{\rho}_2, z)$ and applying the generalized Huygens-Fresnel integral (5.18) we at once obtain the formula

$$
\begin{aligned}
W_{\alpha\beta}(\boldsymbol{\rho}_1, \boldsymbol{\rho}_2, z; \omega) = & \left(\frac{k}{2\pi B(z)}\right)^2 \iint W_{\alpha\beta}(\boldsymbol{\rho}_1', \boldsymbol{\rho}_2'; \omega) \\
& \times \exp\left[-\frac{ik}{2B(z)}\left\{A(z)(\boldsymbol{\rho}_1'^2 - \boldsymbol{\rho}_2'^2)\right.\right. \\
& \left.\left. - 2(\boldsymbol{\rho}_1' \cdot \boldsymbol{\rho}_1 - \boldsymbol{\rho}_2' \cdot \boldsymbol{\rho}_2) + D(z)(\boldsymbol{\rho}_1^2 - \boldsymbol{\rho}_2^2)\right\}\right] d^2\rho_1' d^2\rho_2'.
\end{aligned}
\tag{5.19}
$$

For some models of the cross-spectral density matrices of random beams it is possible to evaluate the double integral in (5.19) in the closed form. For example, for the electromagnetic Gaussian Schell-model source defined by Eq. (4.172) the cross-spectral density matrix elements of the beam after passage through the $ABCD$ system take the form [8]

$$
\begin{aligned}
W_{\alpha\beta}(\boldsymbol{\rho}_1, \boldsymbol{\rho}_2, z; \omega) = & \frac{A_\alpha A_\beta B_{\alpha\beta}(\omega)}{\widehat{\Delta}_{\alpha\beta}(z)} \exp\left[-\frac{\rho_1^2 + \rho_2^2}{4\widehat{\Delta}_{\alpha\beta}(z)\sigma^2(\omega)}\right] \\
& \times \exp\left[-\frac{|\boldsymbol{\rho}_1 - \boldsymbol{\rho}_2|^2}{2\widehat{\Delta}_{\alpha\beta}(z)\delta_{\alpha\beta}^2(\omega)}\right] \exp\left[-ik\frac{\rho_1^2 - \rho_2^2}{2\widehat{R}_{\alpha\beta}(z)}\right],
\end{aligned}
\tag{5.20}
$$

where

$$
\widehat{\Delta}_{\alpha\beta}(z) = A^2(z) + \frac{B(z)}{4\sigma^4(\omega)k^2}\left(1 + \frac{4\sigma^2(\omega)}{\delta_{\alpha\beta}^2(\omega)}\right),
\tag{5.21}
$$

and

$$
\widehat{R}_{\alpha\beta}(z) = \frac{B(z)\widehat{\Delta}_{\alpha\beta}(z)}{D(z)\widehat{\Delta}_{\alpha\beta}(z) - A(z)}.
\tag{5.22}
$$

Parameters $\widehat{\Delta}_{\alpha\beta}(z)$ and $\widehat{R}_{\alpha\beta}(z)$ can be viewed as generalized beam expansion coefficient and radius of curvature due to the optical system.

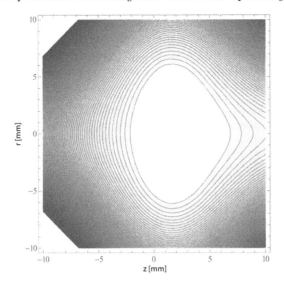

FIGURE 5.4
Contour plots of the refractive index distribution of the crystalline lens.

5.2   Random beams in the human eye

One of the most remarkable optical elements existing in nature is the *crystalline lens* of the human eye, also known as the *aquula* (water, Lat.). The optical structure of this lens appears to be fairly well-known for longer than a century [9], [10]. The first analytical model of the refractive index profile of the lens, $n(\mathbf{r})$, was given in [11]:

$$
\begin{aligned}
n(\rho, z) = {}& 1.406 - 0.0062685(z - z_0)^2 + 0.0003834(z - z_0)^3 \\
& - [0.00052375 + 0.00005735(z - z_0) + 0.00027875(z - z_0)^2]\rho^2 \quad (5.23) \\
& - 0.000066717\rho^4,
\end{aligned}
$$

where ρ, z and z_0 are given in millimeters. The total thickness z of the crystalline lens is about 3.6 mm and the distance z_0 from its entrance plane end to the plane with the maximum refractive index is about 1.7 mm. Figure 5.4 shows contour plots of the refractive index calculated from Eq. (5.23) as a function of z and ρ. Other models for the refractive index distribution within the lens were developed in Refs. [12], [13]. A recent mini-review of the existing models is made in Ref. [14].

Our discussion of light passage in the eye will be based on one of the newer models for the crystalline eye lens [15] (see also [16]–[17]), which relies on the assumption that the variation of the refractive index in the radial direction is parabolic making possible to determine implies the $ABCD$-matrix. Namely, the lens is treated as a GRIN medium limited either by plane-parallel end faces or by curved end faces with a quadratic transverse distribution of refractive index. In the paraxial approximation and for a meridional section of the lens, is given by expression

$$n(\rho, z) = n_0(z)\left[1 - \frac{g_c^2(z)}{2}\rho^2\right],\qquad(5.24)$$

where $n_0(z)$ is the refractive index along the z optical axis and $g(z)$ is the gradient parameter describing the evolution of the transverse parabolic distribution. The expressions for n_0 and g_c depend on the optical modeling of the inhomogeneity of the refractive index within the lens. In [15] the ray transfer matrix elements $A_c(z)$, $B_c(z)$, $C_c(z)$, $D_c(z)$ for the curved-end faces model of the lens were established:

$$A_c(z) = \left[1 - \frac{g_e^2 z^2}{2} + \frac{\dot{g}_e}{2g_e}z\left(1 - \frac{g_e^2 z^2}{6}\right)\right] - \frac{P_f z(1 - \frac{g_e^2 z^2}{6})}{n_e},\qquad(5.25)$$

$$B_c(z) = \frac{n_1}{n_e}z\left(1 - \frac{g_e^2 z^2}{6}\right),\qquad(5.26)$$

$$C_c(z) = -\frac{P_e(z)}{n_1'},\qquad(5.27)$$

$$D_c(z) = \frac{n_1}{n_1'}\left\{-\left[g_e^2 + \left(\frac{\dot{g}_e}{2g_e}\right)^2\right]z\left(1 - \frac{g_e^2 z^2}{6}\right)\right.$$
$$\left. + \frac{P_b\left[g_e^2 + \left(\frac{\dot{g}_e}{2g_e}\right)^2\right]z\left(1 - \frac{g_e^2 z^2}{6}\right)}{n_e}\right\},\qquad(5.28)$$

where the back refractive power or equivalent power $P_e(z)$ of the lens has the form

$$P_e(z) = P_b\left[1 - \frac{g_e^2 z^2}{2} + \frac{\dot{g}_e}{2g_e}z\left(1 - \frac{g_e^2 z^2}{6}\right)\right]$$
$$+ P_f\left(-\left[g_e^2 + \left(\frac{\dot{g}_e}{2g_e}\right)^2\right]z\left(1 - \frac{g_e^2 z^2}{6}\right)\right)$$
$$- n_e\left[1 - \frac{g_e^2 z^2}{2} - \frac{\dot{g}_e}{2g_e}z\left(1 - \frac{g_e^2 z^2}{6}\right)\right] - \frac{P_b P_f z\left(1 - \frac{g_e^2 z^2}{6}\right)}{n_e}.\qquad(5.29)$$

Equations (5.25)–(5.29) represent the thickness-dependent $ABCD$ matrix elements of the curved end faces model of the crystalline lens. The typical values of the parameters are the following: g_e (1425.06) is the gradient parameter, $\dot{g}_e$ (-2×10^3) is its slope at z, n_e (1.386) is the edge index, n_1 (1.336) is the refractive index of the media in the object space, n_1' (1.336) is the refractive

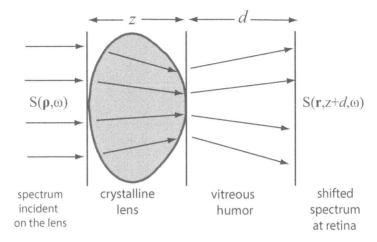

FIGURE 5.5
Illustration of notations used for beam propagation in the human eye. From Ref. [18].

index of the media in the image space, P_f is the power of the front surface of the lens $((n_c - n_1)/R)$, P_b is the power of the back surface of the lens $((n'_1 - n_c)/R)$, n_c (1.406) is the central refractive index and R (0.0004) is the surface radius of curvature.

Since the curved-end model incorporates the edge-refractive index $n_1 = 1.336$ of the vitreous humor the $ABCD$ matrix of the system including the crystalline lens and the vitreous humor followed by the crystalline lens can be represented by a matrix

$$\begin{bmatrix} A(z+d) & B(z+d) \\ C(z+d) & D(z+d) \end{bmatrix} = \begin{bmatrix} 1 & d \\ 0 & 1 \end{bmatrix} \begin{bmatrix} A_c(z) & B_c(z) \\ C_c(z) & D_c(z) \end{bmatrix}, \qquad (5.30)$$

where d is the propagation distance in the vitreous humor from the lens to the retina (see Fig. 5.5).

Human visual recognition of a distant object is made by perceiving and analyzing light scattered from it. The intensity and the spectral (color) signature of the object plays the crucial part in its recognition out of the rest of the environment. Polarization might also play a part but this issue is little studied. The question then arises whether on the way from the object to the human brain the scattered light experiences certain intensity, polarization and spectral changes that restrict us from seeing the objects the way they are, and subjects us to an analysis of somewhat different input information about the object.

In order to illustrate the dependence of the spectral changes of light in the eye numerically we will employ the isotropic Gaussian Schell-model beams with Gaussian intensity and spectral profiles. The cross-spectral density function of such a beam in the source plane $z = 0$ has the form (see also Chapter 3)

$$W(\boldsymbol{\rho}_1', \boldsymbol{\rho}_2'; \lambda) = I_0(\lambda) \exp\left[-\frac{\rho_1'^2 + \rho_2'^2}{4\sigma^2}\right] \exp\left[-\frac{(\boldsymbol{\rho}_1' - \boldsymbol{\rho}_2')^2}{2\delta^2}\right], \tag{5.31}$$

where without loss of generality we assume that the initial spectral composition consists of a single Gaussian spectral line:

$$I_0(\lambda) = \exp\left[-\frac{(\lambda - \lambda_0)^2}{2\Lambda^2}\right], \tag{5.32}$$

with a peak value of one, being centered at wavelength λ_0 and having the root-mean-square width Λ. By adopting the propagation law for the random beam through the $ABCD$ system we find that the spectral density of a Gaussian Schell-model beam takes the form

$$S(\rho, z + d; \lambda) = \frac{I_0(\lambda)}{\widehat{\Delta}(z + d)} \exp\left[-\frac{\rho^2}{2\sigma^2 \widehat{\Delta}(z + d)}\right], \tag{5.33}$$

where

$$\widehat{\Delta}(z + d) = A^2(z + d) + \frac{B^2(z + d)}{4\sigma^4 k^2}\left(1 + \frac{4\sigma^2}{\delta^2}\right). \tag{5.34}$$

We will be interested in evaluation of the normalized spectral density of the beam at distance $d \geq 0$ from the crystalline lens and at any transverse location (x, y), given by the expression

$$S_N(\rho, z + d; \lambda) = \frac{S(\rho, z + d; \lambda)}{\int\limits_0^\infty S(\rho, z + d; \lambda)\, d\lambda}. \tag{5.35}$$

On substituting from Eqs. (5.31) and (5.32) into Eq. (5.33) one can trace the evolution of the spectral density through the vitreous humor of the human eye. Further, the shifted central frequently of the beam can be found from the expression

$$\lambda_1(\rho, z + d) = \frac{\int\limits_0^\infty \lambda S(\rho, z + d; \lambda)\, d\lambda}{\int\limits_0^\infty S(\rho, z + d; \lambda)\, d\lambda}. \tag{5.36}$$

Figure 5.6 (top) shows typical changes in the spectral composition of the incident quasi-monochromatic light beam on propagation through the vitreous humor, right after the interaction with the crystalline lens. In order to compare the spectral changes of the light beam passing through the lens and

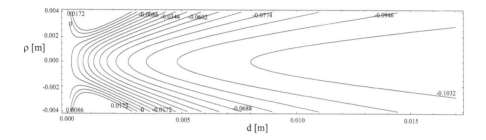

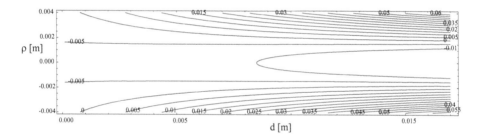

FIGURE 5.6

Contour plots of the normalized spectral shift $\varrho = \frac{\lambda_1 - \lambda_0}{\lambda_0}$ as a function of d (horizontal axis) and ρ (vertical axis); (top) crystalline lens; (bottom) free space. Parameters of the source are $\lambda_0 = 0.633 \ \mu\text{m}$, $\Lambda = 0.3\lambda_0$, $\sigma = 1$ cm, $\delta = 1$ mm.

the vitreous humor with those that would occur on propagation in vacuum at the same distance, we can use the following $ABCD$ matrix:

$$\begin{bmatrix} A(z+d) & B(z+d) \\ C(z+d) & D(z+d) \end{bmatrix} = \begin{bmatrix} 1 & z+d \\ 0 & 1 \end{bmatrix}. \tag{5.37}$$

The spectral changes due to matrix (5.37) are shown in Fig. 5.6 (bottom). Thus, compared to free-space propagation, a random light beam passing through the crystalline lens of the eye, followed by vitreous humor, experiences a considerable blue shift. We note that for free-space propagation there exists a slight shift as well, due to correlations in the beam, but is negligible because of a relatively short propagation distance. Such an optical adjustment of the eye might lead to a significantly different spectral image formed in the human eye and sent to the brain, compared to those obtained with man-made light detectors.

In order to examine polarization changes due to the crystalline lens we must consider the evolution of the cross-spectral density matrix

$$\mathbf{W}(\boldsymbol{\rho}'_1, \boldsymbol{\rho}'_2; \omega) = [W_{\alpha\beta}(\boldsymbol{\rho}'_1, \boldsymbol{\rho}'_2; \omega)], \quad (\alpha, \beta = x, y), \tag{5.38}$$

where, as before, $\boldsymbol{\rho}'_1$ and $\boldsymbol{\rho}'_2$ are the two-dimensional position vectors of points in the plane coinciding with the front plane of the crystalline lens. Further, propagation of each component of the matrix (5.38) through the paraxial optical $ABCD$ system (5.30) is found from the propagation formula [18]:

$$W_{\alpha\beta}(\boldsymbol{\rho}_1, \boldsymbol{\rho}_2, z+d; \omega) = \left(\frac{k}{2\pi(z+d)}\right)^2 \int \int W_{\alpha\beta}(\boldsymbol{\rho}'_1, \boldsymbol{\rho}'_2; \omega)$$
$$\times \exp\left\{ \frac{ik}{2B(z+d)}[A(z+d)(\rho_1'^2 - \rho_2'^2) - 2(\boldsymbol{\rho}_1 \cdot \boldsymbol{\rho}'_1 - \boldsymbol{\rho}_2 \cdot \boldsymbol{\rho}'_2) \right. \tag{5.39}$$
$$+ D(z+d)(\rho_1^2 - \rho_2^2)] \Big\} d\boldsymbol{\rho}'_1 d\boldsymbol{\rho}'_2,$$

where $A(z+d)$, $B(z+d)$, $D(z+d)$ are the elements of the ray transfer 2×2 $ABCD$ matrix (5.30) and $\boldsymbol{\rho} = (x, y)$ is a two-dimensional vector in the plane $z+d > 0$. If as a model for the incident field we take the electromagnetic Gaussian Schell-model beam (4.172) then, on substituting into Eq. (5.39) and performing integrations over the source plane, the elements of the cross-spectral density matrix of such a beam in the plane $z+d > 0$ can be found to be

$$W_{\alpha\beta}(\boldsymbol{\rho}_1, \boldsymbol{\rho}_2, z+d; \omega) = \frac{A_\alpha A_\beta B_{\alpha\beta}}{\widehat{\Delta}_{\alpha\beta}(z+d)} \exp\left[-\frac{\rho_1^2 + \rho_2^2}{4\sigma^2 \widehat{\Delta}_{\alpha\beta}(z+d)} \right]$$
$$\times exp\left[-\frac{(\boldsymbol{\rho}_1 - \boldsymbol{\rho}_2)^2}{2\delta_{\alpha\beta}^2 \widehat{\Delta}_{\alpha\beta}(z+d)} \right] \exp\left[-\frac{ik(\rho_1^2 - \rho_2^2)}{2\widehat{R}_{\alpha\beta}(z+d)} \right], \tag{5.40}$$

where

$$\widehat{\Delta}_{\alpha\beta}(z+d) = A^2(z+d) + \frac{B^2(z+d)}{4\sigma^4 k^2}\left(1 + \frac{4\sigma^2}{\delta_{\alpha\beta}^2(z+d)}\right), \tag{5.41}$$

$$\widehat{R}_{\alpha\beta}(z+d) = \frac{B(z+d)\widehat{\Delta}_{\alpha\beta}(z+d)}{D(z+d)\widehat{\Delta}_{\alpha\beta}(z+d) - A(z+d)}. \quad (5.42)$$

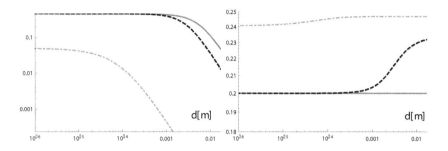

FIGURE 5.7
(left) The spectral density $S(\rho, z+d; \omega)$; (right) the spectral degree of polarization $\wp(\rho, z+d; \omega)$ of the beam vs. propagation distance z [m] after the lens for: coherent beam, $\delta_{xx} = \delta_{yy} = \delta_{xy} = \delta_{yx} \to \infty$ (solid curve); partially coherent beam $\delta_{xx} = \delta_{yy} = 1.125 \times 10^{-4}$ m (dotted curve), $\delta_{xy} = \delta_{yx} = 1.25 \times 10^{-4}$ m, nearly incoherent beam $\delta_{xx} = \delta_{yy} = 1.125 \times 10^{-5}$ m, $\delta_{xy} = \delta_{yx} = 1.25 \times 10^{-5}$ m (dashed curve). Other beam parameters are $\lambda = 0.55$ μm, $A_x = A_y = 1$, $B_{xy} = 0.2$, $\sigma = 0.1$ mm. From Ref. [18].

In Fig. 5.7 the on-axis spectral density and the spectral degree of polarization of the electromagnetic Gaussian Schell-model beam are shown as functions of propagation distance d after the crystalline lens, for several values of source correlation parameters. Evidently, both quantities at the retina are intimately related to the correlation properties of the beam incident onto the lens.

In summary, after we have analyzed how stochastic beams propagate through the human crystalline eye lens it became apparent that the spectral composition, the spatial intensity distribution and polarization of light are drastically modified. Hence, the images formed by the eye are very sensitive to correlation properties of illumination.

5.3   Random beams in negative phase materials

The interaction of electromagnetic radiation with negative phase materials (NPM) as opposed to positive phase materials (PPM) was first considered in [19]. The history and the recent advancements relating to this subject can be found in an excellent review [20]. The PPM and NPM materials are defined as those in which $\mathbf{S}_p \mathbf{k} > 0$ and $\mathbf{S}_p \mathbf{k} < 0$, respectively, $\mathbf{S}_p$ being the pointing vector and $\mathbf{k}$ being the wave vector of the electromagnetic field [19]. The NPM materials were successfully experimentally realized at microwave frequencies [21] and the attempts are currently being made to extend the range to optical frequencies [22]. The problem of anisotropy of such materials is also shown to be resolved (cf. [23]). These achievements carry a significant potential for wave propagation [24]. Until recently, most of the theoretical work on wave interaction with NPM involved plane wave or a Gaussian beam [25] in a slab of NPM except for Refs. [26], [27] where propagation of various properties of stochastic beams was explored.

The second-order properties of stochastic beams were found to envolve in the NPM the same way they do in PPM. However, the substantial qualitative changes occur at the interfaces between the PPM and NPM. We will therefore consider the beam interaction with the stratified media composed by several alternating layers.

In order to derive the $ABCD$ matrices for several illustrative combinations of PPM and NPM we recall, that for passage through a uniform medium of thickness L, and for interaction with a boundary between media with indexes of refraction n_1 (from which the wave is incident) and n_2, the $ABCD$ matrices have forms:

$$\begin{bmatrix} A & B \\ C & D \end{bmatrix}_a = \begin{bmatrix} 1 & L \\ 0 & 1 \end{bmatrix}, \tag{5.43}$$

$$\begin{bmatrix} A & B \\ C & D \end{bmatrix}_b = \begin{bmatrix} 1 & 0 \\ 0 & n_1/n_2 \end{bmatrix}. \tag{5.44}$$

Using matrices (5.43) and (5.44) the matrices that characterize any combinations of PPM and NPM can be readily evaluated with the help of left matrix multiplication. For instance, we obtain for:
1) a single PPM or NPM of length L_1:

$$\begin{bmatrix} A & B \\ C & D \end{bmatrix}_1 = \begin{bmatrix} 1 & L_1 \\ 0 & 1 \end{bmatrix}; \tag{5.45}$$

2) a system of PPM and NPM of lengths L_1 and L_2, respectively:

$$\begin{bmatrix} A & B \\ C & D \end{bmatrix}_2 = \begin{bmatrix} 1 & L_1 + L_2\frac{n_p}{n_n} \\ 0 & \frac{n_p}{n_n} \end{bmatrix}; \tag{5.46}$$

3) a system of PPM of length L_1, NPM of length L_2 and PPM of length L_3:

$$\begin{bmatrix} A & B \\ C & D \end{bmatrix}_3 = \begin{bmatrix} 1 & L_1 + L_3 + L_2\frac{n_p}{n_n} \\ 0 & 1 \end{bmatrix}. \tag{5.47}$$

If $L_2 = 2L_1 = 2L_3$ the arrangement describes the *perfect lens* discussed by Pendry [28].

We will now establish the propagation laws for stochastic electromagnetic beams in an arbitrary combination of PPM and NPM. Suppose the beam is generated in the plane $z = 0$ coinciding with the entrance plane of the first layer of PPM/NPM, and propagates close to the positive z direction. Suppose, the half-space $z > 0$ into which the beam passes is filled with single PPM, NPM or their layers, all confined within planes transverse to the direction of propagation of the beam, altogether characterized by an $ABCD$ matrix. Recall that if a beam propagates in a medium with an arbitrary refractive index $n = n_{(r)} + in_{(i)}$ then the elements of the cross-spectral density matrix of the propagating beam obey the formula (4.146). This result can be readily generalized to the case when the beams pass in the $ABCD$ system in which a medium with index of refraction n fills the regions between the elements. Hence, one obtains the expression

$$
W_{\alpha\beta}(x_1, y_1, x_2, y_2, z; \omega) = \frac{|k|^2}{(2\pi B)^2} \exp(2k_{(i)}z)
$$

$$
\times \iiiint \exp\left\{ -\frac{iA}{2B}[k(x_1'^2 + y_1'^2) - k^*(x_2'^2 + y_2'^2)] \right\}
$$

$$
\times \exp\left\{ \frac{i}{B}[k(x_1 x_1' + y_1 y_1') - k^*(x_2 x_2' + y_2 y_2')] \right\} \tag{5.48}
$$

$$
\times \exp\left\{ -\frac{iD}{2B}[k(x_1^2 + y_1^2) - k^*(x_2^2 + y_2^2)] \right\}
$$

$$
\times W_{\alpha\beta}(x_1', y_1', x_2', y_2', 0; \omega) dx_1' dy_1' dx_2' dy_2'.
$$

Here $(\alpha, \beta = x, y)$, $W_{\alpha\beta}(x_1', y_1', x_2', y_2', 0; \omega)$ and $W_{\alpha\beta}(x_1, y_1, x_2, y_2, z; \omega)$ are the cross-spectral density matrices of the beam in the entrance plane to the first layer and at distance z from it, respectively; A, B, C and D are the elements of the $ABCD$ matrix, and $k = k_{(r)} + ik_{(i)}$ is the wave number in the medium such that $k_{(r)} = k_0 n$, k_0 being wave number in free space, and $k_{(i)}$ is the parameter characterizing gain $(k_{(i)} > 0)$ or absorbtion $(k_{(i)} < 0)$.

Suppose that the beam at the entrance to the layered medium has the form of the Gaussian Schell-model in the waist plane:

$$
W_{\alpha\beta}(x_1', y_1', x_2', y_2', 0; \omega) = A_\alpha A_\beta B_{\alpha\beta}
$$

$$
\times \exp\left[-\frac{x_1'^2 + y_1'^2 + x_2'^2 + y_2'^2}{4\sigma^2} \right] \exp\left[-\frac{(x_1' - x_2')^2 + (y_1' - y_2')^2}{2\sigma_{\alpha\beta}^2} \right].
$$

$$
\tag{5.49}
$$

Then, on substituting from Eq. (5.49) into Eq. (5.48) and performing the integrations one obtains the expression [26]

$$W_{\alpha\beta}(x_1', y_1', x_2', y_2', z; \omega) = \frac{A_\alpha A_\beta B_{\alpha\beta} |k|^2 \delta_{\alpha\beta}^4 \exp(2kzi)}{B^2(4|g_{\alpha\beta}|^2 - 1)}$$

$$\times \exp\left[-\frac{iD}{2B}[k(x_1^2 + y_1^2) - k^*(x_2^2 + y_2^2)]\right] \exp\left[-\frac{\delta_{\alpha\beta}^2 k^2 g_{\alpha\beta}^*(x_1^2 + y_1^2)}{B^2(4|g_{\alpha\beta}|^2 - 1)}\right]$$

$$\times \exp\left[\frac{\delta_{\alpha\beta}^2 |k|^2 (x_1 x_2 + y_1 y_2)}{B^2(4|g_{\alpha\beta}|^2 - 1)}\right] \exp\left[-\frac{\delta_{\alpha\beta}^2 k^{*2} g_{\alpha\beta}(x_2^2 + y_2^2)}{B^2(4|g_{\alpha\beta}|^2 - 1)}\right],$$

$$(5.50)$$

where

$$g_{\alpha\beta} = \frac{1}{2} + \frac{\delta_{\alpha\beta}^2}{4\sigma^2} + i\frac{Ak\delta_{\alpha\beta}^2}{2B}. \tag{5.51}$$

From expression (5.50) all second-order statistics of the beam in the layered medium can be determined.

In Figs. 5.8, 5.9 and 5.10 the spectral density, the spectral degree of coherence and the spectral degree of polarization, respectively, of a typical electromagnetic Gaussian-Schell-model beam propagating through several combinations of PPM and NPM are shown: (a) PPM; (b) NPM; (c) PPM if $0 \leq z < 1$ m and NPM if $1 \text{ m} \leq z \leq 2$ m; (d) ("perfect" lens setup) PPM if $0 \leq z < 0.5$ m, NPM if $0.5 \text{ m} \leq z < 1.5$ m, PPM if $1.5 \text{ m} \leq z \leq 2$ m. The following parameters are used for the numerical curves: $n_p/n_n = -1$, $\sigma_x = \sigma_y = \sigma = 3 \times 10$ mm, $A_x = A_y = 1$, $\delta_{xx} = \delta_{yy} = 10^{-5}$ m, $\delta_{xy} = \delta_{yx} = 1.2 \times 10^{-5}$ m, $B_{xy} = B_{yx} = 0.1$, $k_{(r)} = 5 \times 10^6$ m^{-1}, $k_{(i)} = 0$, $\rho_d = |\rho_1 - \rho_2| = 10^{-5}$ m.

These examples illustrate that the second-order statistical properties of the beam propagating through single layers of PPM and NPM exhibit the same evolution [see curves (a) and (b)]. However, on the boundary between PPM and NPM they can change direction. Hence the boundary acts as a focusing device, similar to a lens. The other turning point occurring for all the considered statistical properties, in curves (d) at $z = 1$ m, is also of interest: at this propagation distance the beam becomes the same as in the source plane $z = 0$. Thus, the same evolution scenario takes place in the interval $z \in [1 \text{ m}, 2 \text{ m}]$ as in $z \in [0 \text{ m}, 1 \text{ m}]$.

We will now turn to demonstration of the spectral switches that may occur with random beams in PPM/NPM layers [27]. The following values of the parameters are assumed: $n_p = 1$, $n_n = n_{(r)} + in_{(i)} = -1$, $\sigma = 3 \times 10^{-3}$ m, $\delta = 10^{-5}$ m, $\omega_0 = 10^{15}$ rad/s, $\bar{\omega} = 0.1\omega_0$. To illustrate the typical switches in the spectrum we first show, in Fig. 5.11, the spectra of the propagating beam, on optical axis, at five transverse planes: $z = 0$ m, 0.5 m, 1 m, 1.5 m and 2 m from the source corresponding to set up (5.47). The alternation of blue-shift and red-shift of the original spectrum takes place after the beam passes through the interface between the PPM and the NPM layers. Note that here lossless materials were considered ($n_{(i)} \equiv 0$).

Since in practice the materials that induce negative refraction of light are

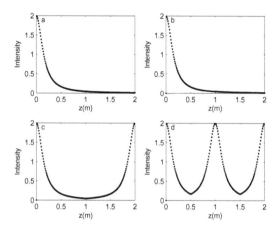

FIGURE 5.8

The spectral density (intensity) of the beam (on-axis) as a function of propagation distance z with in the layers. From Ref. [26].

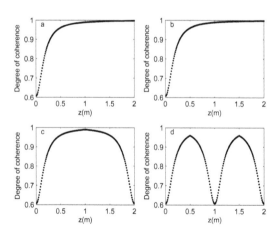

FIGURE 5.9

The degree of coherence of the beam at two points with separation distance $\rho_d = 10^{-5}$ m as a function of propagation distance z from the source. From Ref. [26].

218

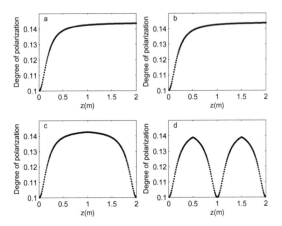

FIGURE 5.10
The degree of polarization of the beam (on-axis) as a function of propagation distance z from the source. From Ref. [26].

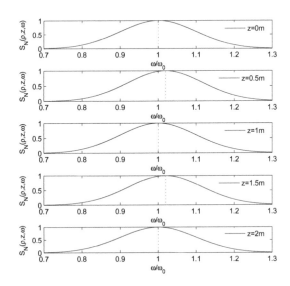

FIGURE 5.11
Spectral changes in Gaussian Schell-model beams on propagation in lossless layers of PPM and NPM. From Ref. [27].

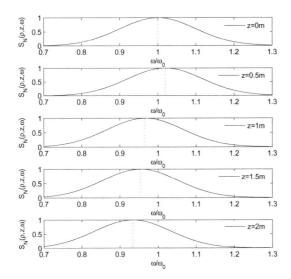

FIGURE 5.12
Spectral changes in Gaussian Schell-model beams on propagation in layers of
PPM and NPM with absorption. From Ref [27].

highly absorptive, it is necessary to include such effects on propagating beams.
Indeed, it can be found that the absorption in the NPM material induces
stronger red shifts compared to a lossless NPM material with the same $n_{(r)}$
[27].

Figure 5.12 shows the on-axis spectra at the same transverse planes: $z = 0$
m, 0.5 m, 1 m, 1.5 m and 2 m from the source, for the case when $n_{(i)} = -10^{-6}$.
In the presence of absorption only a single spectral switch (from blue-shift to
red-shift) is present, while the other one is suppressed. This result is in striking
difference with that of Fig. 5.11 where the beam passes through the lossless
NPM layer and its spectrum's direction switches twice, having the ability to
reconstruct.

5.4 Imaging by twisted random beams

The general subject of image formation by means of partially coherent light
belongs to famous texts [29]–[32] and will not be discussed here. We rather
point to a recent study that explores the effect of the twist phase on the image
resolution.

220

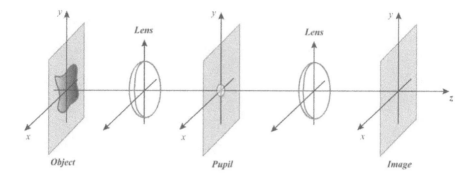

Object   Pupil   Image

FIGURE 5.13
A two-lens imaging system.

Let us confine our attention to the image formation by an isoplanatic, axially symmetric optical systems with low angular aperture, with the illumination being a stochastic, scalar light beam. In particular, consider a telecentric system, consisting of two thin Gaussian lenses with focal lengths F_G and a pupil (see Fig. 5.13), where the object, the pupil and the image are located in the focal planes of the lenses.

Suppose the object is characterized by the amplitude transparency $o(\boldsymbol{\rho};\omega)$, $\boldsymbol{\rho}=(x,y)$ being the two-dimensional position vector in the object plane. It is illuminated by a beam with cross-spectral density function $W^{(ob)}(\boldsymbol{\rho}_1,\boldsymbol{\rho}_2;\omega)$. Then the cross-spectral density function $W^{(im)}(\mathbf{r}_1,\mathbf{r}_2,\omega)$ of the beam in the image plane is given by the well-known relation [30]:

$$W^{(im)}(\mathbf{r}_1,\mathbf{r}_2,\omega)=\int_{-\infty}^{\infty} o^*(\boldsymbol{\rho}_1;\omega)o(\boldsymbol{\rho}_2;\omega)W^{(ob)}(\boldsymbol{\rho}_1,\boldsymbol{\rho}_2;\omega)$$
$$\times h_a^*(\mathbf{r}_1-\boldsymbol{\rho}_1;\omega)h_a(\mathbf{r}_2-\boldsymbol{\rho}_2;\omega)d\boldsymbol{\rho}_1 d\boldsymbol{\rho}_2, \tag{5.52}$$

where $h_a(\mathbf{r}-\boldsymbol{\rho};\omega)$ is the *amplitude spread function* of the optical system, defined as a two-dimensional Fourier transform of the pupil. In particular, if the pupil is a circular aperture with radius R the amplitude spread function reduces to the expression

$$h_a(\mathbf{r}-\boldsymbol{\rho};\omega)=\frac{\pi R^2}{\lambda F_G}\left[2\frac{J_1(2\pi R|\mathbf{r}-\boldsymbol{\rho}|/\lambda F_G)}{2\pi R|\mathbf{r}-\boldsymbol{\rho}|/\lambda F_G}\right], \tag{5.53}$$

where J_1 is the Bessel function of the first kind and the first order.

We will now demonstrate the possibility of overcoming the Rayleigh diffraction limit (the diffraction limit of resolving two point objects in vacuum) by means of twisted, scalar Gaussian Schell-model beams [33], [34], i.e., the beams with the cross-spectral density matrix of the form

$$W^{(ob)}(\boldsymbol{\rho}_1, \boldsymbol{\rho}_2; \omega) = \exp\left[-\frac{\rho_1^2 + \rho_2^2}{4\sigma^2}\right] \exp\left[-\frac{(\boldsymbol{\rho}_1 - \boldsymbol{\rho}_2)^2}{2\delta^2}\right]$$
$$\times \left[\frac{ik\tau_\delta}{2}(\boldsymbol{\rho}_1 - \boldsymbol{\rho}_2)^T \mathbf{J}(\boldsymbol{\rho}_1 + \boldsymbol{\rho}_2)\right], \tag{5.54}$$

where τ_δ is a scalar real-valued twist factor, limited by inequalities $0 \le \tau_\delta^2 \le k^{-2}\delta^{-4}$, and

$$\mathbf{J} = \begin{bmatrix} 0 & 1 \\ -1 & 0 \end{bmatrix}. \tag{5.55}$$

The object will be modeled as an opaque screen with two pinholes separated by distance d and located symmetrically about the y-axis, at distance y_0 from the x-axis. The amplitude transmittance of such an object can be well approximated by two Dirac delta-functions:

$$o(\boldsymbol{\rho}; \omega) = \delta(x - d/2, y - y_0) + \delta(x + d/2, y - y_0). \tag{5.56}$$

On substituting from Eqs. (5.53), (5.54), and (5.56) into Eq. (5.52), we obtain, at $\mathbf{r}_1 = \mathbf{r}_2 = \mathbf{r}$ the image of the object represented by the spectral density function:

$$S^{(im)}(\mathbf{r}; \omega) = W^{(im)}(\mathbf{r}, \mathbf{r}; \omega)$$
$$= \left(\frac{\pi R^2}{\lambda f}\right)^2 \exp\left(-\frac{d^2}{8\sigma^2} - \frac{y_0^2}{2\sigma^2}\right) \tag{5.57}$$
$$\times \left[S_-^2 + S_+^2 + 2\exp\left(-\frac{d^2}{2\delta^2}\right)\cos(\phi_t)S_-S_+\right],$$

where

$$S_\pm = 2\frac{J_1\left(2\pi R\sqrt{(x \pm d/2)^2 + (y - y_0)^2}/\lambda f\right)}{2\pi R\sqrt{(x \pm d/2)^2 + (y - y_0)^2}/\lambda f}, \tag{5.58}$$

and

$$\phi_t = k\tau_\delta dy_0. \tag{5.59}$$

Here the term $\exp\left(-\frac{d^2}{2\delta^2}\right)\cos(\phi_t)$ is the real part of the spectral degree of the coherence of light at two pinholes.

We will first discuss the classic case when the pinholes are imaged with an incoherent illumination, i.e., when $\delta \to 0$ implying that the last (interference) term in the square brackets in Eq. (5.57) vanishes. Under such circumstances the quality of the image may be evaluated on the basis of the Rayleigh resolution criterion [31]: the first zero of the Airy pattern $J_1(x)/x$, produced by one of the pinholes should coincide with the maximum image intensity, produced by the other pinhole. Then the minimum resolvable separation between the images of the pinholes becomes

$$d_R = 0.61\lambda f/R. \tag{5.60}$$

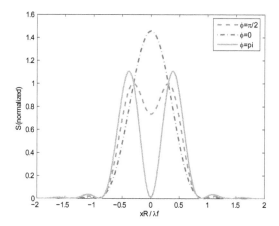

FIGURE 5.14

Image of two illuminated pinholes, plotted in normalized coordinates, for different phases ϕ_t. The other parameters are: $\delta = 0.1$ mm, $\lambda = 0.59$ μm, $R = 1$ cm, $f = 0.5$ m, $\sigma = 1$ mm, $d = 0.61\lambda f/R$, $y_0 = 5$ mm. From Ref. [34].

If illumination is random, without twist phase, i.e., when $0 < \delta \ll \sigma$, $\phi_t = 0$, the notrivial interference term in Eq. (5.57) implies the increase in the minimum resolvable separation of the pinhole's images. Our main task here is to examine how the non-zero values of the twist phase ϕ_t may affect the image quality. Restricting the attention to cases: $\phi_t = \pi/2$ and $\phi_t = \pi$. The former case is equivalent to the image formation with incoherent light, since the interference term vanishes (just like when $\delta \to 0$). In the later case the pinholes are illuminated by light beams that is out of phase, resulting in the substantial reduction of intensity in the central part of the image (see Fig. 5.14). Moreover, since the separation between the two pinholes is $d = 0.61\lambda f/R$, exactly the minimum resolvable distance for two incoherently illuminated pinholes, the out of phase condition is crucial when the pinhole separation is below the Rayleigh diffraction limit.

As is seen from Fig. 5.14 the two intersection points of three curves (closest to the midpoint) are the true positions of the pinholes ($\pm d/2$), but for the curve of interest ($\phi_t = \pi$) the two peak positions are located farther from the center. In order to find the relation between the position of the pinhole and the measured peak position, we calculate the derivative of $S(x, y; \omega)$ at $y = y_0$ with respect to x, and set it to zero. Since

$$\frac{d}{dx}J_1(x) = -J_2(x) + \frac{1}{x}J_1(x), \qquad (5.61)$$

where J_2 is the Bessel function of order 2, we obtain the necessary condition for the extremum:

$$S_{1,-}S_{2,-} + S_{1,+}S_{2,+} + \exp\left(-\frac{d^2}{2\delta^2}\right)\cos\phi_t(S_{1,-}S_{2,-} + S_{1,+}S_{2,+}) = 0, \quad (5.62)$$

where

$$S_{i,\pm} = \frac{J_i(2\pi R|x \pm d/2|/\lambda f)}{2\pi R|x \pm d/2|/\lambda f}, \quad (i = 1, 2). \tag{5.63}$$

These formulas imply that the measured peak position of the image is the first positive or negative value of x that satisfies (5.62). If the separation is greater than the Rayleigh limit and either $\phi_t = \pi/2$ or $\delta \to 0$ the interference term in (5.62) vanishes just like in the incoherent illumination case, which gives the true positions of the two pinholes. When $\phi_t = \pi$ Eq. (5.62) reduces to the form

$$\frac{S_{1,-}S_{2,-} + S_{1,+}S_{2,+}}{S_{1,-}S_{2,-} + S_{1,+}S_{2,+}} = \exp\left(-\frac{d^2}{2\delta^2}\right), \tag{5.64}$$

and can be solved for x numerically. Due to the constraint on the twist phase, $\tau_\delta \le k_{-1}\delta^{-2}$, the corresponding upper bound for ϕ_t becomes

$$\phi_t \le \frac{dy_0}{\delta^2}, \tag{5.65}$$

implying that δ must be small enough to guarantee that ϕ_t is as large as π, precluding from using fairly coherent beams. On the other hand, δ should not be too small, which would lead to the results for incoherent illumination. Therefore, only a random beam with correlation width δ varying between certain limits is suitable for resolving two pinholes with separation below the Rayleigh diffraction limit. Figure 5.15 presents the normalized spectral density profiles of the image for different values of δ, with $\delta = 10^{-4}$ being the optimal choice.

In order to resolve two points with sub-Rayleigh separation, the phase difference between the two pinholes does not need to be exactly π, but can be in its close vicinity. The greater the difference between the Rayleigh limit and the actual separation, the closer the phase must be to π. When the pinhole separation is far below the Rayleigh limit, say, one order of magnitude smaller, only the phase difference of π is capable of resolving them. Therefore, the phase difference is a ruler for determining whether two sub-Rayleigh pinholes can be resolved.

In the best case scenario, when the twist factor takes its maximum value, say $\tau'_\delta = k^{-1}\delta^{-2}$, the corresponding twist phase becomes $\phi'_t = dy_0\delta^{-2}$. Figure 5.16 demonstrates that the two pinholes, undistinguishable for an incoherent source, become resolved if imaged by a partially coherent twisted beam. The chosen phase difference between the two pinholes in the situation considered here is $\phi_t = 3.1417 \approx \pi$. If neither the value of δ nor the separation d cannot be adjusted, then one can change the off-axis locations of the pinholes, y_0 such that the condition $y_0 = \pi\delta^2/d$ is met.

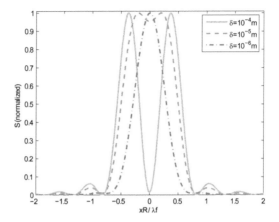

FIGURE 5.15
Image of two illuminated pinholes, plotted in normalized coordinates, for different values of coherence width δ, with $\phi_t = \pi$, $d = 0.24\lambda f/R$. The rest of the parameters are as in Fig. 5.14. From Ref. [34].

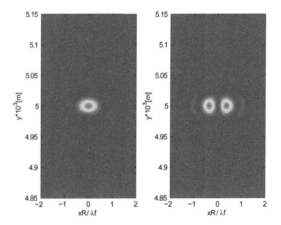

FIGURE 5.16
Contour image of two pinholes with completely incoherent illuminating source (left) and partially coherent illuminating source $\delta = 0.1$ mm (right) with $d = 0.213\lambda f\, R$. The rest of the parameters are as in Fig. 5.14. From Ref. [34].

It is reasonable to assume that the maximum value of y_0 must be no larger than the radius of the pupil R, restrained by the sensitivity of the detector. Then the minimum resolved separation between the two pinholes is determined by the formula $d_{min} = \pi\delta^2/R$. If the twist phase factor τ_δ does not reach its maximum value $k^{-1}\delta^{-2}$ then the minimum separation becomes

$$d_{min} = \frac{\pi}{k\tau_\delta R}. \qquad (5.66)$$

For the parameters used in Fig. 5.16 $d_{min} \approx 0.1746 d_R$, i.e., about one order of magnitude smaller than the Rayleigh diffraction limit.

Thus, the presence of the twist phase in random illumination can substantially improve the resolving quality of imaging. Imaging of random electromagnetic beams has been recently considered in Ref. [35] where the analysis of polarization changes of the beam passing in the telescopic system has been investigated.

5.5   Tensor method for random beam interaction with astigmatic $ABCD$ systems

A convenient tensor method has been developed specifically for passage of scalar and electromagnetic Gaussian Schell-model beams through the $ABCD$ optical systems, which always leads to results in the closed form [36]–[40]. In addition, according to this method more general astigmatic systems can also be treated for which instead of the 2×2 $ABCD$ matrices one employs the 4×4 $ABCD$ matrices formed with four 2×2 submatrices relating to each of the A, B, C and D elements.

The anisotropic electromagnetic Gaussian Schell-model beam (without the twist phase) incident onto the $ABCD$ system can be expressed as

$$W_{\alpha\beta}(\widetilde{\boldsymbol{\rho}'}) = A_\alpha A_\beta B_{\alpha\beta} \exp\left[-\frac{ik}{2}\widetilde{\boldsymbol{\rho}'}^T \widetilde{\mathbf{M}}_{0\alpha\beta}^{-1}\widetilde{\boldsymbol{\rho}'}\right], \quad (\alpha, \beta = x, y), \qquad (5.67)$$

where $\widetilde{\boldsymbol{\rho}'}^T = (\boldsymbol{\rho}_1'^T \ \boldsymbol{\rho}_2'^T)$ is a four-dimensional vector, and

$$\widetilde{\mathbf{M}}_{0\alpha\beta}^{-1} = \begin{bmatrix} \frac{1}{ik}\left(\frac{1}{2\sigma_\alpha^2} + \frac{1}{\delta_{\alpha\beta}^2}\right)\mathbf{I} & \frac{i}{k\delta_{\alpha\beta}^2}\mathbf{I} \\ \frac{i}{k\delta_{\alpha\beta}^2}\mathbf{I} & \frac{1}{ik}\left(\frac{1}{2\sigma_\beta^2} + \frac{1}{\delta_{\alpha\beta}^2}\right)\mathbf{I} \end{bmatrix}, \qquad (5.68)$$

is the 4×4 tensor, $\mathbf{I}$ being the 2×2 identity matrix. The propagation of the elements of the cross-spectral density matrix (5.67) through a general astigmatic $ABCD$ optical system can be expressed with the tensor form of the generalized Collins formula [40]:

$$W_{\alpha\beta}(\tilde{\boldsymbol{\rho}};\omega) = \frac{k^2}{4\pi^2\sqrt{\det(\tilde{\mathbf{B}})}} \int_0^\infty \int_0^\infty \int_0^\infty \int_0^\infty W_{\alpha\beta}(\tilde{\boldsymbol{\rho}}';\omega)$$
$$\times \exp\left[-\frac{ik}{2}(\tilde{\boldsymbol{\rho}}^T\tilde{\mathbf{B}}^{-1}\tilde{\mathbf{A}}\tilde{\boldsymbol{\rho}} - 2\tilde{\boldsymbol{\rho}}^T\tilde{\mathbf{B}}^{-1}\tilde{\boldsymbol{\rho}}'^T + \tilde{\boldsymbol{\rho}}'^T\tilde{\mathbf{D}}\tilde{\mathbf{B}}^{-1}\tilde{\boldsymbol{\rho}}')\right]d\tilde{\boldsymbol{\rho}}'. \tag{5.69}$$

Here, the 4×4 matrix $\tilde{\mathbf{A}}$, $\tilde{\mathbf{B}}$, $\tilde{\mathbf{C}}$ and $\tilde{\mathbf{D}}$ have forms:

$$\tilde{\mathbf{A}} = \begin{bmatrix} \mathbf{A} & 0\mathbf{I} \\ 0\mathbf{I} & \mathbf{A}^* \end{bmatrix}, \qquad \tilde{\mathbf{B}} = \begin{bmatrix} \mathbf{B} & 0\mathbf{I} \\ 0\mathbf{I} & -\mathbf{B}^* \end{bmatrix}, \tag{5.70}$$

$$\tilde{\mathbf{C}} = \begin{bmatrix} \mathbf{C} & 0\mathbf{I} \\ 0\mathbf{I} & -\mathbf{C}^* \end{bmatrix}, \qquad \tilde{\mathbf{D}} = \begin{bmatrix} \mathbf{D} & 0\mathbf{I} \\ 0\mathbf{I} & \mathbf{D}^* \end{bmatrix}. \tag{5.71}$$

While for free-space propagation $\mathbf{A}$, $\mathbf{B}$, $\mathbf{C}$, and $\mathbf{D}$ are real, implying that the conjugation sign "*" is not needed, for a general optical system with loss or gain (e.g., dispersive media, a Gaussian aperture, helical gas lenses, etc.) the elements of these matrices can be complex and conjugation is then required. Also, $\mathbf{A}$, $\mathbf{B}$, $\mathbf{C}$, and $\mathbf{D}$ satisfy the following well-known Luneburg relations that describe the symplecticity of a general astigmatic optical system [41]:

$$(\mathbf{B}^{-1}\mathbf{A})^T = \mathbf{B}^{-1}\mathbf{A}, \quad (-\mathbf{B}^{-1})^T = (\mathbf{C} - \mathbf{D}\mathbf{B}^{-1}\mathbf{A}), \quad (\mathbf{D}\mathbf{B}^{-1})^T = \mathbf{D}\mathbf{B}^{-1}. \tag{5.72}$$

On substituting from Eq. (5.67) into Eq. (5.69) and after some tensor operations we find that

$$W_{\alpha\beta}(\tilde{\boldsymbol{\rho}};\omega) = \frac{k^2 A_\alpha A_\beta B_{\alpha\beta}}{4\pi^2\sqrt{\det(\tilde{\mathbf{B}})}}$$
$$\times \exp\left[-\frac{ik}{2}\tilde{\boldsymbol{\rho}}^T\tilde{\mathbf{D}}\tilde{\mathbf{B}}^{-1}\tilde{\boldsymbol{\rho}} + \frac{ik}{2}\tilde{\boldsymbol{\rho}}^T\tilde{\mathbf{B}}^{-1T}(\mathbf{M}_{0\alpha\beta}^{-1} + \tilde{\mathbf{B}}^{-1}\tilde{\mathbf{A}})^{-1}\tilde{\mathbf{B}}^{-1T}\tilde{\boldsymbol{\rho}}\right]$$
$$\times \int_0^\infty \int_0^\infty \int_0^\infty \int_0^\infty \exp\left[-\frac{ik}{2}\left|(\mathbf{M}_{0\alpha\beta}^{-1} + \tilde{\mathbf{B}}^{-1}\tilde{\mathbf{A}})^{1/2}\tilde{\boldsymbol{\rho}}'\right.\right.$$
$$\left.\left. - (\mathbf{M}_{0\alpha\beta}^{-1} + \tilde{\mathbf{B}}^{-1}\tilde{\mathbf{A}})^{-1/2}\tilde{\mathbf{B}}^{-1}\tilde{\boldsymbol{\rho}}\right|^2\right]d\tilde{\boldsymbol{\rho}}'. \tag{5.73}$$

With the help of the formula

$$\int_{-\infty}^{\infty} \exp(-qx^2)dx = \sqrt{\frac{\pi}{q}}, \tag{5.74}$$

expression (5.73) reduces to

$$W_{\alpha\beta}(\tilde{\boldsymbol{\rho}};\omega) = \frac{k^2 A_\alpha A_\beta B_{\alpha\beta}}{4\pi^2 \sqrt{\det(\tilde{\mathbf{B}})\det(\mathbf{M}_{0\alpha\beta}^{-1} + \tilde{\mathbf{B}}^{-1}\tilde{\mathbf{A}})}}$$

$$\times \exp\left[-\frac{ik}{2}\tilde{\boldsymbol{\rho}}^T\tilde{\mathbf{D}}\tilde{\mathbf{B}}^{-1}\tilde{\boldsymbol{\rho}} + \frac{ik}{2}\tilde{\boldsymbol{\rho}}^T\tilde{\mathbf{B}}^{-1T}(\mathbf{M}_{0\alpha\beta}^{-1} + \tilde{\mathbf{B}}^{-1}\tilde{\mathbf{A}})^{-1}\tilde{\mathbf{B}}^{-1T}\right].$$

$$(5.75)$$

By noting that

$$\det\tilde{\mathbf{B}}\det(\mathbf{M}_{0\alpha\beta}^{-1} + \tilde{\mathbf{B}}^{-1}\tilde{\mathbf{A}}) = \det(\tilde{\mathbf{A}} + \tilde{\mathbf{B}}\mathbf{M}_{0\alpha\beta}^{-1}),\qquad(5.76)$$

$$\tilde{\mathbf{D}}\tilde{\mathbf{B}}^{-1} - \tilde{\mathbf{B}}^{-1T}(\mathbf{M}_{0\alpha\beta}^{-1} + \tilde{\mathbf{B}}^{-1}\tilde{\mathbf{A}})$$

$$= \left[\tilde{\mathbf{D}}\tilde{\mathbf{B}}^{-1}(\tilde{\mathbf{A}} + \tilde{\mathbf{B}}\mathbf{M}_{0\alpha\beta}^{-1}) - \tilde{\mathbf{B}}^{-1}\right](\tilde{\mathbf{A}} + \tilde{\mathbf{B}}\mathbf{M}_{0\alpha\beta}^{-1})^{-1} \qquad (5.77)$$

$$= (\tilde{\mathbf{C}} + \tilde{\mathbf{D}}\mathbf{M}_{0\alpha\beta}^{-1})(\tilde{\mathbf{A}} + \tilde{\mathbf{B}}\mathbf{M}_{0\alpha\beta}^{-1})^{-1}$$

and after setting

$$\mathbf{M}_{1\alpha\beta}^{-1} = (\tilde{\mathbf{C}} + \tilde{\mathbf{D}}\mathbf{M}_{0\alpha\beta}^{-1})(\tilde{\mathbf{A}} + \tilde{\mathbf{B}}\mathbf{M}_{0\alpha\beta}^{-1})^{-1},\qquad(5.78)$$

we express Eq. (5.75) as

$$W_{\alpha\beta}(\tilde{\boldsymbol{\rho}};\omega) = A_\alpha A_\beta B_{\alpha\beta}[\det(\tilde{\mathbf{A}} + \tilde{\mathbf{B}}\mathbf{M}_{0\alpha\beta}^{-1})]^{-1/2}$$

$$\times \exp\left[-\frac{ik}{2}\tilde{\boldsymbol{\rho}}^T\widetilde{\mathbf{M}}_{1\alpha\beta}^{-1}\tilde{\boldsymbol{\rho}}\right].$$

$$(5.79)$$

As we will see from Chapter 7 this tensor method can be easily modified to include a random medium existing between the optical elements without too many complications for the evaluation of the resulting cross-spectral density matrix.

5.6 Electromagnetic random beams in optical resonators

The interaction of monochromatic beams with laser cavities was formulated a long time ago [42], [43]. The transverse modes of the resonator, known as the *Fox-Li modes*, were found to be related to the geometry of the resonator and the characteristics of the light source. The generalization of the Fox-Li theory to random fields was made in Refs. [44]–[46]. The effects of the resonators on the statistical properties of light, e.g., its spectral density and its state of coherence, were also extensively studied [47], [48]. In [49] the theory of resonator modes was developed for electromagnetic random fields (see also

228

[50] for an alternative theory), where the transverse electromagnetic modes have been related to the classic Fox-Li modes [42].

The interaction of the random electromagnetic beams with laser resonators [51]–[54] will be of our primery interest in this section. This topic is of importance because of the fact that the laser cavity is capable of efficiently modulating the polarization properties of incident radiation. Essentially, there are only a few ways of highly controllable modulation of polarimetric properties. The most famous among them is passage of the wave through devices of polarization optics, such as polarizers, rotators, absorbers, compensators, etc. Usually all these methods are associated with certain power losses. On the other hand a laser resonator is capable of either preserving the initial power level of radiation or amplifying it.

A laser resonator is a region in space confined between two reflectors (mirrors or retro-reflectors) that might be vacuum space or filled by a medium. The mirrors are shaped in such a way that once entering the cavity the radiation remains there oscillating between the mirrors until being released. The two mirrors may be having the same or different size and shape. For practical purposes the surfaces of the mirrors are perfectly smooth and typically one of them has negligible absorption. Originally, the beam of light is sent to one of the mirrors in such a way that it then travels back and forth in the cavity. After being bounced between the mirrors for sufficiently many times the radiation acquires certain physical properties, such as intensity and phase curvature profiles, polarization state, depending on the geometry of the resonator and the material properties of the reflectors and medium inside. Figure 5.17 (a) illustrates a typical laser resonator while Fig. 5.17 (b) shows its unfolded version more suitable for analytic study of cavity-wave interaction.

Let us assume that both mirrors, M_1 and M_2 are spherical, with a radius of curvature R_M, mirror spot size η_M, and distance between the centers of the mirrors L_R. The interaction of light with such a resonator is equivalent to its propagation through a sequence of thin spherical lenses with focal lengths $F_G = R_M/2$ combined with filters with a Gaussian amplitude transmission function [47], [48]. The stability parameter $g_R = 1 - L_R/R_M$ leads to the classification of cavities as stable ($0 \le g_R < 1$) or unstable ($g_R \ge 1$) [55].

The 4×4 $ABCD$ matrix for the laser resonator was introduced in Ref. [56] and has the form

$$\begin{bmatrix} \mathbf{A} & \mathbf{B} \\ \mathbf{C} & \mathbf{D} \end{bmatrix} = \begin{bmatrix} \mathbf{A}_1 & \mathbf{B}_1 \\ \mathbf{C}_1 & \mathbf{D}_1 \end{bmatrix}^N, \tag{5.80}$$

with

$$\begin{bmatrix} \mathbf{A}_1 & \mathbf{B}_1 \\ \mathbf{C}_1 & \mathbf{D}_1 \end{bmatrix} = \begin{bmatrix} \mathbf{I} & L_R\mathbf{I} \\ \left(-\frac{2}{R_M} - i\frac{\lambda}{\pi\eta_R^2}\right)\mathbf{I} & \left(1 - \frac{2L_R}{R_M} - i\frac{\lambda L_R}{\pi\eta_R^2}\right)\mathbf{I} \end{bmatrix}, \tag{5.81}$$

where N is the number of passages between the two mirrors. The matrix with elements having subscript "1" describes the single pass between the mirrors.

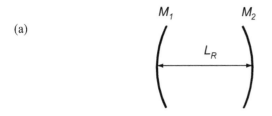

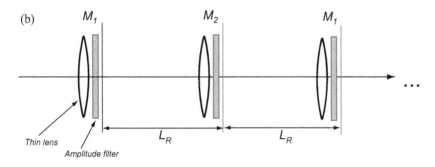

FIGURE 5.17
A typical laser cavity: (a) Actual arrangement; (b) Unfolded version.

The initial beam parameters for the passage in resonators are chosen to have values: $\lambda = 0.590\ \mu$ m, $A_x = A_y = 0.707$, $B_{xy} = B_{yx} = 0.2$ and $\sigma_x = \sigma_y = 1$ mm. Such choice guaranties that the polarization properties are uniform across the source plane with $\wp = 0.2$. Figure 5.18 includes the numerical curves for the on-axis degree of polarization versus the number of passages N, for different values of g_R and the initial correlation coefficients, provided $\epsilon_R = 0.8$ mm. Figues 5.18 (a) and (b) imply that the degree of polarization increases with increasing N, its value approaching different saturation values for different cavities when N is large enough ($N > 30$). The degree of polarization grows showing an oscillatory behavior in stable resonators ($g_R \geq 1$) while growth is monotonic for unstable resonators ($0 \leq g_R < 1$). In addition, in unstable cavities the degree of polarization decreases for higher values of g_R. One finds from Fig. 5.18 (c) and 5.18 (d) that the degree of polarization decreases as the correlation coefficients in the input plane take larger values both in stable and unstable resonators.

The effects of the mirror spot size ϵ_R on the on-axis degree of polarization are illustrated in Fig. 5.19 as N grows, for different values of ϵ_R, for $g_R = 1$. The growth of the degree of polarization becomes more pronounced in response to higher values of ϵ_R for sufficiently large values of N.

In the case when the cavity is lossless ($\epsilon_R \to \infty$) we examine the on-axis degree of polarization versus N for different values of parameter g_R.

230

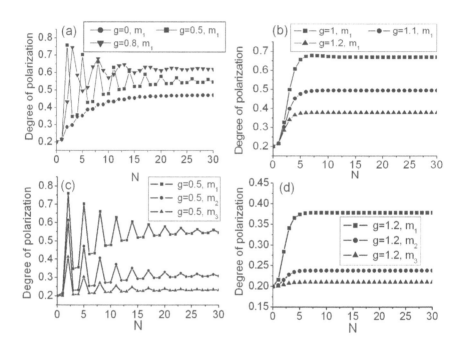

FIGURE 5.18

On-axis degree of polarization versus number of passages N for different values of parameter g_R and the source correlation coefficients. m_1: $\delta_{xx} = \delta_{yy} = 0.1$ mm, $\delta_{xy} = \delta_{yx} = 0.2$ mm; m_2: $\delta_{xx} = \delta_{yy} = 0.25$ mm, $\delta_{xy} = \delta_{yx} = 0.5$ mm; m_3: $\delta_{xx} = \delta_{yy} = 0.5$ mm, $\delta_{xy} = \delta_{yx} = 1$ mm. From Ref. [51].

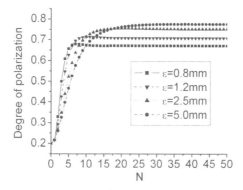

FIGURE 5.19
On-axis degree of polarization versus N for different values of mirror spot size ϵ_R in a Gaussian plane-parallel cavity $(g_R = 1)$ with $\delta_{xx} = \delta_{yy} = 0.1$ mm, $\delta_{xy} = \delta_{yx} = 0.2$ mm. From Ref. [51].

Figure 5.20 refers to stable and unstable lossless cavities. In the former case the degree of polarization exhibits an oscillatory behavior, and its value does not saturate even for large values of N. In unstable cavities, the degree of polarization grows rapidly reaching its saturation value after several passages, this value decreasing as g_R increases. In Fig. 5.21, the degree of polarization versus dimension x is shown for different values of the mirror spot size ϵ_R and the source correlation coefficients in a Gaussian plane-parallel cavity $(g_R = 1)$ with $N = 200$. Figure 5.21 implies that after making N trips in the cavity, an initially uniformly polarized beam acquires different values of the degree of polarization at the transverse plane of the output mirror while its distribution becomes Gaussian with the width being much larger for smaller ϵ_R and for lower values of the source correlation coefficients.

A more complex case of a beam-resonator interaction when the cavity is filled with a gain medium and the beam possesses a twist phase is treated in [54], however, with little changes to the main results of this section.

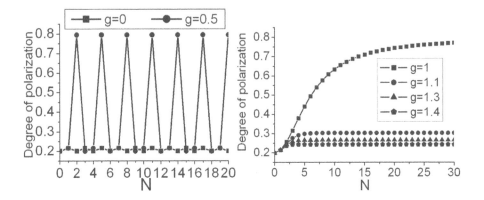

FIGURE 5.20

On-axis degree of polarization versus N for different values of g_R in a lossless cavity ($\epsilon_R \to \infty$) with $\delta_{xx} = \delta_{yy} = 0.1$ mm, $\delta_{xy} = \delta_{yx} = 0.2$ mm. From Ref. [51].

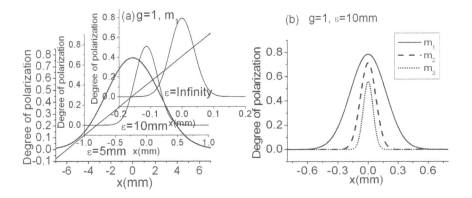

FIGURE 5.21

The degree of polarization versus one transversal dimension x for different values of the mirror spot size and the source correlation coefficients in a Gaussian plane-parallel cavity ($g_R = 1$). m_1, m_2 and m_3 take the same values as in Fig. 5.18. From Ref. [51].

Bibliography

[1] S. Wang and L. Ronchi, "Principles and design of optical arrays," *Progress in Optics*, Vol. XXV, E. Wolf, Ed., North-Holland, 1988, pp. 279–348.

[2] S. Wang, D. Zhao, *Matrix Optics*, Springer-Verlag, 2000.

[3] H. Kogelnik, "Imaging of optical modes-resonators with internal lenses," *Bell Sys. Techn. J.* **44**, 455–494 (1965).

[4] J. A. Arnaud, "Hamiltonian theory of beam mode propagation," in *Progress in Optics*, Vol. XI, E. Wolf, Ed. North-Holland, 1973, pp. 247–304.

[5] A. E. Siegman, *Lasers*, University Science Books, 1986.

[6] C. Gmez-Reino, M. V. Prez, and C. Bao, *Gradient-Index Optics: Fundamentals and Applications*, Springer, 2002.

[7] S. A. Collins, Jr., "Lens-system diffraction integral written in terms of matrix optics," *J. Opt. Soc. Am. A* **60**, 1168–1177 (1970).

[8] B. Lu, and L. Pan, "Propagation of vector Gaussian Schell-model beams through a paraxial optical ABCD system," *Opt. Commun.* **205**, 7–16 (2002).

[9] L. Matthiessen, "Untersuchungen über den Aplanatismus und die Periscopie der Krystalllinsen in den Augen der Fische," *Pfluegers Ar-Chapter Gesamte Physiol. Menschen Tiere* **21**, 287–307 (1880).

[10] L. Mattheissen, "Ueber die Beziehungen, welche zwischen dem Brechungsindex des Kerncentrums der Krystalllinse und den Dimensionen des Auges bestehen," *Pflagers Archiv* **27**, 510–523 (1882).

[11] A. Gullstrand, *Helmholtz's Handbuch der Physiologischen Optik*, 3rd ed., Vol. 1, Appendix II, pp. 301–358, English translation edited by J. P. Southall, Optical Society of America, 1924.

[12] G. Smith, D.A. Atchison, and B. K. Piercionek, "Modeling the power of the aging human eye," *J. Opt. Soc. Am. A* **9**, 2111–2117 (1992).

[13] Y. Huang and D. T. Moore, "Human eye modeling using a single equation of gradient index crystalline lens for relaxed and accomodated states," *Proc. SPIE* **6342**, 63420D (2006).

[14] J. A. Diaz, "ABCD matrix of the human lens gradient-index profile: applicability of the calculation methods," *Applied Optics* **47**, 195–205 (2008).

[15] M. T. Flores-Arias, M. V. Perez, C. Bao, A. Castelo, and C. Gomez-Reino, "Gradient-index human lens as a quadratic phase transformer," *J. Mod. Opt.* **53**, 495–506 (2006).

[16] M. V. Perez, C. Bao, M. T. Flores-Arias, M. A. Rama, and C. G. Reino, "Description of gradient-index crystalline lens by a first-order optical system," *J. Opt. A: Pure Appl. Opt.* **7**, 103–110 (2005).

[17] M. A. Rama, M. V. Perez, C. Bao, M. T. Flores-Arias, and C. G. Reino, "Gradient-index crystalline lens model: A new method for determining the paraxial properties by the axial and field rays," *Optics Communications* **249**, 595–609 (2005).

[18] S. Sahin and O. Korotkova, "Crystalline human eye lens' response to stochastic light," *Opt. Lett.* **36**, 2970–2972 (2011).

[19] V. G. Veselago, "Electrodynamics of substances with simultaneously negative electrical and magnetic permeabilities," *Sov. Phys. Usp.* **10**, 509–514 (1968).

[20] A. D. Boardman, N. King, and L. Velasco, "Negative refraction in perspective," *Electromagnetics* **25**, 365–389 (2005).

[21] R. A. Shelby, D. R. Smith, and S. Schultz, "Experimental verification of a negative index of refraction," *Science* **292**, 77–79 (2001).

[22] B.-J. Seo, T. Ueda, T. Itoh, and H. Fetterman, "Isotropic left handed material at optical frequency with dielectric spheres embedded in negative permittivity medium," *Appl. Phys. Lett.* **88**, 161122 (2006).

[23] J. Zhang, H. Chen, L. Ran, Y. Luo, B. Wu, and J. A. Kong, "Experimental characterization and cell interaction of a two-dimensional isotropic left-handed metamaterial," *Appl. Phys. Lett.* **92**, 084108 (2008).

[24] G. V. Eleftheriades and K. G. Balmain, Eds. *Negative Refraction Metamaterials: Fundamental Principles and Applications*, John Wiley and Sons Inc., 2005.

[25] J. Zhou, H. Luo, S. Wen, and Y. Zeng, "ABCD matrix formalism for propagation of Gaussian beam through left-handed material slab system," *Opt. Commun.* **282**, 2670–2675 (2009).

[26] Z. Tong and O. Korotkova, "Stochastic electromagnetic beams in negative refractive index materials," *Opt. Lett.* **35**, 175–177 (2010).

[27] Z. Tong and O. Korotkova, "Spectral shifts and spectral switches in random fields on interaction with negative phase materials," *Phys. Rev. A* **82**, 013829 (2010).

[28] J. B. Pendry, "Negative refraction makes a perfect lens," *Phys. Rev. Lett.* **85**, 3966–3969 (2000).

[29] M. Born and E. Wolf, *Principles of Optics*, Cambridge University Press, 7th Edition, 1999.

[30] J. W. Goodman, *Statistical Optics*, Wiley, 1985.

[31] J. W. Goodman, *Introduction to Fourier Optics*, McGraw-Hill, 1996.

[32] A. T. Friberg and J. Turunen, "Imaging of Gaussian Schell-model sources," *J. Opt. Soc. Am. A* **5**, 713–720 (1988).

[33] Y. Cai and L. Hu, "Propagation of partially coherent twisted anisotropic Gaussian Schell-model beams through an apertured astigmatic optical system," *Opt. Lett.* **31**, 685–687 (2006).

[34] Z. Tong and O. Korotkova, "Overcoming rayleigh diffraction limit with partially coherent twisted beams," *Opt. Lett.* **37**, 2595–2597 (2012).

[35] A. S. Ostrovsky, M. Á. Olvera-Santamaría, amd P. C. Romero-Soría, "Effect of coherence and polarization on resolution of optical imaging system," *Opt. Lett.* **36**, 1677–1679 (2011).

[36] A. Gerrard and J. M. Burch, *Introduction to Matrix Methods in Optics*, John Wiley and Sons, 1975.

[37] H. T. Yura and S. G. Hanson, "Optical beam wave propagation through complex optical systems," *J. Opt. Soc. Am. A* **4**, 1931–1948 (1987).

[38] H. T. Yura and S. G. Hanson, "Second-order statistics for wave propagation through complex optical systems," *J. Opt. Soc. Am. A* **6**, 564–575 (1989).

[39] Q. Lin, S. Wang, J. Alda, and E. Bernabeu, "Transformation of non-symmetric Gaussian beam into symmetric one by means of tensor *ABCD* law," *Optik* **85**, 67–72 (1990).

[40] Q. Lin and Y. Cai, "Tensor ABCD law for partially coherent twisted anisotropic Gaussian Schell-model beams," *Opt. Lett.* **27**, 216–218 (2002).

[41] R. K. Luneburg, *Mathematical Theory of Optics*, University of California Press, 1964.

[42] A. G. Fox and T. Li, "Resonate modes in a maser interferometer," *Bell Syst. Tech. J.* **40**, 453-488 (1961).

[43] G. D. Boyd and J. P. Gordon, "Confocal multimode resonator for millimeter through optical wavelength masers," *Bell Syst. Tech. J.* **40**, 489-508 (1961).

[44] E. Wolf, "Spatial coherence of resonant modes in a maser interferometer," *Phys. Lett.* **3**, 166-168 (1963).

[45] E. Wolf and G. S. Agarwal, "Coherence theory of laser resonator modes," *J. Opt. Soc. Am. A* **1**, 541-546 (1984).

[46] F. Gori, "Propagation of the mutual intensity through a periodic structure," *Atti Fond. Giorgio Ronchi* **35**, 434-447 (1980).

[47] P. DeSantis, A. Mascello, C. Palma, and M. R. Perrone, "Coherence growth of laser radiation in Gaussian cavities," *IEEE J. Quantum Electron.* **32**, 802-812 (1996).

[48] C. Palma, G. Cardone, and G. Cincotti, "Spectral changes in Gaussian-cavity lasers," *IEEE J. Quantum Electron.* **34**, 1082-1088 (1998).

[49] E. Wolf, "Coherence and polarization properties of electromagnetic laser modes," *Opt. Commun.* **265**, 60-62 (2006).

[50] T. Saastamoinen, J. Turunen, J. Tervo, T. Setälä, and A. T. Friberg, "Electromagnetic coherence theory of laser resonator modes," *J. Opt. Soc. Am. A* **22**, 103-108 (2005).

[51] Y. Min, Y. Cai, H. T. Eyyuboglu, Y. Baykal, and O. Korotkova, "Evolution of the degree of polarization of an electromagnetic Gaussian Schell-model beam in a Gaussian cavity," *Opt. Lett.* **33**, 2266-2268 (2008).

[52] O. Korotkova, M. Yao, Y. Cai, H. T. Eyyuboglu, and Y. Baykal, "State of polarization of a stochastic electromagnetic beam in an optical resonator," *J. Opt. Soc. Am. A* **25**, 2710–2720 (2008).

[53] Z. Tong, O. Korotkova, Y. Cai, H. T. Eyyuboglu and Y. Baykal, "Correlation properties of random electromagnetic beams in laser resonators," *Appl. Phys. B* **97**, 849–857 (2009).

[54] S. Zhu and Y. Cai, "Degree of polarization of a twisted electromagnetic Gaussian Schell-model beam in a Gaussain cavity filled with gain media," *Prog. in Electromag. Res. B* **21**, 171–187 (2010).

[55] H. Kogelnik and T. Li, "Laser beams and resonators," *Appl. Opt.* **5**, 1550-1567 (1966).

[56] W. Casperson and S. D. Lunnam, "Gaussian modes in high loss laser resonators," *Appl. Opt.* **14**, 1193-1199 (1975).

6

Random beams in linear random media

CONTENTS

6.1 Natural random media: turbulence

Turbulent media are very frequently occurring in nature and are the most difficult ones to deal with. Turbulent flows are found in the Earth's atmosphere, jet streams, cumulous clouds, the photosphere of stars, smoke plumes, air flows around ships and airplanes, in rivers, and in oceanic flows [1]. Turbulence can be defined as a fluid regime exhibiting chaotic changes in its characteristics, and such changes are diffusive, dissipative and the flow may possess vortexes [2]. Diffusivity is the mechanism for spreading velocity fluctuations through the flow, or increased rates of mixing, momentum, heat and mass transfer. Turbulence is also characterized by high fluctuations in vorticity, the manifestation of the three-dimensional rotational feature of the flow. Turbulence is difficult to characterize and predict, as observations are sparse and the turbulence theory remains one of the last major unsolved problems in classical physics [3].

The optical turbulence is well explained by the presence of irregularities in the refractive index or, so-called, "turbulent eddies," appearing due to fluctuations in various physical properties of matter, such as temperature, pressure and concentration of inhomogeneous chemical content. Such eddies are created in different types of matter through certain physical/chemical/biological mechanisms. Shearing and mixing of different parts of the irregular structures under influence of winds in atmosphere, currents in the ocean, cell growth and fluid transfer in bio-tissues lead to a mechanism of energy transfer among eddies of different sizes. The largest possible size of an eddy in the turbulent process is taken as the definition of the *outer scale* L_0 of turbulence. Larger eddies break down further into smaller ones with energy until the size of the eddy reaches the lower limit when the energy dissipates. The size of the smallest eddy before dissipation defines the *inner scale* l_0. The interval of eddy sizes between the inner and the outer scales is called the *inertial range*. All turbulent eddies in the inertial range affect the propagating beam and can be simply viewed as the act of diffracting and refracting lenses [4]. The cumulative effect of all such lenses encountered by the wave along the propagation path results in its random phase and amplitude distribution.

The random distribution (in time and space) of the index of refraction in the turbulent medium requires statistical description of the propagating beam, i.e., must be based on the theory of random processes. The presence of time dependence in the statistical characterization can be frequently suppressed with the help of the Taylor's hypotheses of frozen turbulence [4].

The most important statistical characteristics of the refractive index in the three-dimensional space are the first two moments: the mean value of a field

$$n_0(\mathbf{r}) = \langle n(\mathbf{r}) \rangle_M, \tag{6.1}$$

and its covariance function

$$B_n(\mathbf{r}_1; \mathbf{r}_2) = \langle [n(\mathbf{r}_1) - n_0(\mathbf{r}_1)][n(\mathbf{r}_2) - n_0(\mathbf{r}_2)] \rangle_M, \tag{6.2}$$

where the angular brackets with subscript M denote the ensemble average over the realizations of the medium.

The interaction of the electromagnetic fields with the turbulent media is a very complex process in the general case when the latter are anisotropic and inhomogeneous. In the inertial range of scales the random media are often assumed to be homogeneous (statistical moments of the field are translation-invariant). Under such circumstances the relation between the spatial covariance function $B_n(\mathbf{r})$ and the power spectrum $\Phi_n(\boldsymbol{\kappa})$ (see Chapter 1), which determines the distribution of energy among the eddies of different sizes, has the form of the three-dimensional Fourier transform pair:

$$B_n(\mathbf{r}) = \int\!\!\!\int\!\!\!\int_{-\infty}^{\infty} \exp(i\boldsymbol{\kappa} \cdot \mathbf{r}) \Phi_n(\boldsymbol{\kappa}) d^3\kappa, \tag{6.3}$$

$$\Phi_n(\boldsymbol{\kappa}) = \left(\frac{1}{2\pi}\right)^3 \int\!\!\!\int\!\!\!\int_{-\infty}^{\infty} \exp(i\boldsymbol{\kappa} \cdot \mathbf{r}) B_n(\mathbf{r}) d^3r. \tag{6.4}$$

Here $\boldsymbol{\kappa} = (\kappa_x, \kappa_y, \kappa_z)$ is the three-dimensional vector, whose components have the units m^{-1}, representing spatial frequencies.

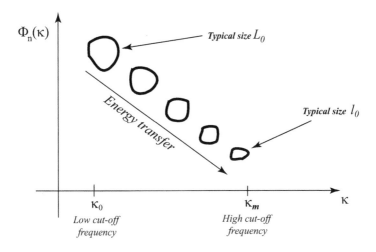

FIGURE 6.1
Turbulent cascade.

The assumption that in the inertial range of scales the atmosphere is also isotropic (statistical moments of the field are rotation invariant) leads to the spectral representation

$$B_n(r) = \frac{4\pi}{r} \int_0^\infty \sin(\kappa r)\Phi_n(\kappa)\kappa d\kappa, \qquad (6.5)$$

$$\Phi_n(\kappa) = \frac{1}{2\pi^2\kappa} \int_0^\infty \sin(\kappa r)B_n(r)r dr, \qquad (6.6)$$

where r and κ are the magnitudes of vectors $\mathbf{r}$ and $\boldsymbol{\kappa}$, respectively. For typical natural turbulent media the power spectrum in Eq. (6.6) has the form of the power law. If presented on a logarithmic scale it returns a line with a negative slope. In Fig. 6.1 the basic representation of a turbulent medium via its one-dimensional spatial power spectrum in the inertial range of scales $1/L_0 << \kappa << 1/l_0$ is shown.

6.1.1 Atmospheric turbulence

For propagation in the classic atmospheric turbulence several power spectra of the refractive index are frequently used in the literature subject to the choice: analytic tractability of the results versus their accuracy. The main cause of the refractive index fluctuations in the atmosphere is the temperature variation. The common feature of these power spectra is the $-11/3$ power law in the inertial range of scales. The very first analytical model for the power spectrum was derived by Kolmogorov [5], [6] by means of non-dimensional analysis:

$$\Phi_n(\kappa) = 0.033 C_n^2 \kappa^{-11/3}, \quad 1/L_0 << \kappa << 1/l_0, \tag{6.7}$$

where C_n^2 is the refractive index structure parameter, having typical values in the range $10^{-15} - 10^{-13} m^{-2/3}$. This parameter is assumed to be constant along the horizontal paths near the ground but can vary with altitude [4].

The Tatarskii's spectrum, which extends the Kolmogorov's model for large values of κ, is expressed as [7]

$$\Phi_n(\kappa) = 0.033 C_n^2 \kappa^{-11/3} \exp\left[-\frac{\kappa^2}{\kappa_m^2}\right], \quad 1/L_0 << \kappa, \tag{6.8}$$

where $\kappa_m = 5.92/l_0$.

The von Kármán's spectrum, which extends the Tatarskii's spectrum to the region of small κ has the form [4]

$$\Phi_n(\kappa) = 0.033 C_n^2 \kappa^{-11/3} \frac{\exp\left[-\frac{\kappa^2}{\kappa_m^2}\right]}{(\kappa^2 + \kappa_0^2)^{11/6}}, \quad \kappa > 0, \tag{6.9}$$

where $\kappa_0 = 1/L_0$.

The model that also takes into account the characteristic bump at high spatial frequencies was suggested by Andrews [8] (based on data of Hill [9]). It has the form

$$\Phi_n(\kappa) = 0.033 C_n^2 \kappa^{-11/3} \frac{\exp\left[-\frac{\kappa^2}{\kappa_m^2}\right]}{(\kappa^2 + \kappa_0^2)^{11/6}}$$
$$\times [1 + 1.802(\kappa/\kappa_l) - 0.254(\kappa/\kappa_l)^{7/6}], \quad 0 \leq \kappa < \infty, \tag{6.10}$$

where $\kappa_l = 3.3/l_0$.

Recently, another mathematical model for the atmospheric power spectrum was obtained specifically for paths in the marine environment [10], based on the data acquired by Friehe et al. [11] and Hill [9] and has the form

$$\Phi_n(\kappa) = 0.033 C_n^2 \kappa^{-11/3} \frac{\exp\left[-\frac{\kappa^2}{\kappa_h^2}\right]}{(\kappa^2 + \kappa_0^2)^{11/6}}$$
$$\times [1 - 0.061(\kappa/\kappa_h) + 2.836(\kappa/\kappa_h)^{7/6}], \quad 0 \leq \kappa < \infty, \tag{6.11}$$

where $\kappa_h = 3.41/l_0$.

It has been experimentally shown that generally atmospheric fluctuations might possess a structure different from the classic Kolmogorov turbulence, i.e., they can have other energy distribution law among the differently sized turbulent eddies, and exhibit non-homogeneity and/or anisotropy (cf. [12]–[16]). Such deviations are usually pertinent to higher atmospheric layers, being

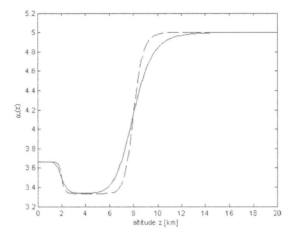

FIGURE 6.2
Non-Kolmogorov atmospheric turbulence: three-layered model. Solid and dashed curves refer to two different data fits. From Ref. [18].

caused by gravity waves, the jet stream, etc., and can substantially affect the statistics of electromagnetic waves, especially at optical frequencies, as can be shown by direct data application in propagation equations (cf. [17]).

The measurements of the atmospheric power spectrum at different heights above the ground have recently revealed that it has the three-layered structure, being of the Kolmogorov's type only in the boundary layer [18] (see Fig. 6.2).

While it is generally impossible to characterize all the features of the non-Kolmogorov's atmosphere, several analytical models have been recently suggested for taking into account the slope variation of the atmospheric power spectrum [19], [20]. In particular, it was assumed that instead of the classic power law $-11/3$ the power spectrum has a generalized form, defined by parameter α, varying in the range $3 < \alpha < 4$ [21].

In the simplest of the suggested non-Kolmogorov models, the power spectrum $\Phi_n(\kappa)$ has the von Kármán form, but with the slope generalized to arbitrary parameter α, i.e., [19]:

$$\Phi_n(\kappa) = A(\alpha)\tilde{C}_n^2 \frac{\exp\left[-\kappa^2/\kappa_m^2\right]}{\left(\kappa^2 + \kappa_0^2\right)^{\alpha/2}}, \quad 0 \leq \kappa < \infty, \quad 3 < \alpha < 4, \quad (6.12)$$

where $\kappa_0 = 2\pi/L_0$, as above, $\kappa_m = c(\alpha)/l_0$, and

$$c(\alpha) = \left[\Gamma\left(\frac{5-\alpha}{2}\right) A(\alpha)\frac{2}{3}\pi\right]^{1/(\alpha-5)}. \quad (6.13)$$

The term $\tilde{C}_n^2$ is the generalized refractive-index structure parameter having

242

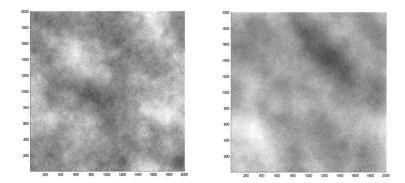

FIGURE 6.3
Computer simulation of turbulent phase screens for $\alpha = 3.10$ (left) and $\alpha = 3.67$ (right).

units $m^{3-\alpha}$, and

$$A(\alpha) = \frac{1}{4\pi^2}\Gamma(\alpha - 1)\cos\left(\frac{\alpha\pi}{2}\right), \tag{6.14}$$

where $\Gamma(x)$ is the Gamma function.

In Fig. 6.3 the turbulent screens are simulated for values of $\alpha = 3.01$ (non-Kolmogorov case) and $\alpha = 3.67$ (Kolmogorov case), for $C_n^2 = 10^{-13}m^{3-\alpha}$, $L_0 = \infty$, $l_0 = 0$ by means of the following MATLAB® code:

```
clear all;
sz = 1000; Cn2=10^(-13); deltaG=2; cx = (-sz:sz);
mx = (ones(2*sz+1,1)*cx).^2; mr = sqrt(mx+transpose(mx));
psd = 0.23*Cn2*mr.^(-alpha); psd(sz+1,sz+1) = 0;
randomcoeffs = randn(2*sz+1)+i*randn(2*sz+1);
phasescreen = real(fft2(fftshift(sqrt(psd).*randomcoeffs)));
figure() imagesc(phasescreen); colormap(gray);
```

Visual comparison between the two screens in Fig. 6.3 clearly shows that the slope of the power spectrum determines the relative weight of its crude and fine features.

6.1.2 Oceanic turbulence

The model for the power spectrum of the refractive-index fluctuations in the turbulent ocean waters can be obtained as the linearized polynomial of two variables: the temperature fluctuations and the salinity fluctuations [22]. This model is valid under the assumption that the fluctuations are isotropic and homogeneous and, hence, requires only of the spectrum depending on κ, which has the form

$$\Phi_n(\kappa) = 0.388 \times 10^{-8} \varepsilon_K^{-1/3} \kappa^{-11/3} [1 + 2.35(\kappa \eta_K)^{2/3}] f(\kappa, w_K, \chi_T), \quad (6.15)$$

where ε_K is the rate of dissipation of turbulent kinetic energy per unit mass of fluid, which may vary in the range from 10^{-4} m$^2/s^3$ to 10^{-10} m$^2/s^3$, $\eta_K = 10^{-3}$m is the Kolmogorov micro-scale (inner scale), and

$$f(\kappa, w, \chi_T) = \frac{\chi_T}{w_K^2} \left(w_K^2 e^{-A_T \delta_K} + e^{-A_S \delta_K} - 2w_K e^{-A_{TS} \delta_K} \right), \quad (6.16)$$

with χ_T being the rate of dissipation of mean-square temperature, $A_T = 1.863 \times 10^{-2}$, $A_S = 1.9 \times 10^{-4}$, $A_{TS} = 9.41 \times 10^{-3}$, and $\delta_K = 8.284(\kappa \eta_K)^{4/3} + 12.978(\kappa \eta_K)^2$, w_K being the relative strength of temperature and salinity fluctuations, which in the ocean waters can vary in the interval $[-5; 0]$, attaining the upper bound for the maximum temperature-induced optical turbulence. All the parameters entering the model (6.15)–(6.16) can be directly measured [24].

While parameters χ_T and ε_K primarily influence the height of the power spectrum, the balance parameter w_K affects its shape. To illustrate typical dependence of the spectrum (6.15)–(6.16) on w_K we show it in Fig. 6.4 as a function of wave number κ, for several values of w_K.

6.1.3 Biological tissues

The optical properties of different bio-tissues are determined by local refractive index variations, typically in the range $0.04 - 0.01$ with the background index of ≈ 1.34 for extracellular fluid and ≈ 1.36 for cytoplasm, as well as by the presence of micro-structures of different sizes (50 nm for protein macromolecules to 1 mm for thick blood vessels) and shapes (spheres, cylinders, etc.). Perhaps the first successful attempt to characterize the random variation in the refractive index on the continuum of scales was made by Schmitt and Kumar [25], [26], where the (two-dimensional) fractal power spectrum resembling the von Karman model for the atmospheric turbulence

$$\Phi_n(\kappa) = \frac{4\pi C_n^2 L_0^2 (\alpha - 1)}{(1 + \kappa^2 L_0^2)^{\alpha}} \quad (6.17)$$

was fitted to the experimental data. Here the slope α belongs to the range between 1.28 and 1.41, C_n^2 takes values from 0.001 to 0.005 in most soft tissues, L_0 significantly depends on the tissue type but is typically on the order of a few micrometers. Figure 6.5 presents a measured and fitted power spectrum of the mouse liver tissue.

Other mathematical models for fluctuations in optical properties of bio-tissues have later been proposed, which may prove to be more suitable for

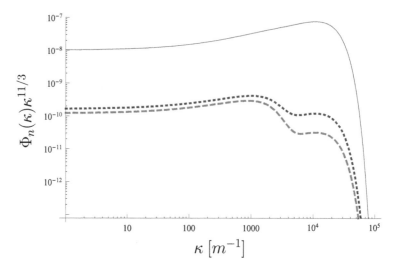

FIGURE 6.4

Log-log plot of the oceanic power spectrum $\Phi_n(\kappa)$, calculated from Eqs. (6.15)-(6.16) and normalized by the Kolmogorov power-law $\kappa^{-11/3}$, for $w_K = -0.1$ (solid curve), $w_K = -2.5$ (dotted curve), $w_K = -4.9$ (dashed curve). From Ref. [23].

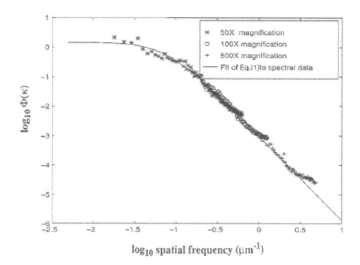

FIGURE 6.5

The structure of the mouse liver tissue power spectrum in the logarithmic scale. From Ref. [25].

analytic calculations. The introduction of all these models begins with the correlation function (rather than with the power spectrum)

$$C_\epsilon(r) = \langle \delta\epsilon_m(r')\delta\epsilon_m(r'+r)\rangle, \tag{6.18}$$

of weakly fluctuating dielectric permittivity. If the fluctuations are weak then $\delta\epsilon_m(\mathbf{r}) = 2n_0^2(\delta n(\mathbf{r}) - 1)$, with n_0 being the background value of the refractive index and $\delta n(\mathbf{r})$ being the relative refractive index at position $\mathbf{r}$.

For instance, in a two-parametric model known as the Booker-Gordon formula [27], [28] the correlation function of an isotropic tissue has the form:

$$C_\epsilon(r) = C_\epsilon^2 \exp\left(-\frac{r}{l_c}\right), \tag{6.19}$$

leading to the power spectrum

$$\Phi_\epsilon(\kappa) = \frac{C_\epsilon^2 l_c^3}{\pi^2(1+\kappa^2 l_c^2)^2}. \tag{6.20}$$

Here C_ϵ^2 and l_c denote the relative strength of fluctuations and the typical correlation width.

The model developed by Xu and Alfano [29] is based on the assumption that the refractive index variations caused by underlying microscopic structures can be treated as a collection of spheres with the radii distributed as the power law [26]

$$\eta(l) = \eta_0 l^{3-D_f}, \tag{6.21}$$

where $\eta(l)$ is the volume fraction of spherical particles with radii $0 \le l \le L_0$, η_0 being a constant and D_f being the particle size distribution, which can be readily fitted for a given tissue sample [29]. Then the correlation function of an isotropic tissue volume is a weighted average of exponential functions given by the integral

$$C_\epsilon(r) = C_\epsilon^2 \int\limits_0^{L_0} \exp\left(-\frac{r}{l}\right) \eta(l) dl$$
$$= C_\epsilon^2 \eta_0 L_0^{4-D_f} E_{5-D_f}\left(\frac{r}{L_0}\right), \tag{6.22}$$

where $C_\epsilon = 2n_0^2(\alpha - 1)$, L_0 is the outer scale, and

$$E_n(z) = \int_0^1 \exp\left(-\frac{z}{\xi}\right)\xi^{n-2}d\xi, \tag{6.23}$$

ξ being the integration variable. The corresponding power spectrum then takes the form:

$$\Phi_\epsilon(\kappa) = \int\limits_0^{L_0} \frac{C_\epsilon^2 l^3}{\pi^2}(1+\kappa^2 l^2)^2 \eta(l) dl. \tag{6.24}$$

For $D_f = 3, 4, 5$ integral in Eq. (6.24) reduces to analytical expressions [29].

Structurally similar model to that in [29] has been developed in [30] with which analytic calculations can be readily performed for all values of parameter D_f. The correlation function is expressed as

$$C_\epsilon(r) = \frac{2}{\Gamma(\alpha - 3/2)} \left(\frac{r}{2L_0}\right)^{\alpha - 3/2} K_{\alpha - 3/2}\left(\frac{r}{L_0}\right), \qquad (6.25)$$

where $K_m(x)$ is the modified Bessel function of the second kind. The corresponding power spectrum then becomes

$$\Phi_\epsilon(\kappa) = C_\epsilon^2 \left(\frac{2}{L_0}\right)^{\alpha - 3/2} \frac{\Gamma(\alpha)}{\Gamma(\alpha - 3/2)[1 + (\kappa L_0)^2]^\alpha}, \qquad (6.26)$$

where $\alpha = 3.5 - 0.5 D_f$.

The Whittle-Matern correlation family

$$C_n(r) = C_n^2 \frac{2^{5/2 - \alpha}}{|\Gamma(\alpha - (3/2))|} \left(\frac{r}{L_0}\right)^{\alpha - 3/2} K_{\alpha - 3/2}\left(\frac{r}{L_0}\right), \qquad (6.27)$$

being a generalized function of the refractive index fluctuations, reduces to other models for the selected values of parameter α. It has been applied by Rogers et al. [31] for light scattering. For instance, it reduces to Gaussian for $\alpha \to \infty$, and exponential for $\alpha = 2$. The corresponding power spectrum of the refractive index then has the form

$$\Phi_n(\kappa) = \frac{C_n^2 L_0^3 \Gamma(\alpha)}{\pi^{3/2} |\Gamma(\alpha - (3/2))| (1 + \kappa^2 L_0^2)^\alpha}. \qquad (6.28)$$

While the examples relating to propagation of beams in the atmosphere and in the ocean will be considered in the following sections their interaction with biological tissues are left until Ch. 8 where the theory of beams scattered from deterministic and random media is considered. In the case of biological tissues the statistics of beams scattered from thin slices, rather than those for continuous propagation within the slices, are of most practical importance.

6.2   Scalar random beam interaction with random media

In this section we will first outline two methods for evaluating the statistics of scalar beams propagating in the turbulent media. The first method based on the extended Huygens-Fresnel principle has been, perhaps, the most popular one for analytical calculations. The second method has been specifically developed for deterministic and random beams for which the angular spectrum representation is known and has a simple form. Yet another method based on coherent mode decomposition has also been developed [32] but is not considered here since it has been little employed. In the second part of the section we will discuss some of the effects of random media on the propagating beams, such as fractional power loss, spectral shifts and polarimetric changes. In this chapter we only consider the case of the clean-air or clean-water turbulence, i.e., we assume that the light wave is not affected by particles. Hence, the optical turbulence, i.e., continuous temporal and spatial random variations in the index of refraction is assumed to be the only mechanism affecting the beam on propagation. In Chapter 8 we will outline the procedure for combining optical turbulence and particulate scattering effects.

6.2.1 Extended Huygens-Fresnel principle

Suppose a scalar beam-like field $U(\rho'; \omega)$, where $\rho' = (x', y')$ is a two-dimensional vector, is generated by a planar source located in the plane $z = 0$ and propagates into the half-space $z > 0$, close to the positive z direction, which is filled with a random medium (see Fig. 6.6).

Under the assumption that the fluctuations in the turbulent medium are very mild, i.e., if $\delta n/n << 1$ over the distances compared to the wavelength of light, the electric field $U(\mathbf{r}; \omega)$ at a point with position vector $\mathbf{r} = (\rho; z)$ satisfies the wave equation of the form

$$\nabla^2 U(\mathbf{r}; \omega) + k^2 n^2(\mathbf{r}; \omega)U(\mathbf{r}; \omega) = 0. \tag{6.29}$$

The solution of Eq. (6.29) can be found for the field $U(\mathbf{r}; \omega)$ in the form of the *extended Huygens-Fresnel integral* [4]:

$$U(\mathbf{r}; \omega) = -\frac{ik\exp(ikz)}{2\pi z} \int U(\rho', \omega)\exp\left[ik\frac{(\rho' - \rho)^2}{2z}\right]\exp[\Psi(\rho', \mathbf{r}; \omega)]d^2\rho',$$
$$\tag{6.30}$$

where $\Psi(\rho', \mathbf{r}; \omega)$ is the complex phase perturbation caused by the random distribution of the index of refraction in the medium.

Let the second-order correlation properties of such a beam at points ρ_1' and ρ_2' be described by the cross-spectral density function

$$W(\rho_1', \rho_2'; \omega) = \langle U^*(\rho_1'; \omega)U(\rho_2'; \omega)\rangle, \tag{6.31}$$

where the average over the ensemble of monochromatic realizations is implied (see Chapter 3). If we also characterize the fluctuations in the propagating field by the cross-spectral density function

$$W(\mathbf{r}_1, \mathbf{r}_2; \omega) = \langle U^*(\mathbf{r}_1; \omega)U(\mathbf{r}_2; \omega)\rangle, \tag{6.32}$$

then on substituting from Eq. (6.30) into Eq. (6.32) we find that the it takes the form:

$$W(\mathbf{r}_1, \mathbf{r}_2; \omega) = \int\int W(\rho_1', \rho_2'; \omega)K(\mathbf{r}_1, \mathbf{r}_2; \rho_1', \rho_2'; \omega)d^2\rho_1' d^2\rho_2', \tag{6.33}$$

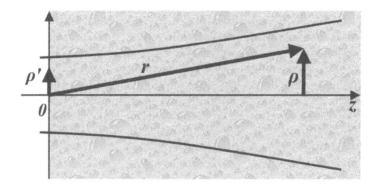

FIGURE 6.6
Illustrating notations relating to beam propagation in turbulence.

where propagator K is given by the expression

$$K(\mathbf{r}_1, \mathbf{r}_2; \boldsymbol{\rho}'_1, \boldsymbol{\rho}'_2; \omega) = \left(\frac{k}{2\pi z}\right)^2 \exp\left[-ik\frac{(\boldsymbol{\rho}'_1 - \boldsymbol{\rho}_1)^2 - (\boldsymbol{\rho}'_2 - \boldsymbol{\rho}_2)^2}{2z}\right]$$
$$\times \langle \exp[\Psi^*(\boldsymbol{\rho}'_1, \mathbf{r}_1; \omega) + \Psi(\boldsymbol{\rho}'_2, \mathbf{r}_2; \omega)]\rangle_M. \tag{6.34}$$

Here the term in the angular brackets represents the correlation function of the complex phase perturbed by the random medium, and subscript M denotes the average over the medium realizations. For the homogeneous and isotropic turbulence the phase correlation function is related to the power spectrum $\Phi_n(\kappa)$ by the formula [4]

$$\langle \exp[\Psi^*(\boldsymbol{\rho}_1, \mathbf{r}_1; \omega) + \Psi(\boldsymbol{\rho}_2, \mathbf{r}_2; \omega)]\rangle_M$$
$$= \exp\left[-4\pi^2 k^2 z \int_0^1 \int_0^\infty \kappa \Phi_n(\kappa)[1 - J_0[|(1 - \xi)(\mathbf{r}_1 - \mathbf{r}_2) + \xi(\boldsymbol{\rho}_1 - \boldsymbol{\rho}_2)|\kappa]]d\kappa d\xi\right],$$
$$\tag{6.35}$$

where $J_0(x)$ stands for the 0th order Bessel function. If the points are located in the region sufficiently close to the optical axis, it is possible to show that after approximating the Bessel function by its two first terms:

$$J_0(x) \approx 1 - \frac{1}{4}x^2, \tag{6.36}$$

the expression (6.35) reduces to the form [33] (see also [34])

$$K(\mathbf{r}_1, \mathbf{r}_2, \boldsymbol{\rho}'_1, \boldsymbol{\rho}'_2; \omega) = \left(\frac{k}{2\pi z}\right)^2 \exp\left[-ik\frac{(\boldsymbol{\rho}'_1 - \boldsymbol{\rho}_1)^2 - (\boldsymbol{\rho}'_1 - \boldsymbol{\rho}_2)^2}{2z}\right]$$

$$\times \exp\left[-\frac{\pi^2 k^2 z}{3}[(\boldsymbol{\rho}_1 - \boldsymbol{\rho}_2)^2 + (\boldsymbol{\rho}_1 - \boldsymbol{\rho}_2)(\boldsymbol{\rho}'_1 - \boldsymbol{\rho}'_2) + (\boldsymbol{\rho}'_1 - \boldsymbol{\rho}'_2)^2]\right.$$

$$\left.\times \int_0^\infty \kappa^3 \Phi_n(\kappa) d\kappa\right].$$

(6.37)

For example, the Gaussian Schell-model beam propagating in a turbulent medium with power spectrum $\Phi_n(\kappa)$ has a particularly simple analytical form of the cross-spectral density function. On substituting from the expression for the cross-spectral density function of such source and propagator (6.37) into Eq. (6.33), one can find that for the propagating beam [34]

$$W(\mathbf{r}_1, \mathbf{r}_2; \omega) = \frac{A_0^2}{\Delta^2(z; \omega)} \exp\left[-\frac{(\boldsymbol{\rho}_1 + \boldsymbol{\rho}_2)^2}{8\sigma^2 \Delta^2(z; \omega)}\right]$$

$$\times \exp\left\{-\left[\frac{1}{2\Delta^2(z; \omega)}\left(\frac{1}{4\sigma^2} + \frac{1}{\delta^2}\right) + M_T(1 + \sigma^2)\right.\right.$$

(6.38)

$$\left.\left.-\frac{M_T^2 z^2 \lambda^2}{8\pi^2 \sigma^2 \Delta^2(z; \omega)}\right](\boldsymbol{\rho}_1 - \boldsymbol{\rho}_2)^2\right\} \exp\left(\frac{i\pi(\rho_2^2 - \rho_1^2)}{\lambda R(z; \omega)}\right).$$

Here

$$\Delta^2(z; \omega) = 1 + \left(\frac{z\lambda}{2\pi\sigma}\right)^2 \left(\frac{1}{4\sigma^2} + \frac{1}{\delta^2}\right) + \frac{M_T z^2 \lambda^2}{2\pi^2 \sigma^2},$$

(6.39)

and

$$R(z; \omega) = \frac{4\pi^2 \sigma^2 \Delta^2(z; \omega)z}{4\pi^2 \sigma^2 \Delta^2(z; \omega) + M_T z^2 \lambda^2 - 4\pi^2 \sigma^2}$$

(6.40)

are the generalized expansion coefficient and the radius of curvature of the beam in the presence of random medium. In these equations factor M_T is given by the expression

$$M_T = \frac{4\pi^4 z}{3\lambda^2} \int_0^\infty \kappa^3 \Phi_n(\kappa) \, d\kappa.$$

(6.41)

On substituting from Eq. (6.41) into Eq. (6.39) we find at once that on propagation in a random medium at large distances from the source under approximation (6.36) a beam expands as z^3, rather than as z^2 for the case of free-space propagation, i.e., when $M_T = 0$ [33], [35].

For some models of the atmospheric power spectrum the integral in Eq.

(6.41) and, hence, the cross-spectral density function of the beam can be evaluated analytically. For instance, for the non-Kolmogorov spectrum (6.12) we find that

$$
I = \int_0^\infty \kappa^3 \Phi_n(\kappa) \, d\kappa
$$

$$
= \frac{A(\alpha)\tilde{C}_n^2}{2(\alpha - 2)} \left[\kappa_m^{2-\alpha} \beta \exp\left(\frac{\kappa_0^2}{\kappa_m^2}\right) \Gamma\left(2 - \frac{\alpha}{2}, \frac{\kappa_0^2}{\kappa_m^2}\right) - 2\kappa_0^{4-\alpha} \right],
$$

(6.42)

where $\beta = 2\kappa_0^2 - 2\kappa_m^2 + \alpha\kappa_m^2$ and $\Gamma(\cdot, \cdot)$ denotes the incomplete Gamma function.

6.2.2 Angular spectrum method

An alternative technique for the description of optical beam propagation in random media is based on its representation by the angular spectrum [36]. The advantage of this method stems from the fact that the main procedure for calculation of the statistics of the propagating beams is the same for all incident fields and the change in their properties only leads to modification of the few last steps. The basic idea behind this technique is that once the incident field is decomposed into the spectrum of plane waves traveling at different directions they can be individually propagated through the given medium. The statistics of the resulting field (after interaction with the medium) can be determined by combining the contributions of the individual plane waves. Because the propagation characteristics of plane waves in random media are typically better known than those of arbitrarily shaped waves, such an approach reduces to breaking of the beam into the plane waves and its subsequent recombination.

Suppose, as before, that a scalar beam-like field $U(\mathbf{r}; \omega)$ propagates in a random medium that fills the half-space $z > 0$. Then it may be represented by the integral [36]

$$
U(\mathbf{r}; \omega) = \iint a(\mathbf{u}; \omega) P_{\mathbf{u}}^M(\mathbf{r}; \omega) d\mathbf{u}_\perp,
$$

(6.43)

where $P_{\mathbf{u}}^M(\mathbf{r}; \omega)$ is a plane wave at point $\mathbf{r} = (\boldsymbol{\rho}, z)$ traveling in direction $\mathbf{u}$, distorted by the random medium. The cross-spectral density function of the beam propagating to the points with position vectors $\mathbf{r}_1$ and $\mathbf{r}_2$ then takes the form

$$
W(\mathbf{r}_1, \mathbf{r}_2; \omega) = \langle U^*(\mathbf{r}_1; \omega) U(\mathbf{r}_2; \omega) \rangle
$$

$$
= \iiiint a^*(\mathbf{u}_1; \omega) a(\mathbf{u}_2; \omega) \langle P_{\mathbf{u}_1}^{*M}(\mathbf{r}_1; \omega) P_{\mathbf{u}_2}^M(\mathbf{r}_2; \omega) \rangle_M d\mathbf{u}_{1\perp} d\mathbf{u}_{2\perp},
$$

(6.44)

subscript M denoting the average taken over the realizations of the random medium.

For a stochastic incident field characterized by the cross-spectral density $W(\boldsymbol{\rho}_1', \boldsymbol{\rho}_2'; \omega)$ in the source plane, $z = 0$, Eq. (6.44) takes the form

$$W(\mathbf{r}_1, \mathbf{r}_2; \omega) = \iiiint A(\mathbf{u}_1, \mathbf{u}_2; \omega) \langle P_{\mathbf{u}_1}^{*M}(\mathbf{r}_1; \omega) P_{\mathbf{u}_2}^M(\mathbf{r}_2; \omega) \rangle_M d\mathbf{u}_{1\perp} d\mathbf{u}_{2\perp}, \tag{6.45}$$

where

$$A(\mathbf{u}_1, \mathbf{u}_2; \omega) = \langle a^*(\mathbf{u}_1; \omega) a(\mathbf{u}_2; \omega) \rangle \tag{6.46}$$

is the angular correlation function given by the formula

$$\begin{aligned} A(\mathbf{u}_1, \mathbf{u}_2; \omega) = \frac{1}{(2\pi)^4} \iiiint & W(\boldsymbol{\rho}_1', \boldsymbol{\rho}_2'; \omega) \\ & \times \langle P_{\mathbf{u}_1}^{M*}(\boldsymbol{\rho}_1'; \omega) P_{\mathbf{u}_2}^M(\mathbf{r}_2; \omega) \rangle_M d\boldsymbol{\rho}_{1\perp}' d\boldsymbol{\rho}_{2\perp}'. \end{aligned} \tag{6.47}$$

This formula incorporates two ensemble averages: one over fluctuations in the field and the other over fluctuations of the random medium. It is commonly assumed that these two random processes are statistically independent. In general, this assumption is not valid. For instance, if a beam with high intensity interacts with the medium it is capable of modifying its statistics leading to nonlinear coupling. It also follows from Eq. (6.47) that the statistical properties of the field are contained entirely in the function $A(\mathbf{u}_1, \mathbf{u}_2; \omega)$ which is independent from the statistical properties of the medium. Hence, once the statistical properties of a given medium are known, the propagation characteristics of a beam of any type can be evaluated in a straightforward manner by carrying out the integral in (6.47).

The plane wave $P_{\mathbf{u}}^M(\mathbf{r}'; \omega)$ propagating through a weakly scattering medium can be represented by the Rytov series

$$P_{\mathbf{u}}^M(\mathbf{r}; \omega) = P_{\mathbf{u}}(\mathbf{r}; \omega) \exp[\Psi_{\mathbf{u}}^{(1)}(\mathbf{r}; \omega) + \Psi_{\mathbf{u}}^{(2)}(\mathbf{r}; \omega) + ...], \tag{6.48}$$

where $P_{\mathbf{u}}(\mathbf{r}; \omega)$ is the plane wave in the absence of the medium while $\Psi^{(1)}(\mathbf{r}; \omega)$ and $\Psi^{(2)}(\mathbf{r}; \omega)$ are the complex phase perturbations of the first and the second order, respectively. Keeping terms only to the second order in the Rytov approximation, we may express the cross-spectral density function $W_{\mathbf{u}_1, \mathbf{u}_2}^M$ for two plane waves propagating in turbulence as

$$\begin{aligned} W_{\mathbf{u}_1, \mathbf{u}_2}^M(\mathbf{r}_1, \mathbf{r}_2; \omega) = & P_{\mathbf{u}_1}^*(\mathbf{r}_1; \omega) P_{\mathbf{u}_2}(\mathbf{r}_2; \omega) \\ & \times \langle \exp[\Psi_{\mathbf{u}_1}^{(1)}(\mathbf{r}_1; \omega) + \Psi_{\mathbf{u}_1}^{(2)}(\mathbf{r}_1; \omega) \\ & + \Psi_{\mathbf{u}_2}^{(1)}(\mathbf{r}_2; \omega) + \Psi_{\mathbf{u}_2}^{(2)}(\mathbf{r}_2; \omega) + ...] \rangle. \end{aligned} \tag{6.49}$$

Equation (6.49) can be simplified by using the method of cumulants: keeping terms to the second order we may approximate the average of an exponential function as

$$\langle \exp[\Psi] \rangle \approx \exp\left[\langle \Psi \rangle + \frac{1}{2}\left(\langle \Psi^2 \rangle - \langle \Psi \rangle^2\right)\right]. \tag{6.50}$$

Following this approach one can then show that the cross-spectral density function (6.49) for two tilted plane waves reduces to expression [36]

$$W^M_{u_1,u_2}(\mathbf{r}_1,\mathbf{r}_2;\omega) = P^*_{u_1}(\mathbf{r}_1;\omega)P_{u_2}(\mathbf{r}_2;\omega)$$
$$\times \exp[2E^{(1)}_{u_1,u_2}(\mathbf{r}_1,\mathbf{r}_2;\omega) + E^{(2)}_{u_1,u_2}(\mathbf{r}_1,\mathbf{r}_2;\omega)], \quad (6.51)$$

where

$$E^{(1)}_{u_1,u_2}(\mathbf{r}_1,\mathbf{r}_2;\omega) = -\pi k^2 \int_0^z d\xi \iint d^2\kappa \Phi_n(\xi,\boldsymbol{\kappa}), \quad (6.52)$$

and

$$E^{(2)}_{u_1,u_2}(\mathbf{r}_1,\mathbf{r}_2;\omega) = 2\pi k^2 \int_0^z d\xi \iint d^2\kappa \Phi_n(\xi,\boldsymbol{\kappa})$$
$$\times \exp[-i(z-\xi)(\mathbf{u}_1-\mathbf{u}_2)\cdot\boldsymbol{\kappa}]\exp[-i(\mathbf{r}_2-\mathbf{r}_1)\cdot\boldsymbol{\kappa}]. \quad (6.53)$$

Here ξ is the variable of integration, $\Phi_n(z,\boldsymbol{\kappa})$ is the power spectrum of refractive index fluctuations, $\boldsymbol{\kappa} = (\kappa_x,\kappa_y,\kappa_z)$ is the spatial frequency vector. Similar expressions have been also derived for the special cases of a single, normally incident plane wave, a spherical wave, and a Gaussian beam [4]. In the case when the atmospheric fluctuations are isotropic, Eqs. (6.52) and (6.53) reduce to the expressions

$$E^{(1)}_{u_1,u_2}(\mathbf{r}_1,\mathbf{r}_2;\omega) = -2\pi^2 k^2 \int_0^z d\xi \int_0^\infty \kappa d\kappa \Phi_n(\xi,\kappa), \quad (6.54)$$

and

$$E^{(2)}_{u_1,u_2}(\mathbf{r}_1,\mathbf{r}_2;\omega) = 4\pi^2 k^2 \int_0^z d\xi \int_0^\infty \kappa d\kappa \Phi_n(\xi,\kappa)$$
$$\times J_0[\kappa|(\mathbf{r}_{2\perp}-\mathbf{r}_{1\perp})-(z-\xi)(\mathbf{u}_{1\perp}-\mathbf{u}_{2\perp})|], \quad (6.55)$$

where J_0 is the Bessel function of the first kind of zero order and $\kappa = |\boldsymbol{\kappa}|$. In this case the complex phase correlation function given by the exponential term in Eq. (6.51), expressed by Eqs. (6.54) and (6.55), reduces to Eq. (6.35).

If the power spectrum of atmospheric fluctuations is z-independent (for instance in the case when the beam propagates in the atmosphere close to the ground along a horizontal path) Eqs. (6.54) and (6.55) can be further simplified. In particular, the moment $E^{(1)}_{u_1,u_2}$ then becomes

$$E^{(1)}_{u_1,u_2}(\mathbf{r}_1,\mathbf{r}_2;\omega) = -2\pi^2 k^2 z \int_0^\infty \kappa d\kappa \Phi_n(\kappa). \quad (6.56)$$

The moment $E^{(2)}_{\mathbf{u}_1,\mathbf{u}_2}$ can be evaluated by expanding the Bessel function in the form of infinite series as:

$$J_0[\kappa|(\mathbf{r}_{2\perp} - \mathbf{r}_{1\perp}) - (z - \xi)(\mathbf{u}_{1\perp} - \mathbf{u}_{2\perp})|]$$
$$= \sum_0^\infty J_m[\kappa|\mathbf{r}_{2\perp} - \mathbf{r}_{1\perp}|]J_m[\kappa\xi|\mathbf{u}_{2\perp} - \mathbf{u}_{1\perp}|]\exp[im(\phi_r - \phi_u)],$$

$$(6.57)$$

where J_m is the Bessel function of the first kind of order m, ϕ_r and ϕ_u are the angles that vectors $\mathbf{r}_{2\perp} - \mathbf{r}_{1\perp}$ and $\mathbf{u}_{2\perp} - \mathbf{u}_{1\perp}$ make with the x-axis, respectively. Then, according to the result in [37] (11.1.3 and 11.1.4) for the integral over z one obtains:

$$\int_0^z d\xi[\kappa\xi|\mathbf{u}_2 - \mathbf{u}_1|] = \frac{1}{\kappa}|\mathbf{u}_2 - \mathbf{u}_1|\left[\int_0^{\kappa z|\mathbf{u}_2 - \mathbf{u}_1|} J_0(\xi)d\xi\right.$$

$$(6.58)$$

$$\left. - 2\sum_{k=0}^{m-1} J_{2k+1}(\kappa z|\mathbf{u}_2 - \mathbf{u}_1|)\right],$$

if m is even and

$$\int_0^z d\xi[\kappa\xi|\mathbf{u}_2 - \mathbf{u}_1|] = \frac{1}{\kappa}|\mathbf{u}_2 - \mathbf{u}_1|\left[1 - J_0(\kappa z|\mathbf{u}_2 - \mathbf{u}_1|)\right.$$

$$(6.59)$$

$$\left. - 2\sum_{k=0}^{m} J_{2k}(\kappa z|\mathbf{u}_2 - \mathbf{u}_1|)\right],$$

if m is odd. Therefore, the integral for $E^{(2)}_{\mathbf{u}_1,\mathbf{u}_2}$ can be expressed via the sum of a series of one-dimensional integrals in κ. On summing up these results, the propagating cross-spectral density of a general beam-like field may be calculated with the help of the angular correlation function as

$$W(\mathbf{r}_1, \mathbf{r}_2; \omega) = \iiiint A(\mathbf{u}_1, \mathbf{u}_2; \omega)\exp\left[2E^{(1)}_{\mathbf{u}_1,\mathbf{u}_2}(\mathbf{r}_1, \mathbf{r}_2; \omega)\right.$$

$$(6.60)$$

$$\left. + E^{(2)}_{\mathbf{u}_1,\mathbf{u}_2}(\mathbf{r}_1, \mathbf{r}_2; \omega)\right]d\mathbf{u}_{1\perp}d\mathbf{u}_{2\perp}.$$

Once the exponential terms in Eq. (6.60) are calculated for a particular medium, i.e., for the given propagation distance z and refractive index power spectrum Φ_n, these partial results can be used for different sources. In comparison with the expression based on the extended Huygens-Fresnel integral of the previous section, the angular spectrum method is more effective if the angular correlation function has a particularly simple form. Perhaps, this fact can be illustrated at best for the J_0-Bessel beam. Since the angular spectrum of such a beam is just a delta-function, the number of integrals involved in the

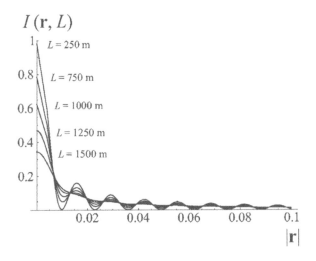

FIGURE 6.7
Radial distribution of the spectral density of a Bessel beam propagating in atmospheric turbulence. From Ref. [36].

calculation can be significantly reduced. Figure 6.7 shows the typical behavior of the transverse spectral density profile for a narrow Bessel beam propagating at several distances in the atmospheric turbulence with the von Karman power spectrum. As the propagation distance increases the Bessel beam looses the oscillatory behavior of its tails and assumes the Gaussian profile.

Figure 6.8 shows the radial intensity profiles of typical Gaussian and Gaussian Schell-model beams at the propagation distance $z = 1500$ m from the source plane. The solid curve is the angular spectrum based result, while the dashed curve is obtained with the help of the extended Huygens-Fresnel method. The dashed-dotted curve shows the profile of the beam in the absence of turbulence. It is seen from the curves that for both coherent beams and partially coherent beams the results based on the two methods are in good agreement with each other.

6.2.3 Fractional power changes

One of the major effects of the random medium on the propagating beams is the spatial spreading occurring in the cross-sections perpendicular to the direction of propagation. As we have seen from the example relating to the Gaussian Schell-model beam propagating in the turbulent atmosphere (see Sections 6.2.1 and 6.2.2), the spreading is well-pronounced at large distances from the source, having in this limit the z^3-dependence. Such turbulence-induced spreading results in the more drastic power loss compared to that for beams in free-space propagation. The degree of coherence of the beam in

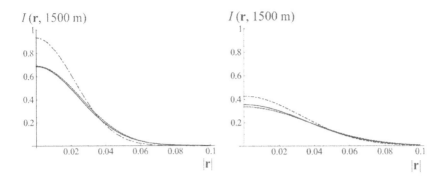

FIGURE 6.8
Radial distribution of the spectral density (intensity) of a Gaussian beam (left) and a Gaussian Schell-model beam (right) propagating in free space (dash-dotted curves) and in atmospheric turbulence (solid and dashed curves) with the von Karman power spectrum, at different distances z from the source. From Ref. [36].

the source plane influences the fractional power as well, just like in the free-space propagation case (see Chapter 3) [38]. In Fig. 6.9 the contour plots of fractional power of a beam propagating in the Kolmogorov-type atmosphere are shown as functions of propagation distance z from the source and of the radius $\bar{\rho}$ of the circular aperture of detector. While in the case of an almost coherent beam most of the power can be still captured by an aperture on the order of 1m at 6–8 kilometers from the source, for nearly incoherent beams it is not achievable already at distances shorter than 1 km.

6.2.4 Correlation-induced spectral changes

Spectral composition of beam-like light fields is sometimes employed as a carrier of a signal or an image, whether for communication or remote sensing [39, 40]. It is well known that for information transfer in extended random media, e.g., turbulent atmosphere or ocean, it is often preferable to use stochastic beams rather than deterministic [41]. As we will show in Chapter 7, in the case of stochastic beams the signal-to-noise ratio of the detected signal after transmission through a random medium can be controlled to some extent [42]. Hence, it is of importance to explore how the spectra of random beams can be modified by random media. On the other hand, the correlation-induced spectral changes in light fields can play the crucial part for solving inverse problems of determining the properties of media with which the beam has interacted [43].

After the seminal work showing the possibility for optical fields generated by random sources and propagating in vacuum to exhibit changes in their spectral composition [44], [45], the spectral shifts of light waves were also

256

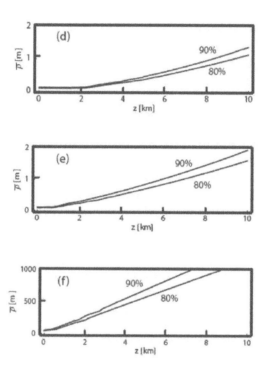

FIGURE 6.9
Contours of the fractional power in cross-sections of a Gaussian Schell-model beam with $\lambda = 0.633 \ \mu$ m, $\sigma = 1$ cm, on propagation in turbulent amosphere with Kolmogorov power spectrum and $C_n^2 = 10^{-13}$ m$^{-2/3}$: almost coherent source (top), partially coherent source with $\delta/\sigma = 0.1$ (middle) and nearly incoherent source (bottom). From Ref. [38].

observed for beams scattered from media, confined to some regions in space, being of whether a continuous [46] or particulate [47] nature, and also on propagation in media with a variety of refractive properties [48]. In addition in some of the recent studies the phenomenon of switching in the direction of the spectral shift after passage of a light beam through an aperture [49], a convergent lens [50], or an interface between media with positive and negative phase materials [51] (see Chapter 5) was discovered. In what follows we will show that similar switches are also pertinent to propagation in the turbulent media.

In spite of the fact that an enormous body of literature discusses evolution of various beams characteristics in the atmospheric turbulence, only a few separate studies examine specifically the spectral changes [52]–[58]. Recall that in order to characterize the correlation-induced spectral changes in scalar beams propagating in random media it suffices to apply the expressions given in Sections 3.2.1 or 3.2.2 in the definitions of the normalized spectrum and the spectral shift (3.13)–(3.15). We will now follow Ref. [58] for demonstration of the typical spectral shifts and switches in Gaussian Schell-model beams propagating in the atmospheric turbulence with the power spectrum (6.12). Since the central frequency $\lambda_1(\mathbf{r})$ and, hence, the shift $\varrho(\mathbf{r})$ generally vary with propagation distance z and wavelength λ, the spectral composition is expected to be modified. Because the spectral shifts are difficult to assert qualitatively we will illustrate the possibilities by numerical examples. The following parameters for the source and the atmosphere have been chosen: $\lambda_0 = 0.5435$ μm, $\Lambda = \lambda_0/3$; $\sigma = 1$ cm; $\delta = 1$ mm; $\tilde{C}_n^2 = 10^{-13}\text{m}^{3-\alpha}$; $L_0 = 1$ m; $l_0 = 1$ mm.

In figure 6.10 the changes in the on-axis behavior of the normalized spectral density S_N are shown at the source and in the field, calculated at three different ranges: 0.5, 1 and 1.5 km. For the power spectra with low values of the slope ($\alpha = 3.01$ and $\alpha = 3.1$) the spectral density gradually recovers with growing distance from the source. This effect appears for the Kolmogorov's power spectrum ($\alpha = 3.67$) and completely vanishes in free space.

Figure 6.11 includes the contour plot of the normalized shift ϱ as a function of z and ρ, for several values of slope α (6.12). In the region close to the optical axis the blue shift of the spectrum is well pronounced in free space, but is somewhat suppressed by the atmosphere. The best reconstruction of the oroginal spectrum occurs when $\alpha \approx 3.1$, while for the Kolmogorov's turbulence it is partial.

Figure 6.12 gives a slightly different perspective than Fig. 6.11 on the changes in the spectral composition of the beam. Here we show the contours of the spectral shift ϱ as a function of z and α on the axis of the beam. Note again that the value $\alpha \approx 3.1$ of the atmospheric spectrum corresponds to the strongest effects of the atmospheric turbulence on the beam's spectral composition.

Thus, we have shown that in comparison with the classic Kolmogorov's turbulence ($\alpha_K = 11/3$) in which the spectral changes are known to be par-

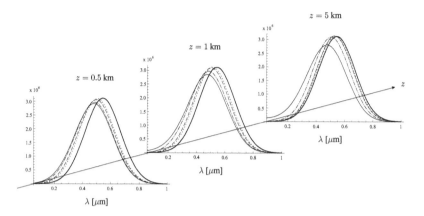

FIGURE 6.10
The normalized spectral density $\mathcal{S}_N$ [unitless] of a random beam propagating in atmospheric turbulence as a function of λ for $\rho = 0$ and different α: $\alpha = 3.01$ (dashed curve), $\alpha = 3.1$ (dotted curve), $\alpha = 3.67$ (dot-dashed curve) and free space (solid, thin curve), at $z = 0.5$ km, $z = 1$ km, and $z = 5$ km. Solid thick curve shows the normalized spectral density in the source plane. From Ref. [58].

tially suppressed, they are mitigated much stronger for $\alpha < \alpha_K$ and weaker for $\alpha_K < \alpha$. The atmosphere reconstructs the original spectral composition of the beam at best for $\alpha \approx 3.1$. While along the optical axis the original blue-shift can only get suppressed, resulting in blue-red spectral switch, for the off-axis positions the initially induced red-shift turns to the blue-shift at sufficiently large distance, resulting in the red-blue switch.

The formation of light colors in oceanic waters was a subject of extensive investigations for a number of decades [59]–[63]. However, the majority of such studies were devoted to the interaction of natural light with waters and only explored the spectral changes due to absorption and scattering by molecules and particles. In several studies, the propagation of laser light beams in the ocean was explored [64]–[66], where the accent again was made on various effects due to scattering and absorption. In these cases, i.e., either for an unbounded sunlight wave, with a very wide spectrum, or for a bounded laser beam, the spectral composition can not be modified due to the source correlations. As we have pointed out earlier, in order to have such a modification in the spectrum the generated radiation must remain highly directional, have a narrow initial spectrum, and be partially coherent. Since recently the interest in the active optical underwater communications, imaging and sensing appeared [67]–[71] it becomes important to investigate how oceanic turbulence affects spectra of optical stochastic beams.

Recall that the oceanic power spectrum (6.15)–(6.16) has a very complex

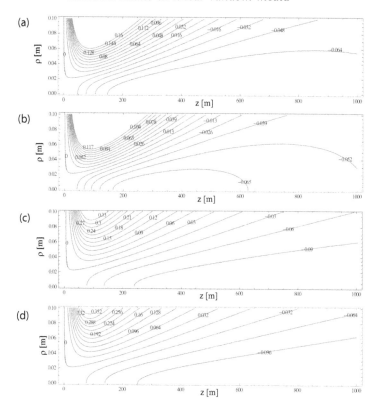

FIGURE 6.11
Contour plots of normalized spectral shift $\varrho_\lambda = \frac{\lambda_1 - \lambda_0}{\lambda_0}$ of a random beam propagating in atmospheric turbulence as a function of z (horizontal axis, in meters) and ρ (vertical axis, in meters) for (a) $\alpha = 3.01$; (b) $\alpha = 3.10$, (c) $\alpha = 3.67$; (d) free space. From Ref. [58].

structure. Hence, the dependence of the spectral shifts in random beams on the oceanic parameters can only be approached by numerical integration. Following Ref. [23] we assume that the following parameters for the source are used: $\lambda_0 = 0.5435 \ \mu$ m, $\Lambda = \sqrt{2}\lambda_0/6$; $\sigma = 1$ cm and $\delta = 0.1$ mm. In Fig. 6.13 the dependence of the spectral shift on the position in the propagating beam is illustrated as a set of contour plots. In the four considered cases different values of the dissipation rate of the mean-square temperature χ_T are chosen, varying from $\chi_T = 10^{-10} \ K^2/s$, corresponding to a very weak oceanic turbulence to $\chi_T = 10^{-2} \ K^2/s$, the limit pertaining to fairly strong fluctuations. Out of several parameters of oceanic turbulence, χ_T has the strongest influence on the beam. We find from Fig. 6.13 (b)-(d) that unlike for free-space propagation [see Fig. 6.13 (a)] where the spectrum of the beam undergoes a

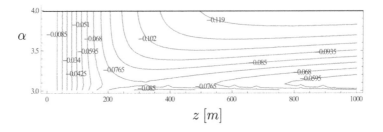

$$z\ [m]$$

FIGURE 6.12

Contour plots of the normalized spectral shift $\varrho_\lambda = \frac{\lambda_1 - \lambda_0}{\lambda_0}$ of a random beam propagating in atmospheric turbulence as a function of z (horizontal axis, in meters) and α (vertical axis) for $\rho = 0$ (on the beam axis).

blue-shift, for larger values of χ_T it is gradually suppressed, leading to the spectral switch. From Figs. 6.13 (b)–6.13 (d) we see that for larger values of χ_T the location of the switch moves towards the source plane, occurring within the first several meters of propagation. In Fig. 6.13 (d) the turbulence is so strong that there is no change in the central wavelength: the source-induced spectral shift is effectively mitigated very close to the source.

Figure 6.14 shows the contours of the on-axis shift ϱ_λ versus propagation distance z, and three parameters of oceanic turbulence: (a) temperature mean-square dissipation rate χ_T; (b) temperature-salinity balance parameter w_K and (c) energy dissipation rate ε. Figure 6.14 (a) implies that the value for χ_T at which the turbulence becomes effective is about $10^{-5}\ K^2/s$. On the other hand, Fig. 6.14 (b) shows that temperature-salinity parameter w_K must be close to zero in order to suppress the induced spectral change. Hence the salinity-induced optical turbulence must be minimal for fast reconstruction of the initial beam spectrum. Figure 6.14 (c) illustrates that the spectral composition is also sensitive to the kinetic energy dissipation rate ε_K, but this parameter appears to be not that important as χ_T and w_K.

In summary, the random beams exhibits spectral switches on propagation in ocean turbulence at relatively short distances, on the order of tens of meters. As we have seen earlier, a similar effect occurs on propagation in atmospheric turbulence, however for much longer ranges, on the order of tens of kilometers [58]. The explanation of the spectral recovery can be made by comparing the source and the turbulence correlations: while the former modifies the spectrum at some short propagation range, the later affects it stronger at greater propagation distances.

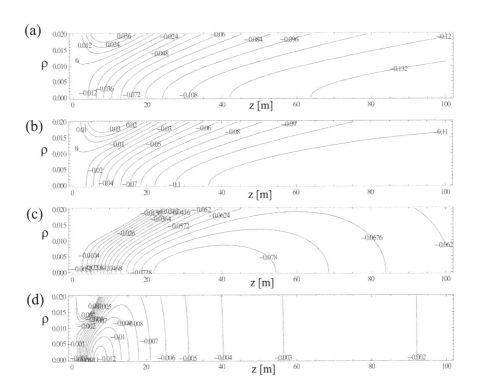

FIGURE 6.13
Contour plots of normalized spectral shift $\varrho_\lambda = \frac{\lambda_1 - \lambda_0}{\lambda_0}$ as a function of z (horizontal axis, in meters) and ρ (vertical axis, in meters) for (a) $\chi_T = 10^{-10}$ K^2/s; (b) $\chi_T = 10^{-5}$ K^2/s, (c) $\chi_T = 10^{-4}$ K^2/s; (d) $\chi_T = 10^{-2}$ K^2/s; $\varepsilon_K = 10^{-4}$ m^2/s^3, $w_K = -4.5$. From Ref. [23].

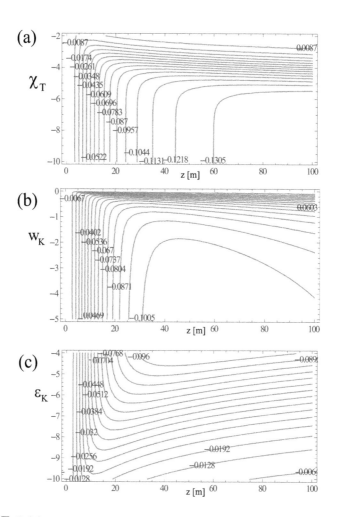

FIGURE 6.14
Contour plots of normalized spectral shift $\varrho_\lambda = \frac{\lambda_1 - \lambda_0}{\lambda_0}$ as a function of z (horizontal axis, in meters) and (a) χ_T, on a log scale for $\varepsilon_K = 10^{-4}$ m^2/s^3, $w_K = -2.5$, (b) w_K for $\varepsilon_K = 10^{-4}$ m^2/s^3, $\chi_T = 10^{-4.5}$ K^2/s, (c) ε_K, on a log scale (vertical axis) for $\chi_T = 10^{-4.5}$ K^2/s, $w_K = -2.5$; $\rho = 0$. From Ref. [23].

6.3 Electromagnetic random beam interaction with random media

6.3.1 General theory

We will now assume that a random electromagnetic beam travels in linear medium with fluctuating index of refraction $n(\mathbf{r}; \omega)$, in the absence of absorption and gain, from the plane $z = 0$ into the half-space $z > 0$, close to the z-axis. As we have stated in Chapter 1, it follows from Maxwell's equations that the space-dependent part $\mathbf{E}(\mathbf{r}; \omega)$ of a monochromatic electric field vector $\mathbf{E}(\mathbf{r}; \omega)exp(-i\omega t)$ propagating in the medium satisfies equation ([4], Section 13.1.1)

$$\nabla^2 \mathbf{E}(\mathbf{r}; \omega) + k^2 n^2(\mathbf{r}; \omega)\mathbf{E}(\mathbf{r}; \omega) + \nabla[\mathbf{E}(\mathbf{r}; \omega)\nabla \ln n^2(\mathbf{r}; \omega)] = 0. \qquad (6.61)$$

Only the third term in this equation couples the Cartesian components of the electric field. Elementary arguments show that this term may be neglected if the refractive index varies slowly with position, i.e., if $\Delta n/n << 1$, over distances of the order of a wavelength. Under these circumstances, which are typical for propagation in atmospheric and oceanic turbulence as well as in some biological tissues, we may replace Eq. (6.61) by equation [72]

$$\nabla^2 \mathbf{E}(\mathbf{r}; \omega) + k^2 n^2(\mathbf{r}; \omega)\mathbf{E}(\mathbf{r}; \omega) = 0. \qquad (6.62)$$

Hence, just like in the case of free-space propagation the transverse Cartesian components E_x and E_y of the electric field propagate independently of each other, obeying the equation

$$\nabla^2 E_\alpha(\mathbf{r}; \omega) + k^2 n^2(\mathbf{r}; \omega)E_\alpha(\mathbf{r}; \omega) = 0, \quad (\alpha = x, y). \qquad (6.63)$$

Consequently, the changes in the properties of electromagnetic beams (such as polarization) propagating in linear random media are not due to the coupling of the field components by the medium but due to the correlations set at the source plane. The inclusion of the field component E_z must be made if the propagating field is considered beyond the paraxial region [73].

It follows from Eq. (6.63) that in the paraxial case the electric field components at any point in the half-space $z > 0$ to which the beam propagates can be expressed with the help of the electromagnetic version of the extended Huygens-Fresnel principle [72]:

$$E_\alpha(\boldsymbol{\rho}, z; \omega) = -\frac{ik\exp(ikz)}{2\pi z} \iint E_\alpha(\boldsymbol{\rho}', z; \omega)$$
$$\times \exp\left[-ik\frac{(\boldsymbol{\rho} - \boldsymbol{\rho}')^2}{2z}\right]\exp[\Psi(\boldsymbol{\rho}, \boldsymbol{\rho}', z; \omega)]d^2\rho', \quad (\alpha = x, y). \qquad (6.64)$$

Here, just like in the scalar case, Ψ is a random, complex phase factor that represents the effect of the random medium on a monochromatic spherical wave. On substituting from Eq. (6.64) into the definition of the cross-spectral density matrix we obtain the following propagation law:

$$W_{\alpha\beta}(\mathbf{r}_1, \mathbf{r}_2; \omega) = \iint W_{\alpha\beta}(\boldsymbol{\rho}_1', \boldsymbol{\rho}_2'; \omega) K(\mathbf{r}_1, \mathbf{r}_2; \boldsymbol{\rho}_1, \boldsymbol{\rho}_2; \omega) d^2\rho_1' d^2\rho_2',$$

$$(\alpha, \beta = x, y),$$

(6.65)

where propagator K was introduced in Section 6.2.1. On using any of the source models introduced in Chapter 4 and any of the random media of Section 6.1 in Eq. (6.65) the evolution of the cross-spectral density matrix and, hence, all the second-order electromagnetic properties of the beam, such as spectrum, coherence and polarization can be predicted [57], [72], [74]–[77]. The angular spectrum-based method has also been extended to the electromagnetic domain by using in Eq. (6.60) the 2×2 cross-spectral density elements instead of the scalar cross-spectral density function [78].

6.3.2 Polarization changes in random media

Until very recently little attention has been paid to investigations about the effects of turbulence on the polarimetric properties of optical beams. This could be due to the fact that for monochromatic electromagnetic beams the estimation has been made that showed that the turbulence-induced polarization changes occuring due to the third coupling term in Eq. (6.61) must be negligible, at least on passage through an isotropic and homogeneous turbulence [7]. However, polarization of such beams does not change in free space either. On the other hand, in the case when an electromagnetic beam is of a random nature, its polarization properties can evolve even in vacuum, as we have seen in Chapter 4. Hence, it is expected that polarization of random electromagnetic beams can change also in turbulence, subject to both the random nature of the source and the fluctuations in the medium. In fact, as was recently shown [74]–[77] the degree and the state of polarization can undergo changes that are fundamentally different from the ones the beams exhibit in free space.

We will now derive expressions for the polarimetric changes in electromagnetic Gaussian Schell-model beams propagating in linear random media. By a straightforward generalization of Eq. (6.38) to the electromagnetic domain we find that the elements of the cross-spectral density matrix of such a beam evaluated at the coinciding arguments $\boldsymbol{\rho}_1 = \boldsymbol{\rho}_2 = \boldsymbol{\rho}$ take the form

$$W_{\alpha\beta}(\boldsymbol{\rho}, \boldsymbol{\rho}, z; \omega) = \frac{A_\alpha A_\beta B_{\alpha\beta}}{\Delta_{\alpha\beta}^2(z; \omega)} \exp\left[-\frac{\rho^2}{2\sigma^2 \Delta_{\alpha\beta}^2(z; \omega)}\right],$$

(6.66)

where

$$\Delta_{\alpha\beta}^2(z; \omega) = 1 + \left(\frac{1}{4\sigma^2} + \frac{1}{\delta_{\alpha\beta}^2}\right)\frac{z^2}{k^2\sigma^2} + M_T \frac{z^2}{k^2\sigma^2},$$

(6.67)

with M_T given in Eq. (6.41). Without loss of generality we will use the Tatarskii's model of the power spectrum, for which this parameter takes the form

$$M_T = 1.098 C_n^2 l_0^{-1/3} k^2 z. \tag{6.68}$$

We note here that similarly to the case of a scalar stochastic beam, the turbulence-induced spreading, being represented by the last term in Eq. (6.67), has the cubic dependence on the propagation distance and, hence, is more effective than the source-induced diffraction for larger propagation distances. In the case when power spectrum Φ_n is given by other models the power of z in expansion coefficient (6.67) can slightly deviate from the value of three.

On substituting from Eqs. (6.66)–(6.68) into the formula for the spectral degree of polarization (4.31) one can analyze the behavior of this quantity when the beam propagates in the atmosphere with the Tatarskii's spectrum. More specifically,

$$\wp(\boldsymbol{\rho}, z; \omega) = \frac{\sqrt{F(\boldsymbol{\rho}, z; \omega)}}{G(\boldsymbol{\rho}, z; \omega)}, \tag{6.69}$$

where

$$
F(\boldsymbol{\rho}, z; \omega) = \left(\frac{A_x^2}{\Delta_{xx}^2(z; \omega)} \exp\left[-\frac{\rho^2}{2\sigma^2 \Delta_{xx}^2(z; \omega)} \right] \right.
$$
$$
\left. - \frac{A_y^2}{\Delta_{yy}^2(z; \omega)} \exp\left[-\frac{\rho^2}{2\sigma^2 \Delta_{yy}^2(z; \omega)} \right] \right)^2
$$
$$
+ \frac{4 A_x^2 A_y^2 |B_{xy}|^2}{\Delta_{xy}^2(z; \omega)} \exp\left[-\frac{\rho^2}{2\sigma^2 \Delta_{xy}^2(z; \omega)} \right], \tag{6.70}
$$

and

$$
G(\boldsymbol{\rho}, z; \omega) = \frac{A_x^2}{\Delta_{xx}^2(z; \omega)} \exp\left[-\frac{\rho^2}{2\sigma^2 \Delta_{xx}^2(z; \omega)} \right]
$$
$$
+ \frac{A_y^2}{\Delta_{yy}^2(z; \omega)} \exp\left[-\frac{\rho^2}{2\sigma^2 \Delta_{yy}^2(z; \omega)} \right]. \tag{6.71}
$$

These formulas are valid at any distance from the source but for small values of ρ, due to the small argument approximation (6.36) to the J_0-Bessel function used in derivation of the extended Huygens-Fresnel integral.

Perhaps, the most interesting fact about polarimetric changes in random beams on propagation in (isotropic and homogeneous) random media is that their initial polarization properties can self-reconstruct as sufficiently large distances from the source plane. We will now prove this statement by considering the atmospheric turbulence with the Tatarskii's spectrum and a Gaussian Schell-model beam with uniform polarization [79]. By means of asymptotic analysis we will show that sufficiently far from the source the degree of polarization returns to exactly the same value it had in the source plane. Indeed, in the asymptotic regime, as $kz \to \infty$, the elements of the cross-spectral density matrix (6.66) tend to values

$$W_{\alpha\beta}(\boldsymbol{\rho}, \boldsymbol{\rho}, z; \omega) = \frac{A_\alpha A_\beta B_{\alpha\beta} \varsigma_{\alpha\beta}}{M_T} z^{-3} - \frac{A_\alpha A_\beta B_{\alpha\beta}}{M_T^2} z^{-4} + O(z^{-5}), \quad (6.72)$$

where O denotes the order of magnitude and

$$\varsigma_{\alpha\beta} = \left(\frac{1}{4\sigma^2} + \frac{1}{\delta_{\alpha\beta}^2} \right) \frac{1}{k^2 \sigma^2}. \quad (6.73)$$

On substituting from Eq. (6.72) into expressions for F and G we find that

$$F(z; \omega) = \left[\frac{A_x^2 - A_y^2}{M_T} z^{-3} - \frac{A_x^2 \varsigma_{xx} - A_y^2 \varsigma_{yy}}{M_T^2} z^{-4} + O(z^{-5}) \right]^2$$

$$+ 4 \left[\frac{A_x^2 A_y^2 B_{xy}}{M_T} z^{-3} - \frac{A_x^2 A_y^2 B_{xy} \varsigma_{xy}}{M_T^2} z^{-4} + O(z^{-5}) \right]^2, \quad (6.74)$$

and

$$G(z; \omega) = \left[\frac{A_x^2 + A_y^2}{M_T} z^{-3} - \frac{A_x^2 \varsigma_{xx} + A_y^2 \varsigma_{yy}}{M_T^2} z^{-4} + O(z^{-5}) \right]^2. \quad (6.75)$$

Finally, on substituting these asymptotic results into Eq. (6.69) and retaining only terms with z^{-3} we find that

$$\wp(\boldsymbol{\rho}, z; \omega) \sim \frac{\sqrt{(A_x^2 - A_y^2)^2 + 4A_x^2 A_y^2 |B_{xy}|^2}}{A_x^2 + A_y^2}, \quad as \quad kz \to \infty. \quad (6.76)$$

It is important to note that atmospheric terms M_T entering the expressions for F and G cancel out as $kz \to \infty$. This implies that atmospheric fluctuations contribute to polarization changes only at intermediate distances from the source. More importantly, on comparing the formula (6.76) and the degree of polarization $\wp(\boldsymbol{\rho}'; \omega)$ in the source plane $z = 0$, readily found from Eqs. (6.66) and (6.67) we are led to the conclusion that

$$\wp(\boldsymbol{\rho}, z; \omega) \sim \wp(\boldsymbol{\rho}'; \omega), \quad as \quad kz \to \infty, \quad (6.77)$$

the expressions on both sides being independent of the transverse variables, $\boldsymbol{\rho}$ and $\boldsymbol{\rho}'$. Thus, the result (6.77) is the mathematical statement of the fact that after a sufficiently long distance of propagation in a linear random medium, such as turbulent atmosphere, the degree of polarization of an initially uniformly polarized electromagnetic Gaussian Schell-model beam returns to its value in the source plane. The same result can be readily shown to hold for other models of the atmospheric and oceanic power spectrum. Moreover, we have only demonstrated the recovery of the degree of polarization but it has also been shown that the state of polarization of the completely polarized portion of the beam also recovers as well [74]. It turns out that for the beams generated by the uniformly polarized sources the polarization ellipse can exhibit drastic changes at the intermediate distances from the source. However, it returns to the original state at sufficiently large distances.

In Fig. 6.15 typical changes in the degree of polarization and the polarization ellipse of a beam propagating through the turbulent atmosphere is illustrated. In Figure 6.16 the changes in the polarization properties of a beam are compared on propagation in free space and in turbulent atmosphere.

The validity of Eq. (6.77) was only proven for the beams generated by the uniformly polarized sources ($\sigma_x = \sigma_y$). In the case when the beam is not initially uniformly polarized the recovery of the polarization properties does not occur [80].

Figure 6.17 illustrates typical changes in the degree of polarization of an electromagnetic Gaussian Schell-model beam, which has (a) and does not have (b) uniform polarization in the source plane.

6.3.3 Propagation in non-Kolmogorov atmospheric turbulence

We will now discuss the changes in basic statistical properties of random electromagnetic beams in atmospheric turbulence with power spectrum (6.12)–(6.14) in a greater detail. If Eqs. (6.38)–(6.40) are generalized to electromagnetic domain we then obtain for the elements of the cross-spectral density matrix of the Gaussian Schell-model beam with the uniform polarization the formulas:

$$W_{\alpha\beta}\left(\mathbf{r}_1, \mathbf{r}_2; \omega\right) = \frac{B_{\alpha\beta} A_\alpha A_\beta}{\Delta^2_{\alpha\beta}(z;\omega)} \exp\left[-\frac{(\mathbf{r}_1 + \mathbf{r}_2)^2}{8\sigma^2 \Delta^2_{\alpha\beta}(z;\omega)}\right] \exp\left[\frac{ik\left(\mathbf{r}_2^2 - \mathbf{r}_1^2\right)}{2R_{\alpha\beta}(z;\omega)}\right]$$

$$\times \exp\left\{-\left[\frac{1}{2\Delta^2_{\alpha\beta}(z;\omega)}\left(\frac{1}{4\sigma^2} + \frac{1}{\delta^2_{\alpha\beta}}\right) + \frac{1}{3}\pi^2 k^2 z I (1 + \sigma^2)\right.\right.$$

$$\left.\left. - \frac{\pi^4 k^2 z^4 I^2}{18\sigma^2 \Delta^2_{\alpha\beta}(z;\omega)}\right](\mathbf{r}_1 - \mathbf{r}_2)^2\right\},$$

$$(6.78)$$

where the spreading coefficients and curvature terms are given respectively by the expressions

$$\Delta^2_{\alpha\beta}(z;\omega) = 1 + \frac{z^2}{k^2\sigma^2}\left(\frac{1}{4\sigma^2} + \frac{1}{\delta^2_{\alpha\beta}}\right) + \frac{2\pi^2 z^3 I}{3\sigma^2}, \qquad (6.79)$$

$$R_{\alpha\beta}(z;\omega) = \frac{\sigma^2 \Delta^2_{\alpha\beta}(z;\omega)z}{\sigma^2 \Delta^2_{\alpha\beta}(z;\omega) + \frac{1}{3}\pi^2 z^3 I - \sigma^2}, \qquad (6.80)$$

268

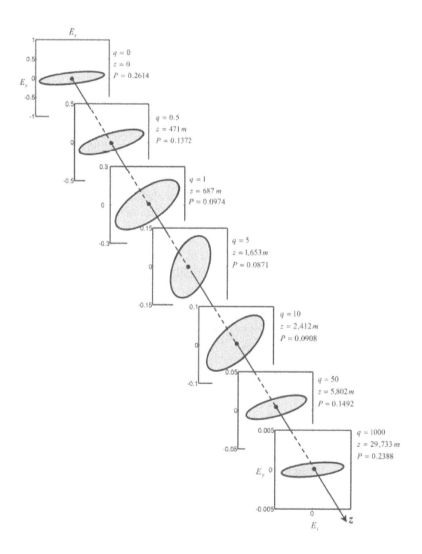

FIGURE 6.15
Propagation of the polarization ellipse of the electromagnetic Gaussian Schell-model beam along the optical axis. Here q denotes the Rytov variance σ_R^2. From Ref. [74].

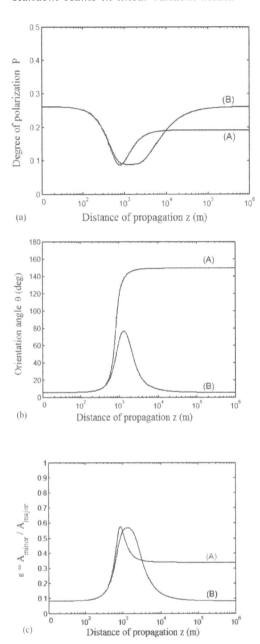

FIGURE 6.16
Comparison of the polarization properties of a typical electromagnetic Gaussian Schell-model beam on propagation in free space and in turbulent atmosphere. From Ref. [74].

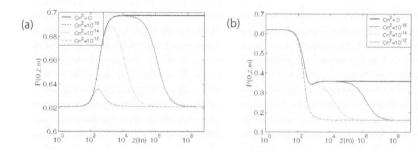

FIGURE 6.17

Propagation of the degree of polarization along the optical axis of the beam produced by an electromagnetic Gaussian Schell-model source with (a) uniform polarization; (b) non-uniform polarization. The Kolmogorov model for the power spectrum was used to characterize the atmospheric fluctuations, values of the structure parameter are given for each curve in $m^{-2/3}$. From Ref. [80].

and I was defined in Eq. (6.42).

Following Ref. [77] we will now illustrate the behavior of a typical electromagnetic Gaussian-Schell-model beam (with diagonal matrix) and analyze its dependence on parameter α. We will set the following values of the parameters for the atmosphere and the beam: $\tilde{C}_n^2 = 10^{-13}$ m$^{3-\alpha}$, $A_y = 1$, $l_0 = 1$ mm, $L_0 = 1$ m, $\lambda = 0.6328$ μm, $\sigma = 2.5$ cm, $\delta_{xx} = 5$ mm, $\delta_{yy} = 0.5$ mm.

Figure 6.18 shows variation of the on-axis spectral density normalized by its value in the source plane $S_N(0, z; \omega)$ and the spectral degree of polarization $\wp(0, z; \omega)$ with slope α at a distance of 1 km from the source. In this figure the illumination beam is uniformly unpolarized across the source, but due to source correlations it becomes nearly polarized at 1 km just like for propagation in vacuum. In this case the atmospheric turbulence substantially modifies these statistics, the strength of the effect being dependent on α. The main trends of both properties are similar, with maximum values at the ends of the interval $3 < \alpha < 4$ and one minimum inside. This figure shows that the turbulence affects both the spectral density and the spectral degree of polarization at most when α is close to its lower limit but does not reach it.

In Fig. 6.19 (left) is indicated the on-axis normalized spectral density as a function of distance z from the source for several values of parameter α. The spectral density remains almost the same up to about a kilometer and then decreases at a rate depending on α.

Figure 6.19 (right) shows the variation in the on-axis spectral degree of polarization $\wp$ as a function of z, for several values of the parameter α. The degree of polarization of the initially unpolarized beam first grows to almost

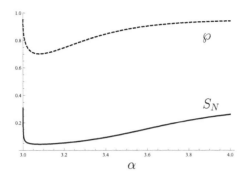

FIGURE 6.18
The normalized spectral density S_N and the spectral degree of polarization $\wp$ as functions of α for $\rho = 0$, $z = 1$ km, $A_x = 1$. From Ref. [77].

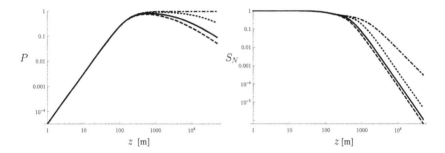

FIGURE 6.19
(left) The normalized spectral density S_N and (right) the spectral degree of polarization $\wp$ as functions of distance z for $\rho = 0$, $A_x = 1$, and different α: $\alpha = 3.01$ (solid curve), $\alpha = 3.1$ (dashed curve), $\alpha = 3.67$ (dotted curve) and free space (dot-dashed curve) on a log scale. From Ref. [77].

unity due to source correlations, independently of α, but decreases to level depending on the value of α at large distances.

Figure 6.20 concerns the changes in the modulus of the spectral degree of coherence η with propagation distance z, the separation distance $\rho_d = |\boldsymbol{\rho}_1 - \boldsymbol{\rho}_2|$ between two points in the transverse plane and the parameter α. In particular, Fig. 6.20 (left) illustrates the dependence of the absolute value of the spectral degree of coherence on z, between two fixed points symmetrically situated about the optical axis. While this quantity reaches high values at distances close to the source, the effect being attributed to source correlations, it then decreases, at about 1 km from the source, to lower values, depending on α. For other statistics the dependence of the degree of coherence on α is non-

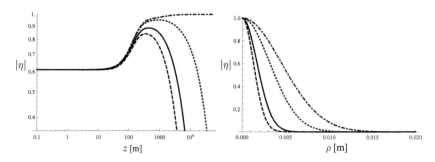

FIGURE 6.20
(left) The absolute value of the spectral degree of coherence as a function of distance z for $\rho = 10^{-3}$ m, $A_x = 1.3$ and different α: $\alpha = 3.01$ (solid curve), $\alpha = 3.1$ (dashed curve), $\alpha = 3.67$ (dotted curve) and free space (dot-dashed curve) on a log scale; (right) The absolute value of the spectral degree of coherence as a function of ρ for $z = 1$ km, $A_x = 1.3$ and the same α. From Ref. [77].

monotonic, the fastest drop corresponding to $\alpha = 3.1$. Figure 6.20 (right) presents the decrease in the modulus of the degree of coherence varying with ρ at fixed distance 1 km from the source, for several values of α. Thus, all the second-order statistics of the beam are affected by non-Kolmogorov turbulence the most in a region about $\alpha = 3.1$.

6.3.4 Propagation in oceanic turbulence

We will now employ the power spectrum model (6.15)–(6.16) of oceanic turbulence [22], which takes into account both temperature and salinity fluctuations for predicting the changes in the second-order statistics of electromagnetic Gaussian Schell-model beams on traveling in the oceanic waters [75]. Due to the complexity of the oceanic spectrum the changes of polarization properties of the propagating beam can be found from Eqs. (6.86) and (6.87) with integral I evaluated numerically. Unless it is specified overwise we will assume the following values of parameters of the source and of the turbulent ocean: $\lambda = 0.633$ μm, $\sigma_x = \sigma_y = \sigma = 1$ cm, $\delta_{xx} = 0.5$ mm, $\delta_{yy} = 4$ mm, $\delta_{xy} = 5$ mm, $A_x = 1.3$, $A_y = 1$ and $B_{xy} = B_{yx} = 0.1$, $B_{xx} = B_{yy} = 1$, $\chi_T = 10^{-6} K^2/s$, $\varepsilon_K = 10^{-7} m^2/s^3$, $w_K = -2.5$.

Figure 6.21 shows the typical evolution of the polarization ellipse of a propagating beam along the optical axis and includes the values of the spectral density (normalized by its value at the source), the degree of polarization, the orientation angle and the degree of ellipticity for four fixed values of propagation distance z. Just like in atmospheric turbulence, the beam can change its polarimetric properties in a non-monotonic manner, as z increases.

Moreover, at sufficiently large distances from the source, the polarization may self-reconstruct.

It is evident from Fig. 6.22 that all three polarization properties behave qualitatively in a very similar way. This figure is organized as follows: spectral density (upper-left); degree of polarization (upper-right); orientation angle (lower-left) and degree of ellipticity (lower-right). All the quantities vary with propagation distance z from the source (in meters, horizontal axis) and show the dependence on the source correlation properties, namely on δ_{xx}: $\delta_{xx} = 0.5$ mm (solid curves), $\delta_{xx} = 1$ mm (dotted curves), $\delta_{xx} = 5$ mm (dashed curves) while the two other root-mean-square correlation widths, δ_{yy} and δ_{xy}, are kept fixed. All polarization properties are appreciably sensitive to the source correlations, with larger changes (in magnitude) occurring for smaller values of δ_{xx}.

Figure 6.23 illustrates how the degree of polarization of the beam depends on the parameters of oceanic turbulence, in particular mean square temperature dissipation rate χ_T, energy dissipation rate per unit mass ϵ_K, and temperature-salinity balance parameter w_K. Figure 6.23 (a) explores the dependence of the beam statistics on propagation distance z for three fixed values of the mean square temperature dissipation rate: $\chi_T = 10^{-2} K^2/s$ (solid curves), $\chi_T = 10^{-6} K^2/s$ (dotted curves), $\chi_T = 10^{-10} K^2/s$ (dashed curves). This figure indicates that larger values of χ_T (stronger turbulence) correspond to earlier drop in the spectral density but smaller changes in the polarization properties of the beam. In particular, for $\chi_T = 10^{-2} K^2/s$ (solid curves) the degree of polarization remains invariant in the considered range values of z. In this case, because of the extremely strong turbulence, the polarization changes due to the source correlations are almost entirely suppressed by the fluctuations in the medium. For weaker turbulence (dotted and dashed curves) the source correlations and the medium compete resulting in several changes in polarization as the beam travels.

Similar analysis is carried out in Fig. 6.23 (b) where the degree of polarization of the beam is shown as a function of propagation distance z, for three values of the energy dissipation rate per unit mass ε_K, namely $\varepsilon_K = 10^{-4} m^2/s^3$ (solid curves), $\varepsilon_K = 10^{-7} m^2/s^3$ (dotted curves), $\varepsilon_K = 10^{-10} m^2/s^3$ (dashed curves). In contrast with Fig. 6.23 (a) the polarization properties start changing sooner and do so less for smaller values of ε_K . Finally, Fig. 6.23 (c) illustrates that the salinity parameter w_K also plays an important part in determining the behavior of a beam's polarization. Three cases are considered: $w_K = -0.1$ (solid curves), $w_K = -2.5$ (dotted curves), $w_K = -4.9$ (dashed curves) the first option leading to the fastest and smallest changes in polar-

274

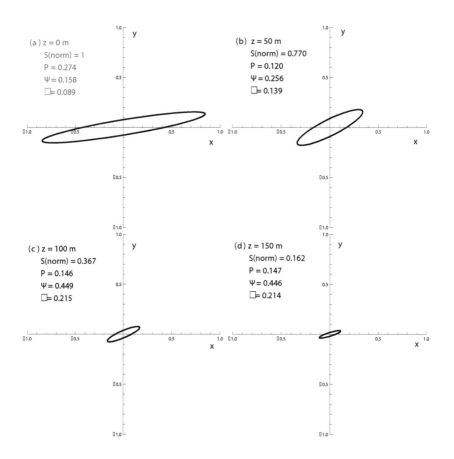

FIGURE 6.21
Typical evolution of the polarization properties of a stochastic beam in oceanic turbulence (on-axis). From Ref [75].

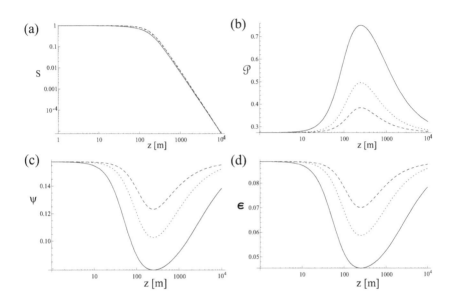

FIGURE 6.22
Variation of the statistical properties of the beam with distance z (horizontal axis, in meters) for several values of the root-mean-square correlation coefficient δ_{xx} of the source. From Ref [75].

276

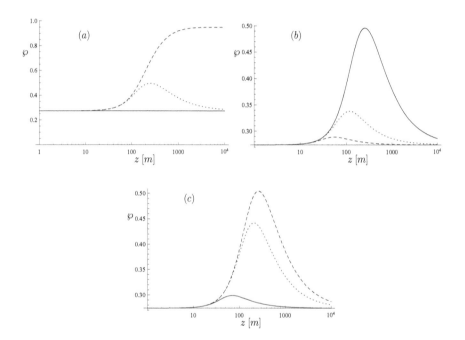

FIGURE 6.23
Variation of the degree of polarization of the beam with distance z (horizontal axis, in meters) for several values of (a) the mean square temperature dissipation rate χ_T; (b) the energy dissipation rate per unit mass ε_K; (c) the temperature-salinity balance parameter w_K. From Ref [75].

ization. This situation takes place when the salinity fluctuations in the ocean substantially balance temperature fluctuations.

Bibliography

[1] H. Tennekes and J. L. Lumley, *A First Course in Turbulence*, The MIT Press, Cambridge, 1972.

[2] P. K. Kundu and I. M. Cohen, *Fluid Mechanics*, Elsevier, 1987.

[3] U. Frisch, *Turbulence: The Legacy of A. N. Kolmogorov*, Cambridge University Press, 1995.

[4] L. C. Andrews and R. L. Phillips, *Laser Beam Propagation in the Turbulent Atmosphere*, 2nd edition, SPIE press, 2005.

[5] A. N. Kolmogorov, "The local structure of turbulence in an incompressible viscous fluid for very large Reynolds numbers," *C. R. Acad. Sci. U.S.S.R.*, **30**, 301–305 (1941).

[6] A. N. Kolmogorov, "Dissipation of energy in the locally isotropic turbulence," *C. R. Acad. Sci. U.S.S.R.* **32**, 16-18 (1941).

[7] V. I. Tatarskii, *Wave Propagation in a Turbulent Medium*, Nauka, 1967.

[8] L. C. Andrews, "An analytical model for the refractive index power spectrum and its application to optical scintillations in the atmosphere," *J. Mod. Opt.* **39**, 1849–1853 (1992).

[9] R. J. Hill, "Spectra of fluctuations in refractivity, temperature, humidity, and the temperature-humidity cospectrum in the inertial and dissipation ranges," *Radio Sci.* **13**, 953–961 (1978).

[10] K. J. Grayshan, F. S. Vetelino, and C. Y. Young, "A marine atmopsperic spectrum for laser propagation," *Waves in Random and Complex Media* **18**, 173–184 (2008).

[11] C. A. Friehe, J. C. La Rue, F. H. Champagne, C. H. Gibson, and G. F. Dreyer, "Effects of temperature and humidity fluctuations on the optical refractive index in the marine boundary layer," *J. Opt. Soc. Am.* **65**, 1502–1511 (1975).

[12] F. Dalaudier, M. Crochet, and C. Sidi, "Direct comparison between in situ and radar measurements of temperature fluctuation spectra: a puzzling results," *Radio Sci.* **24**, 311–324 (1989).

[13] M. S. Belen'kii, J. D. Barchers, S. J. Karis, C. L. Osmon, J. M. Brown, and R. Q. Fugate, "Preliminary experimental evidence of anisotropy of turbulence and the effect of non-Kolmogorov turbulence on wavefront tilt statistics," *Proc. SPIE* **3762**, 396-406 (1999).

[14] B. Joseph, A. Mahalov, B. Nicolaenko, and K. L. Tse, "Variability of turbulence and its outer scales in a model tropopause jet," *J. Atm. Sci.* **61**, 621-643 (2004).

[15] A. Mahalov, B. Nicolaenko, K. L. Tse, and B. Joseph, "Eddy mixing in jet-stream turbulence under stronger stratification," *Geophys. Res. Lett.* **31**, L23111 (2004).

[16] A. Zilberman, E. Golbraikh, N. S. Kopeika, A. Virtser, I. Kupershmidt, and Y. Shtemler, "LIDAR study of aerosol turbulence characteristics in the troposphere: Kolmogorov and non-Kolmogorov turbulence," *Atmos. Res.* **88**, 66–77 (2008).

[17] O. Korotkova, N. Farwell, and A. Mahalov, "The effect of the jet-stream on the intensity of laser beams propagating along slanted paths in the upper layers of the turbulent atmosphere," *Waves in Random Media*, **19**, 692–702 (2009).

[18] A. Zilberman, E. Golbraikh, and N. S. Kopeika, "Propagation of electromagnetic waves in Kolmogorov and non-Kolmogorov atmospheric turbulence: three-layer altitude model," *Appl. Opt.* **47**, 6385–6391 (2008).

[19] I. Toselli, L. C. Andrews, R. L. Phillips, and V. Ferrero, "Free-space optical system performance for laser beam propagation through non-Kolmogorov turbulence," *Opt. Eng.* **47**, 026003 (2008).

[20] C. Rao, W. Jiang, and N. Ling, "Spatial and temporal characterization of phase fluctuations in non-Kolmogorov atmospheric turbulence," *J. Mod. Opt.* **47**, 1111–1126 (2000).

[21] M. M. Charnotskii, "Intensity fluctuations of flat-topped beam in non-Kolmogorov weak turbulence: comment," *J. Opt. Soc. Am. A* **29**, 1838–1840 (2012).

[22] V. V. Nikishov and V. I. Nikishov, "Spectrum of turbulent fluctuation of the sea–water refractive index," *Int. J. Fluid Mech. Res.* **27**, 82–98 (2000).

[23] E. Shchepakina, N. Farwell, and O. Korotkova, "Spectral changes in stochastic light beams propagating in turbulent ocean," *Appl. Phys. B* **105**, 415–420 (2011).

[24] M. Alford, D. Gerdt, and C. Adkins, "An ocean refractometer: resolving millimeter-scale turbulent density fluctuations via the refractive index," *J. Atmos. and Ocean. Tech.* **23**, 121–137 (2006).

[25] J. M. Schmitt and G. Kumar, "Turbulent nature of refractive-index variations in biological tissue," *Opt. Lett.* **21**, 1310–1312 (1996).

[26] J. M. Schmitt and G. Kumar, "Optical scattering properties of soft tissue: a discrete particle model," *Appl. Opt.* **37**, 2788–2797 (1998).

[27] A. Ishimaru, *Wave Propagation and Scattering in Random Media*, IEEE Press, 1997.

[28] M. Moscoso, J. B. Keller, and G. Papanicolaou, "Depolarization and blurring of optical images by biological tissue," *J. Opt. Soc. Am. A* **18**, 948–960 (2001).

[29] M. Xu and R. R. Alfano, "Fractal mechanisms of light scattreing in biological tissue and cells," *Opt. Lett.* **30**, 3051–3053 (2005).

[30] C. J. R. Sheppard, "Fractal model of light scattering in biological tissue and cells," *Opt. Lett.* **32**, 142–144 (2007).

[31] J. D. Rogers, İ. R. Çapoğlu, and V. Backman, "Nonscalar elastic light scattreing from continuous random media in the Born approxiamtion," *Opt. Lett.* **34**, 1891–1893 (2009).

[32] T. Shirai, A. Dogariu, and E. Wolf, "Mode analysis of spreading of partially coherent beams propagating through atmospheric turbulence," *J. Opt. Soc. Am. A* **20**, 1094–1102 (2003).

[33] G. Gbur and E. Wolf, "Spreading of partially coherent beams in random media," *J. Opt. Soc. Am. A* **19**, 1592–1598 (2002).

[34] W. Lu, L. Liu, J. Sun, Q. Yang, and Y. Zhu, "Change in degree of coherence of partially coherent electromagnetic beams propagating through atmospheric turbulence," *Opt. Commun.* **271**, 1–8 (2007).

[35] T. Shirai, A. Dogariu, and E. Wolf, "Directionality of Gaussian Schell-model beams propagating in atmospheric turbulence," *Opt. Lett.* **28**, 610–612 (2003).

[36] G. Gbur and O. Korotkova, "Angular spectrum representation for propagation of arbitrary coherent and partially coherent beams through atmospheric turbulence," *J. Opt. Soc. Am. A* **24**, 745–752 (2007).

[37] M. Abramowitz and I. A. Stegun, Eds. *Handbook of Mathematical Functions with Formulas, Graphs, and Mathematical Tables*, Dover, 1965.

[38] O. Korotkova and E. Wolf, "Beam criterion for atmospheric propagation," *Opt. Lett.* **32**, 2137–2139 (2007).

[39] Jerry D. Gibson, Ed., *The Communications Handbook*, CRC Press, 2002.

[40] J.B. Campbell, *Introduction to Remote Sensing*, Taylor and Francis, 4th Edition, 2007.

[41] Z. Tong, S. Sahin, and O. Korotkova, "Sensing of semi-rough targets embedded in atmospheric turbulence by means of stochastic electromagnetic beams," *Opt. Comm.* **283**, 4512–4518 (2010).

[42] O. Korotkova, L. C. Andrews, and R. L. Phillips, "A model for a partially coherent Gaussian beam in atmospheric turbulence with application in LaserCom," *Opt. Eng.* **43**, 330–341 (2004).

[43] D. Zhao, O. Korotkova, and E. Wolf, "Application of correlation-induced spectral changes to inverse scattering," *Opt. Lett.* **32**, 3483–3485 (2007).

[44] E. Wolf, "Invariance of the spectrum of light on propagation," *Phys. Rev. Lett.* **56**, 1370–1372 (1986).

[45] E. Wolf and D. F. V. James, "Correlation-induced spectral changes," *Rep. Prog. Phys.* **59**, 771–818 (1996).

[46] F. Gori, J. T. Foley, and E. Wolf, "Frequency shifts of spectral lines produced by scattering from spatially random media," *J. Opt. Soc. Am. A* **6**, 1142–1149 (1989); errata, ibid. 7, 173 (1990).

[47] A. Dogariu and E. Wolf, "Spectral changes produced by static scattering on a system of particles," *Opt. Lett.* **23**, 1340–1342 (1998).

[48] H. Roychowdhury, G. P. Agrawal, and E. Wolf, "Changes in the spectrum, in the spectral degree of polarization, and in the spectral degree of coherence of a partially coherent beam propagating through a gradient-index fiber," *J. Opt. Soc. Am. A* **23**, 940–948 (2006).

[49] J. Pu, H. Zhang, and S. Nemoto, "Spectral shifts and spectral switches of partially coherent light passing through an aperture," *Opt. Comm.* **162**, 57–63 (1999).

[50] L. Pan and B. Lu, "Spectral changes and spectral switches of partially coherent beams focused by an aperture lens," *J. Opt. Soc. A* **21**, 140–148 (2004).

[51] Z. Tong and O. Korotkova, "Spectral Shifts and spectral switches in random fields on interaction with negative phase materials," *Phys. Rev. A* **82**, 013829 (2010).

[52] X. Ji, E. Zhang, and B.Lu, "Changes in the spectrum of Gaussian Schell-model beams propagating through turbulent atmosphere," *Opt. Commun.* **259**, 1-6 (2006).

[53] X. Ji, E. Zhang, and B. Lu, "Changes in the spectrum and polarization of polychromatic partially coherent electromagnetic beams in the turbulent atmosphere," *Opt. Commun.* **275**, 292-300 (2007).

[54] M. Alavinejad, B. Ghafarya, and D. Razzaghia, "Spectral changes of partially coherent flat topped beam in turbulent atmosphere," *Opt. Comm. Vol.* **281**, 2173–2178 (2008).

[55] G. Zhang and J. Pu, "Spectral changes of polychromatic stochastic electromagnetic vortex beams propagating through turbulent atmosphere," *J. Mod. Opt.* **55**, 2831–2842 (2008).

[56] H. Wang and X. Lia, "Changes in the spectrum of twist anisotropic Gaussian Schell-model beams propagating through turbulent atmosphere," *Opt. Comm.* **281**, 2337–2341 (2008).

[57] O. Korotkova, J. Pu, and E. Wolf, "Spectral changes of stochastic electromagnetic beams propagating in atmospheric turbulence," *J. Mod. Opt.* **55**, 1199–1208 (2008).

[58] O. Korotkova and E. Shchepakina, "Color changes in stochastic light fields propagating in non-Kolmogorov turbulence," *Opt. Lett.* **35**, 3772–3774 (2010).

[59] N. G. Jerlov, *Marine Optics*, Elsevier, 1976.

[60] C. Mobley, *Light and Water*, Academic Press, 1994.

[61] K. S. Shifrin, *Physical Optics of Ocean Water*, American Institute of Physics, 1988.

[62] R. W. Spinrad, K. L. Carder, M. J. Perry [Eds.], *Ocean Optics*, Oxford University Press, 1994.

[63] R. E. Walker, *Marine Light Field Statistics*, John Wiley and Sons, 1984.

[64] N. Swanson, "Coherence loss of laser light propagated through simulated coastal waters," *Proc. SPIE* **1750**, 397–406 (1992).

[65] K. Arora and E. O. Sheybani, "Measurement of the temporal coherence of an underwater optical scattered field," *Microwave Opt. Techol. Lett.* **6**, 1510–1514 (1993).

[66] A. Perennou, J. Cariou, and J. Lotrian, "Two inteferometric evaluations of the spacial coherence of a laser beam scatterd by turbid water," *Pure Appl. Opt.* **4**, 617–628 (1995).

[67] J. B. Snow, J. P. Flatley, D. E. Freeman, M. A. Landry, C. E. Lindstrom, J. R. Longacre, and J. A. Schwartz, "Underwater propagation of high-data-rate laser communications pulses," *Proc. SPIE* **1750**, 419–427 (1992).

[68] S. Arnon and D. Kedar, "Non-line-of-sight underwater optical wireless communication network," *J. Opt. Soc. Am. A* **26**, 530–539 (2009).

[69] W. Lu, L. Liu, and J. Sun, "Influence of temperature and salinity fluctuations on propagation behaviour of partially coherent beams in oceanic turbulence," *J. Opt. A: Pure Appl. Opt.* **8**, 1052–1058 (2006).

[70] W. Hou, "A simple underwater imaging model," *Opt. Lett.* **34**, 2688–2690 (2009).

[71] F. Hansonand M. Lasher, "Effects of underrwater tubulence on laser beam propagation and coupling into single-mode fiber," *Appl. Optics* **49**, 3224–3230 (2010).

[72] M. Salem, O. Korotkova, A. Dogariu, and E. Wolf, "Polarization changes in partially coherent elecytomagnetic beams propagating through turbulent atmosphere," *Waves in Random Media* **14**, 513–523 (2004).

[73] M. A. Alonso, O. Korotkova, and E. Wolf, "Propagation of the electric correlation matrix and the van Cittert-Zernike theorem for random electromagnetic fields," *J. Mod. Opt.* **53**, 969–978 (2006).

[74] O. Korotkova, M. Salem, A. Dogariu, and E. Wolf, "Changes in the polarization ellipse of random electromagnetic beams propagating through turbulent atmosphere," *Waves in Random and Complex Media* **15**, 353–364 (2005).

[75] O. Korotkova and N. Farwell, "Effect of ocean turbulence on polarization of stochastic beams," *Opt. Commun.* **284**, 1740–1746 (2011).

[76] W. Gao and O. Korotkova, "Changes in the state of polarization of a random electromagnetic beam propagating through tissue," *Opt. Comm.* **270**, 474–478 (2007).

[77] E. Shchepakina and O. Korotkova, "Second-order statistics of stochastic electromagnetic beams propagating through non-Kolmogorov turbulence," *Opt. Express* **18**, 10650–10658 (2010).

[78] O. Korotkova and G. Gbur, "Angular spectrum representation for propagation of random electromagnetic beams in a atmospheric turbulence," *J. Opt. Soc. Am. A* **24**, 2728–2736 (2007).

[79] O. Korotkova, M. Salem, and E. Wolf, "The far-zone behavior of the degree of polarization of partially coherent beams propagating through atmospheric turbulence," *Opt. Comm.* **233**, 225–230 (2004).

[80] X. Du, D. Zhao, and O. Korotkova, "Changes in the statistical properties of stochastic anisotropic electromagnetic beams on propagation in the turbulent atmosphere," *Optics Express* **15**, 16909–16915 (2007).

7

Mitigation of random media effects with random beams

CONTENTS

7.1 Free-space optical communications

One of the major applications that the stochastic beams have already found is the Free-Space Optical (FSO) communications, in situations when the optical channel has to pass through the atmosphere. The efficient optical communications through the atmosphere have been the challenge for the scientific and the engineering communities for more than a century. As a matter of fact one of the first devices for converting sound to the electromagnetic wave is presented in Fig. 7.1 [1]. The first wireless speech communication was achieved at optical wavelengths in 1878, more than 25 years before Reginald Fessenden did the same with radio.

Practically unlimited bandwidth of the signal at optical frequencies is the most desirable feature that distinguishes an optical channel from a classic (radio wave) link. With the advent of the highly directional laser radiation the additional advantage of the FSO communications related to link security also became apparent. However, the atmospheric effects, such as scintillations and beam wander were found to be very severe for deterministic (laser) light beams. It later became apparent that in order to mitigate the turbulent atmosphere several types of diversity can be effectively employed at the signal transmitter as well as at the receiver system. For instance, a source trans-

286

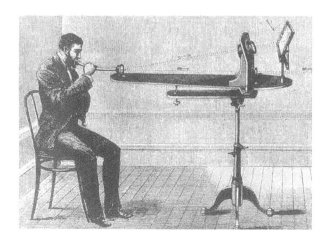

FIGURE 7.1
Photophone: one of the first devices transforming sound to optical wave. From
Ref. [1].

mitting at several frequencies or a collector consisting of an array of sparsely
located detectors can significantly reduce fluctuations in the received signal.
Another method for atmospheric mitigation is the use of partial coherence
and partial polarization of optical sources generating the beams. We will first
discuss the effects of partial polarization and partial coherence on communication systems separately and then demonstrate that the use of both of them
provides the best solution. As we will soon find, while partial coherence of
the source significantly helps with scintillation reduction but leads to overall
power loss, partial polarization of the source is limited in terms of scintillation
mitigation but does not lead to any power loss. Both scintillation reduction
and maximum power collection are crucial for high quality of the communication channels.

A schematic diagram of a typical FSO link is presented in Fig. 7.2. A
generated laser light can be spatially randomized by means of a diffuser or
a Spatial Light Modulator (SLM) and sent through the channel containing
the turbulent atmosphere. It is then collected and focused by a lens onto the
surface of the detector.

7.1.1 Communication link quality criteria

The quantity that characterizes the quality of the communication link at best
is the Bit-Error-Rate (BER). For the simplest On-Off key modulation scheme

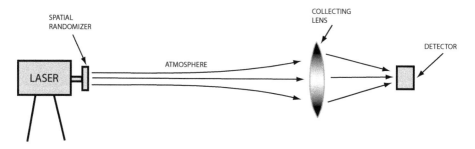

FIGURE 7.2
A typical optical communication link.

it is given by the expression [2]

$$BER = \int_0^\infty \operatorname{erfc}\left(\frac{SNR_A}{2\sqrt{2}}I\right) p_A(I) dI,\qquad(7.1)$$

where $\operatorname{erfc}(x)$ is the complimentary error function, I is the intensity of the fluctuating signal at the receiver and $p_A(I)$ is its probability density function (pdf). We will discuss the models and the experiments for the probability density functions of the intensity in the following section. The BER in Eq. (7.1) returns the ratio of the number of bits of information lost in transmission to the total number of bits sent. Generally, the BER of 10^{-6} and lower is considered as a satisfactory rate and 10^{-9} corresponds to a desired robust level of communication.

The averaged over atmospheric fluctuations signal-to-noise ratio SNR_A in Eq. (7.1) can be expressed as [3]

$$SNR_A = \frac{SNR_c}{\sqrt{\frac{p_{EA}}{p_{E0}} + c_I SNR_c^2}},\qquad(7.2)$$

where SNR_c is the free-space signal-to-noise ratio of the corresponding coherent beam, defined as

$$SNR_c = \frac{\langle i_s \rangle}{\sigma_{DN}}.\qquad(7.3)$$

Here $\langle i_s \rangle$ is the mean value of the received signal current (proportional to the transmitted power) and σ_{DN} is the variance of the detector noise. The term p_{EA}/p_{E0} in Eq. (7.2) is the power loss due to source correlations and turbulence and c_I is the scintillation flux of the beam. It follows at once from Eq. (7.2) that with the increase in the input power ($SNR_c \to \infty$) the signal-to-ratio of the beam propagating in the atmosphere, SNR_A, tends to its limiting value, namely

$$SNR_A = \frac{1}{\sqrt{c_I}}.\qquad(7.4)$$

Relation (7.4) implies that at high input power levels the scintillation index of the beam at the receiver places the ultimate restriction on the quality of the transmitted information. At smaller power levels both the power loss and the scintillations can substantially affect the SNR_A.

7.1.2   The pdf models for beam intensity in the atmosphere

Intuitively, the probability density function (pdf) of a fluctuating quantity shows with which chance it attains a certain level. Practically, from the acquired data of the fluctuating signal one can numerically calculate its statistical moments of any order. However, determination of the pdf from the moments is a very challenging problem which has not been solved exactly so far. In fact, such pdf reconstruction problem is known as the Hausdorff moment problem [4].

The analytic description of the optical signal intensity's pdf in the presence of the turbulent atmosphere is still an open problem, even though it was approached by a number of scientists during the last several decades. The major complication in modeling of the pdf stems from the fact that in various regimes of the refractive-index fluctuations different analytical approaches to characterization of the beam statistics are employed. The matching of the approximate solutions obtained for different regimes leads to ambiguities. Another difficulty stems from the fact that the majority of the analytical models for the intensity pdf have been developed under the assumption that the atmosphere is clear from scatterers (aerosols, water droplets, smoke particles, etc.). In reality, macroscopic scattering can appreciably affect the statistics of the beam. Moreover, as has been recently shown [4], even the power spectrum of the clear-air atmospheric turbulence in maritime environment differers substantially from the ordinary above-ground spectrum. For a thorough review of various intensity pdf models the reader is referred to [2].

The probability density function pdf(I) of the fluctuating intensity I returns the probability with which the beam's intensity attains a certain level. More precisely, if I_N is the intensity, normalized by its mean value, i.e., if

$$I_N = \frac{I}{\langle I \rangle} \tag{7.5}$$

then the probability that the normalized intensity takes on values between I_a and I_b is related to the pdf as

$$Probability(a < I_N < b) = \int_{I_a}^{I_b} \mathrm{pdf}(I_N) dI_N. \tag{7.6}$$

The statistical moment of order n can be obtained from the intensity's pdf

pdf(I_N) by the formula

$$\langle I_N^{(n)} \rangle = \int\limits_0^\infty \text{pdf}(I_N) I_N^n \, dI_N. \tag{7.7}$$

As we have already mentioned, while the calculation of moments $\langle I_N^{(n)} \rangle$ from a given pdf(I_N) is an easy task, the reconstruction of the pdf from the measured moments is considerably more involved. For problems involving light propagation in random media several pdf reconstruction procedures have been suggested. In this work we will only be concerned with two resulting models, one introduced by Barakat [5] suitable for turbulent media containing scatterers and the other by Al-Habash et al. [6] only applicable to clear-air turbulence.

The former approach which we will refer to as the Gamma-Laguerre pdf model, suggests finding several first moments of intensity with the help of the Gamma distribution weighted by the generalized Laguerre polynomials. More precisely, it has the form

$$\text{pdf}^{(GL)}(I_N) = \text{pdf}^{(G)}(I_N) \sum_{n=0}^\infty p_n(I_N) L_n^{(\beta-1)}\left(\frac{\beta I_N}{\mu}\right), \tag{7.8}$$

where pdf$^{(G)}(I_N)$ is the Gamma distribution given by the formula

$$\text{pdf}^{(G)}(I_N) = \frac{1}{\Gamma(\beta)} \left(\frac{\beta}{\mu}\right)^\beta I_N^{(\beta-1)} \exp\left(-\frac{\beta I_N}{\mu}\right), \tag{7.9}$$

with Γ denoting the Gamma-function, and the two parameters of the distribution are defined through the first and second moment:

$$\mu = \langle I_N \rangle, \quad \beta = \frac{\langle I_N \rangle^2}{\langle I_N^2 \rangle - \langle I_N \rangle^2}, \tag{7.10}$$

i.e., the average value and the reciprocal of the scintillation index. Further, $p_n(I_N)$ are the weighing coefficients:

$$p_n(I_N) = n! \Gamma(\beta) \sum_{k=0}^n \frac{(-\beta/\mu)^k \langle I_N^k \rangle}{k!(n-k)! \Gamma(\beta+k)}. \tag{7.11}$$

It turns out that $p_0 = 1$, $p_1 = 0$ and $p_2 = 0$, since the first two moments determine μ and β. The generalized Laguerre polynomials $L_n^{(\beta-1)}(x)$ entering formula (7.8) are given by the expressions

$$L_n^{(\beta-1)}(x) = \sum_{k=0}^n \binom{n+\beta-1}{n-1} \frac{(-x)^k}{k!}. \tag{7.12}$$

In the original paper [5] it is recommended that only five first moments are included for accurate and stable approximation of the pdf. Hence, we would like to note that the Gamma-Laguerre model is only based on the several first statistical moments of the fluctuating intensity, and does not require the knowledge of the atmospheric parameters, characteristics of the source and propagation distance. More importantly, this model is valid everywhere in the cross-section of the beam, while making it possible to account for possible scattering and absorption from particles and aerosols.

In Ref. [6] another model was introduced that became known as the Gamma-Gamma pdf model. It is based on the assumption that only the first and the second moments of the fluctuating intensity are directly used. Gamma-Gamma model also requires specification of several source parameters and atmospheric channel specifications (propagation distance, atmospheric spectrum, etc.). Moreover, in this case the pdf reconstruction can be done only for the coherent Gaussian beam and it is derived under the assumption of the clear-air turbulence (scattering and absorption effects cannot be taken into account). The Gamma-Gamma pdf has the form

$$\text{pdf}^{(GG)}(I_N) = \frac{2(\alpha\beta)^{(\alpha+\beta)/2}}{\Gamma(\alpha)\Gamma(\beta)} I_N^{\frac{\alpha+\beta}{2}-1} K_{\alpha-\beta}(2\sqrt{\alpha\beta I_N}), \qquad (7.13)$$

where $K_m(x)$ is the modified Bessel function of the second kind, and the parameters α and β are defined as follows

$$\alpha = \frac{1}{\exp(\sigma_{\ln x}^2) - 1}, \quad \beta = \frac{1}{\exp(\sigma_{\ln y}^2) - 1}, \qquad (7.14)$$

with $\sigma_{\ln x}^2$ and $\sigma_{\ln y}^2$ being normalized variances of intensity due to perturbations caused by large and small scales of the turbulent medium. For instance, under the assumption of the Kolmogorov power spectrum of atmospheric fluctuations these quantities are given by the formulas [6]

$$\sigma_{\ln x}^2 = \frac{0.49\sigma_B^2}{[1 + 0.56(1 + \theta_D)\sigma_B^{12/5}]^{7/6}}, \quad \sigma_{\ln y}^2 = \frac{0.51\sigma_B^2}{[1 + 0.69\sigma_B^{12/5}]^{5/6}}, \qquad (7.15)$$

with

$$\sigma_B^2 = \frac{\langle I_N^2 \rangle - \langle I_N \rangle^2}{\langle I_N \rangle^2} \qquad (7.16)$$

being the scintillation index in the center of the beam, and

$$\theta_D = \left[1 + \frac{4z^2}{k^2 w_0^4}\right]^{-1} \qquad (7.17)$$

being the divergence angle, depending on propagation distance from the source to the receiver, z, wave number k, and the initial beam radius w_0 of the Gaussian beam.

In order to illustrate how well the two models approximate the intensity

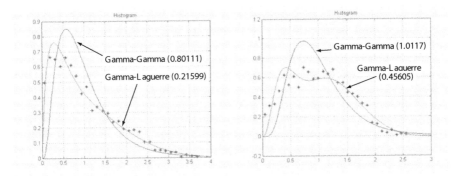

FIGURE 7.3
The probability density function of the normalized intensity reconstructed
by the Gamma-Gamma and the Gamma-Laguerre models. (left) Experiment
above land; (right) Experiment above water. Dots represent actually measured
intensity levels. From Ref. [7].

distributions for propagating laser beams in Fig. 7.3 we include their comparison with the actual histogram of the measured intensity (see [7] for the
technical details of the experiment and postprocessing). The above-ground
link in Fig. 7.3 (left) and the above-water link in Fig. 7.3 (right) imply that
the data (stars) for normalized intensity is distributed differently. In particular, for both links the Gamma-Laguerre model has a somewhat better match
with the data than the Gamma-Gamma model. The double-bump structure
of the Gamma-Laguerre curve in Fig. 7.3 (right) is solely due to the fact that
the finite number of Laguerre polynomials is used. The least square errors of
the histograms are included for a more quantitative illustration of how well
the two analytical models predict the intensity pdfs for each case.

In Fig. 7.4 the comparison of the intensity probability density functions
of a HeNe laser beam and a HeNe laser beam reflected by an SLM propagating in a turbulent atmosphere is shown. The measurements were made by
the authors of Ref. [4] in October 2012 (over the water, 314 meter propagation path, HeNe detector, 10,000 samples/sec, about 1 million data points,
$C_n^2 \approx 10^{14}$ m$^{-2/3}$). It is clear from the two distributions that the intensity
pdf of the partially coherent beam is narrower and taller compared with that
of the corresponding laser beam which implies its better performance in FSO
communications through atmosphere. It is in part the case because the scintillation index being the typical width of the distribution takes on smaller value
for the prerandomized beam.

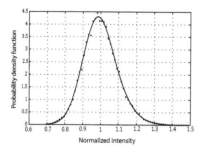 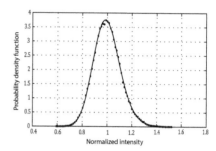

FIGURE 7.4
The probability density function of the normalized intensity of a partially coherent beam beam (left) and a corresponding laser beam (right). Dots represent the data points and the line is produced with the help of the Gamma-Laguerre pdf model.

7.2 Mitigation of scintillations by different randomization schemes

7.2.1 Non-uniform polarization

We will now discuss the possibility of mitigating the scintillations in a deterministic beam propagating in the turbulent atmosphere by introducing a non-uniform polarization pattern of its source without affecting its state of coherence [8]. Hence, we assume that at all points of the source plane the degree of polarization of the beam takes on value one, but all the characteristics of the polarization ellipse can differ from point to point. The reason why such a beam leads to lower levels of scintillation may be understood as follows: a non-uniformly polarized beam can be expressed as the coherent superposition of a pair of mutually orthogonally polarized spatial modes. Because the polarization of monochromatic beams does not change significantly in turbulence [9] the modes remain mutually orthogonal over appreciable propagation distances. Just like a pair of partially coherent modes, the orthogonally polarized modes do not interfere with one another, and their respective interference patterns add by intensity.

Let us consider a monochromatic, non-uniformly polarized beam consisting of a coherent superposition of an x-polarized Laguerre-Gaussian beam $U_{00}^{(LG)}$ in the source plane $z = 0$ (see Chapter 2):

$$U_{00}^{(LG)}(x, y, 0) = \sqrt{\frac{2}{\pi w_0^2}} \exp\left(-\frac{x^2 + y^2}{w_0^2}\right), \qquad (7.18)$$

and a y-polarized Laguerre-Gaussian beam $U_{01}^{(LG)}$:

$$U_{01}^{(LG)}(x, y, 0) = \frac{2}{\sqrt{\pi}w_0^2}(x + iy)\exp\left(-\frac{x^2 + y^2}{w_0^2}\right). \qquad (7.19)$$

Both modes are normalized as

$$\iint |U_{nm}(x, y, 0)|^2 dxdy = 1, \qquad (7.20)$$

where the integration extends over the entire source plane.

In [8] numerical simulations based on the multiple-phase screen method [10] for atmospheric propagation have been carried out for the electromagnetic beam which is a superposition of modes (7.18) and (7.19). The wavelength was taken to be $\lambda = 1.55$ μm, the retractive index structure parameter is $C_n^2 = 10^{-14}$ m$^{-2/3}$, the propagation distance $z = 5$ km, and the width of the beam is taken to be $w_0 = 5$ cm. The two modes are propagated through the same realization of turbulence, and their intensities are added at the detector plane.

The degree to which the scintillation index of the beam can be reduced largely depends on the choice of the amplitudes of the two orthogonally polarized modes in the source plane. In Fig. 7.5 (left), the on-axis scintillation index of such beams is illustrated versus the ratio of the amplitudes of the two modes. For comparison, the scintillation index is also shown for the modes individually and for the modes superimposed with the same polarization. It can be seen that the minimum of the scintillation occurs when the amplitude of the $U_{01}^{(LG)}$ mode is about unity, providing more than 30% reduction as compared to the scintillation of the Gaussian beam alone. Figure 7.5 (right) illustrates the scintillation index of the same four beams, as a function of the Rytov variance, $\sigma_R^2 = 1.23C_n^2 k^{7/6}z^{11/6}$ being the scintillation index of a plane wave [11]. It can be seen that the non-uniformly polarized field outperforms the individual modes as well as the combination of the modes with the same polarization.

Such mitigation is possible because the modes $U_{01}^{(LG)}$ and $U_{00}^{(LG)}$ propagate through the turbulence differently. The interference pattern produced by the $U_{01}^{(LG)}$ mode is therefore different from that of the $U_{00}^{(LG)}$ mode. Instead of the two scalar modes producing a mutual interference pattern, the polarization of the field is scrambled; a realization of this, determined by numerical simulation, is shown in Fig. 7.6. The independence of the two polarized modes is crucial; this reduction of scintillation index would not occur if two Gaussian beams, identical save for their direction of polarization, were coherently superimposed and propagated through turbulence. It is worth noting that, over time scales small compared to the turbulence fluctuations, the field remains essentially fully polarized. However, over time scales long compared with the turbulence fluctuations, degree of polarization decreases. This nonuniform polarization effect leads to relatively straightforward way of reducing the scintillation of a coherent optical beam, as optical elements to convert linear to

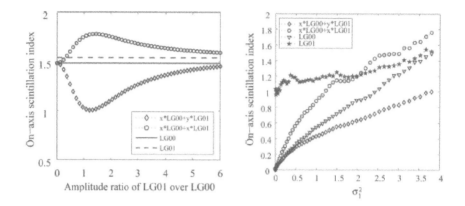

FIGURE 7.5
Simulation of the scintillation index of the non-uniformly polarized beam propagating at 5 km in the turbulent atmosphere of the two Laguerre-Gaussian modes as a function of (left) the amplitude ratio; (right) the Rytov variance. From Ref. [8].

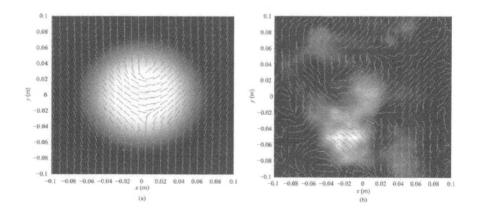

FIGURE 7.6
Simulation of the intensity and polarization state of a non-uniformly polarized beam in atmospheric turbulence. Density plot (gray scale) shows distribution of intensity along the major axis of polarization, x and, lines - polarization state in: (a) the source plane, $z = 0$, (b) in the field $z = 5$ km. From Ref. [8].

nonuniform polarization are now common. For instance, a radially polarized beam can be generated by use of a conical Brewster prism [12].

Before concluding this section we mention that while the non-uniform polarization provides a simple tool for mitigation of scintillations, it does not lead to significant loss of power. Thus, according to Eq. (7.2) it definitely has an advantage for the FSO communication systems.

7.2.2 Partial coherence

It was analytically demonstrated some time ago (see, for instance, [13], [14]) that the scintillation index of a partially coherent field propagating in a turbulent medium can be lower than that of its fully coherent counterpart. This has become a topic of renewed interest in recent years [3], [15]–[17] and can be roughly understood as follows. A partially coherent beam carries its energy in multiple, mutually incoherent, spatial modes, with a larger number of modes corresponding to a lower degree of coherence. Each mode produces its own distinct interference pattern on propagation through turbulence. Because these modes are mutually incoherent, hence, their interference patterns add by intensity, this results in a more uniform intensity pattern (on average) at the detector. Incorporating partial coherence into an optical communications system, however, requires a source whose statistical properties can be easily controlled. This can at best be done with phase diffusers or the Spatial Light Modulators (SLMs), as was discussed in Chapters 3 and 4.

The expression for the on-axis scintillation index of a scalar Gaussian Schell-model beam propagating through weak atmospheric turbulence with Kolmogorov power spectrum was derived in Ref. [14]:

$$
c_I^{(weak)}(0, z; \omega) = 2.67 \sigma_R^2 \Omega^{5/6} \left(\frac{1 + q_1^2}{2q_2^2} \right) \left\{ \frac{11}{16} Re \left[\left(1 + \frac{2i\Omega}{1 + q_1^2} \right)^{5/6} \right. \right.
$$
$$
\left. \left. \times {}_1F_1 \left(-\frac{5}{6}, 1; \frac{q_3 - 2q_1 q_2^2/\Omega}{q_3 + 2q_2^2} \right) \right] - 1 \right\},
$$

(7.21)

where ${}_1F_1$ is the hypergeometric function and $\Omega = \frac{k\sigma^2}{z}$, $q_1^2 = 1 + \frac{\sigma^2}{\delta^2}$, $q_2 = q_1^2 + \Omega^2$, $q_3 = (1 - q_1)^2(q_1 - i\Omega)$, and $\sigma_R^2 = 1.23 C_n^2 k^{7/6} z^{11/6}$ is a Rytov variance being the scintillation index of the unbounded plane wave, calculated for a Kolmogorov power spectrum. For extension of the analysis into moderate and strong regimes of atmospheric fluctuations one can employ the recently formulated theory of spatial filters [2]. For a random beam propagating in such conditions the scintillation index can be approximated as (Ref. [2])

$$
c_I^{(strong)}(0, z) = \exp \left[\frac{0.49 c_I^{(weak)}(0, z)}{[1 + 0.56 c_I^{(weak)6/5}(0, z)]^{7/6}} \right.
$$
$$
\left. + \frac{0.51 c_I^{(weak)}(0, z)}{[1 + 0.69 c_I^{(weak)6/5}(0, z)]^{5/6}} \right] - 1,
$$

(7.22)

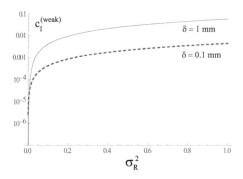

FIGURE 7.7

The scintillation index of a scalar Gaussian Schell-model beam with $\sigma = 1$ cm propagating in the turbulent atmosphere versus the Rytov variance σ_R^2.

where Kolmogorov power spectrum was assumed.

In Fig. 7.7 we show the scintillation index of two partially coherent beams as functions of the Rytov variance.

7.2.3 Combination of non-uniform polarization and partial coherence

After discussing separately the beam scintillation reduction by means of non-uniform polarization and partial coherence we turn to the next possibility of doing this in the class of random electromagnetic beams, i.e., to the possibility of combining these two methods [18].

Recall that the instantaneous intensity $I(\boldsymbol{\rho}, z; \omega)$ of a random electromagnetic beam at transverse coordinate $\boldsymbol{\rho}$ and distance z from the source is defined by the formula

$$I(\boldsymbol{\rho}, z; \omega) = Tr[\mathbf{E}^\dagger(\boldsymbol{\rho}, z; \omega)\mathbf{E}(\boldsymbol{\rho}, z; \omega)], \qquad (7.23)$$

where $\dagger$ stands for the Hermitian adjoint. Then the normalized variance of the intensity fluctuations or the scintillation index, is defined by the formula [20]:

$$c_I(\boldsymbol{\rho}, z; \omega) = \frac{\langle I(\boldsymbol{\rho}, z; \omega)^2 \rangle - \langle I(\boldsymbol{\rho}, z; \omega) \rangle^2}{\langle I(\boldsymbol{\rho}, z; \omega) \rangle^2}, \qquad (7.24)$$

where $\langle I(\boldsymbol{\rho}, z; \omega) \rangle$ is the first moment or average intensity and $\langle I(\boldsymbol{\rho}, z; \omega)^2 \rangle$ is its second moment. Then the scintillation index of the electromagnetic field can be expressed in terms of the scintillation indexes associated with individual field components as:

$$c_I(\boldsymbol{\rho}, z; \omega) = \frac{c_{xx}\langle I_x \rangle^2 + 2c_{xy}\langle I_x \rangle \langle I_y \rangle + c_{yy}\langle I_y \rangle^2}{(\langle I_x \rangle + \langle I_y \rangle)^2}. \qquad (7.25)$$

Here and below the argument $(\boldsymbol{\rho}, z; \omega)$ on the right-hand side is suppressed for brevity, and $\langle I_\alpha(\boldsymbol{\rho}, z; \omega)\rangle$, $(\alpha = x, y)$ are the intensities of the beam components, while

$$c_{\alpha\beta}(\boldsymbol{\rho}, z; \omega) = \frac{\langle I_\alpha I_\beta\rangle - \langle I_\alpha\rangle\langle I_\beta\rangle}{\langle I_\alpha\rangle\langle I_\beta\rangle} \tag{7.26}$$

are the scalar scintillation indexes. Namely, c_{xx} and c_{yy} are the scintillation indexes of $x-$ and y-components of the electric vector and $c_{xy} = c_{yx}$ are their joint scintillation indexes. Thus, the study of propagation of the scintillation index $c_I(\boldsymbol{\rho}, z; \omega)$ of an electromagnetic beam can be reduced to calculation of three scintillation indexes $c_{\alpha\beta}(\boldsymbol{\rho}, z; \omega)$ together with two average intensities $\langle I_x(\boldsymbol{\rho}, z; \omega)\rangle$ and $\langle I_y(\boldsymbol{\rho}, z; \omega)\rangle$ of the components of the electric field.

In order to illustrate the advantage of using partial polarization in the class of random beams we turn to a comparison between two beams generated by a linearly polarized source and an unpolarized source, described by the cross-spectral density matrices:

$$\mathbf{W}_{(LP)}(\boldsymbol{\rho}'_1, \boldsymbol{\rho}'_2, 0; \omega) = \begin{bmatrix} W_{xx}(\boldsymbol{\rho}'_1, \boldsymbol{\rho}'_2, 0; \omega) & 0 \\ 0 & 0 \end{bmatrix}, \tag{7.27}$$

and

$$\mathbf{W}_{(UP)}(\boldsymbol{\rho}'_1, \boldsymbol{\rho}'_2, 0; \omega) = \frac{1}{2}\begin{bmatrix} W_{xx}(\boldsymbol{\rho}'_1, \boldsymbol{\rho}'_2, 0; \omega) & 0 \\ 0 & W_{yy}(\boldsymbol{\rho}'_1, \boldsymbol{\rho}'_2, 0; \omega) \end{bmatrix}, \tag{7.28}$$

where $W_{xx}(\boldsymbol{\rho}'_1, \boldsymbol{\rho}'_2, 0; \omega) = W_{yy}(\boldsymbol{\rho}'_1, \boldsymbol{\rho}'_2, 0; \omega)$ everywhere in the source plane. Let us also assume that $x-$ and y-components of the electric field are mutually independent. This type of light is sometimes referred to as a natural light [21]. On comparing the two sources we stress on the fact that their spectral densities and the degrees of coherence are the same. Hence, the only difference is in their degrees of polarization: it takes on value one for the linearly polarized source and value zero for the unpolarized source. On substituting from Eqs. (7.27) and (7.28) into Eqs. (7.21), (7.22) and then into (7.25) we find that

$$c_{UP}^{(weak)}(0, z; \omega) = \frac{1}{2}c_{LP}^{(weak)}(0, z; \omega), \tag{7.29}$$

and, hence,

$$c_{UP}^{(strong)}(0, z; \omega) = \frac{1}{2}c_{LP}^{(strong)}(0, z; \omega). \tag{7.30}$$

These formulas express simple relations between the on-axis scintillation indexes of two beams produced by a linearly polarized source and a unpolarized, natural source: the scintillation index of the natural beam is twice as low as that of the polarized beam, and this result holds for any propagation distance from the source.

Figure 7.8 shows the behavior of the on-axis scintillation index of the beams generated by sources (7.27) and (7.28) with increasing Rytov variance σ_R^2: (left) in weak turbulence and (right) in moderate and strong regimes of

298

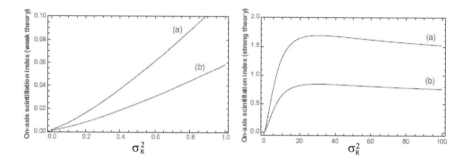

FIGURE 7.8
Evolution of the on-axis scintillation index of two Gaussian Schell-model beams: (a) linearly polarized, (b) unpolarized, in atmospheric turbulence as a function of the Rytov variance: (left) weak turbulence, (right) moderate and strong turbulence. From Ref. [18].

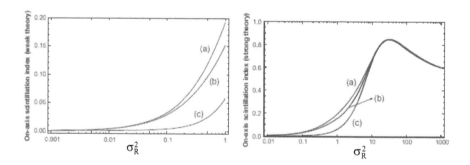

FIGURE 7.9
Propagation of the scintillation index of three unpolarized electromagnetic Gaussian Schell-model beams as a function of Rytov variance: (a) almost coherent source; (b) partially coherent source, (c) almost incoherent source. (left) Weak theory, (right) moderate and strong theory. From Ref. [18].

atmospheric fluctuations. We find, in particular, that unlike a typical scalar beam, the electromagnetic unpolarized beam leads to scintillation level that can be lower than one, in the moderate regime of atmospheric fluctuations. In Fig. 7.9 we show the scintillation indexes of several unpolarized, random beams as functions of the Rytov variance.

In summary, partial coherence and partial polarization can work together for mitigation of the atmospheric effects on a propagating beam. While partial coherence of the source is most effective in weak turbulence and leads to additional loss of power, partial polarization can reduce the scintillation by a factor of two (at most) in all regimes of atmospheric fluctuations and does not affect the power.

7.3   Active LIDAR systems with rough targets

LIDAR (Light Detection And Ranging) is a system that employs the information about reflected and/or scattered light waves for detection, tracking and characterization of a target, which might be either natural (clouds, aerosoles or smoke distributed in the atmosphere) or man-made (an aircraft) [22, 23]. LIDARs can use either continuous or pulsed signals. The major distinction between a LIDAR and a radar is in the choice of the wavelength of the electromagnetic wave. While radars use radio waves, LIDARs can involve wavelengths in ultraviolet, visible, or near infrared ranges. Since, in general, it is possible to image a feature of an object only about the same size as the wavelength or larger, a LIDAR is highly sensitive to aerosols and cloud particles and has many applications in atmospheric and ocean research. While active LIDAR systems use artificial light sources, passive LIDARs rely on the natural light having only the detection component.

In what follows we will confine our attention to a problem of identification of stationary targets embedded in the atmospheric turbulence that have hard surfaces with a given reflection coefficient distribution, size, curvature and roughness [23] [see Fig. 7.10 (top)]. In this case, the working mechanism of an active LIDAR system is the following: a source of illumination generates an electromagnetic wave with controllable properties which propagates to a remote target; after being reflected/scattered from its surface the wave propagates back to the receiver, which is collocated (monostatic LIDAR) or not (bistatic LIDAR) with the source. The return wave possesses the target's signature that results in modification of its statistical properties. Typically, both the amplitude and the phase of the illumination beam are randomly modified by the rough surface of the target that results in complex interference effects in the return wave. The basic problem of the LIDAR is determination of target properties from comparison of the illumination and return waves, the latter is also subject to modification by propagation between the LIDAR and the target. While in the case of free-space propagation only diffraction should be taken into account for propagation in turbulence, the wave-front perturbation caused by inhomogeneities of the medium must also be accounted for. Separation of target and medium perturbations remains a daunting task. We will only restrict our attention to bistatic systems because of the complex backscatter multiplication effects present in the monostatic case scenario [23].

7.3.1 Beam propagation in optical systems in the presence of random medium

We will now analyze various phenomena arising on propagation of random beams through a general paraxial $ABCD$ optical system in the presence of random medium by first deriving general analytic formulas and then applying them to the case when the $ABCD$ matrix represents the bistatic LIDAR system. This problem was previously extensively treated in the framework of scalar theory with deterministic light sources in [23]–[25]. We will largely follow the analysis given in Y. Cai et al. [26], which is capable of employing the theory of electromagnetic random beams.

As we have seen in Chapter 5 within the validity of the paraxial approximation, propagation of an electromagnetic beam through an astigmatic $ABCD$ optical system can be studied with the help of the generalized Huygens-Fresnel integral. Further, as we discussed in Chapter 6, if the wave experiences the phase perturbation due to the (isotropic) random medium its propagation is described by the extended Huygens-Fresnel integral. The combination of two approaches results in the expression

$$
E_\alpha(\boldsymbol{\rho}_1, z = l) = -\frac{i}{\lambda[Det(\mathbf{B})]^{1/2}} \int_{-\infty}^{\infty} E_\alpha(\boldsymbol{\rho}_1', z = 0)
$$

$$
\times \exp\left[-\frac{ik}{2}(\boldsymbol{\rho}_1'^T \mathbf{B}^{-1}\mathbf{A}\boldsymbol{\rho}_1' - 2\boldsymbol{\rho}_1'^T \mathbf{B}^{-1}\boldsymbol{\rho}_1 + \boldsymbol{\rho}_1^T \mathbf{D}\mathbf{B}^{-1}\boldsymbol{\rho}_1) + \Psi(\boldsymbol{\rho}_1', \boldsymbol{\rho}_1)\right] d^2\rho_1',
$$

$$
(\alpha = x, y).
$$

$$(7.31)$$

Here $E_\alpha(\boldsymbol{\rho}_1', z = 0)$ and $E_\alpha(\boldsymbol{\rho}_1, z = l)$ are the realizations of the electric fields in the source plane ($z = 0$) and in the output plane ($z = l$), respectively, $\boldsymbol{\rho}_1'^T = (x_1', y_1')$ and $\boldsymbol{\rho}_1^T = (x_1, y_1)$ with $\boldsymbol{\rho}_1'$ and $\boldsymbol{\rho}_1$ being the position vectors in the source and output planes, $\Psi(\boldsymbol{\rho}_1', \boldsymbol{\rho}_1)$ is the perturbation being the random part of the complex phase of the beam induced by random medium.

We will now develop the relation between the cross-spectral density matrices in the input and output planes, in the case when the propagating beam is random

$$
W_{\alpha\beta}(\boldsymbol{\rho}_1', \boldsymbol{\rho}_2', z = 0) = [\langle E_\alpha(\boldsymbol{\rho}_1', z = 0)E_\beta^*(\boldsymbol{\rho}_2', z = 0)\rangle],
$$

$$
W_{\alpha\beta}(\boldsymbol{\rho}_1, \boldsymbol{\rho}_2, z = l) = [\langle E_\alpha(\boldsymbol{\rho}_1, z = l)E_\beta^*(\boldsymbol{\rho}_2, z = l)\rangle],
$$

$$(7.32)$$

with $(\alpha = x, y; \beta = x, y)$. On substituting from Eq. (7.31) into Eqs. (7.32) we find that the cross-spectral density of a stochastic electromagnetic beam propagating through a general astigmatic optical system is given by the expression

$$W_{\alpha\beta}(\boldsymbol{\rho}_1, \boldsymbol{\rho}_2, z = l) = \frac{1}{\lambda^2 \sqrt{Det(\mathbf{B})} \sqrt{Det(\mathbf{B}^*)}}$$

$$\int_{-\infty}^{\infty} \int_{-\infty}^{\infty} \int_{-\infty}^{\infty} \int_{-\infty}^{\infty} W_{\alpha\beta}(\boldsymbol{\rho}'_1, \boldsymbol{\rho}'_2, z = 0)$$

$$\times \exp\left[-\frac{ik}{2}(\boldsymbol{\rho}'^T_1 \mathbf{B}^{-1}\mathbf{A}\boldsymbol{\rho}'_1 - 2\boldsymbol{\rho}'^T_1\mathbf{B}^{-1}\boldsymbol{\rho}_1 + \boldsymbol{\rho}^T_1\mathbf{DB}^{-1}\boldsymbol{\rho}_1)\right]$$

$$\times \exp\left[\frac{ik}{2}(\boldsymbol{\rho}'^T_2(\mathbf{B}^*)^{-1}\mathbf{A}^*\boldsymbol{\rho}'_2 - 2\boldsymbol{\rho}'^T_2(\mathbf{B}^*)^{-1}\boldsymbol{\rho}_2 + \boldsymbol{\rho}^T_2\mathbf{D}^*(\mathbf{B}^*)^{-1}\boldsymbol{\rho}_2)\right]$$

$$\times \langle \exp[\Psi(\boldsymbol{\rho}'_1, \boldsymbol{\rho}_1) + \Psi^*(\boldsymbol{\rho}'_2, \boldsymbol{\rho}_2)]\rangle d^2\rho'_1 d^2\rho'_2.$$

$$(7.33)$$

Under the quadratic approximation of the wave structure function the expression in the angular brackets of the formula above can be approximated by the product [27]

$$\langle \exp[\Psi(\boldsymbol{\rho}'_1, \boldsymbol{\rho}_1) + \Psi^*(\boldsymbol{\rho}'_2, \boldsymbol{\rho}_2)]\rangle = \exp\left[-\frac{(\boldsymbol{\rho}'_1 - \boldsymbol{\rho}'_2)^2}{\rho_0^2}\right]\exp\left[-\frac{(\boldsymbol{\rho}_1 - \boldsymbol{\rho}_2)(\boldsymbol{\rho}'_1 - \boldsymbol{\rho}'_2)}{\rho_0^2}\right]$$

$$\times \exp\left[-\frac{(\boldsymbol{\rho}_1 - \boldsymbol{\rho}_2)^2}{\rho_0^2}\right].$$

$$(7.34)$$

Here ρ_0 is the coherence length of a spherical wave propagating in the random medium. Under the assumption of the Kolmogorov spectrum it is given by the expression [28, 29]

$$\rho_0 = \sqrt{Det[\mathbf{B}(z)]}\left(1.46k^2C_n^2\int_0^l \det[\mathbf{B}(z)]^{5/6}dz\right)^{-3/5}. \qquad (7.35)$$

Here $\mathbf{B}$ is the submatrix for back propagation from the output plane to propagation distance z, and C_n^2 is the atmospheric structure constant. On substituting from Eq. (7.34) into Eq. (7.33) we find, after some rearrangement, that in the tensor form the elements of the cross-spectral density matrix become

$$
W_{\alpha\beta}(\tilde{\boldsymbol{\rho}}, l) = \frac{k^2}{4\pi^2 \sqrt{det(\tilde{\mathbf{B}})}} \int_{-\infty}^{\infty} \int_{-\infty}^{\infty} \int_{-\infty}^{\infty} \int_{-\infty}^{\infty} W_{\alpha\beta}(\tilde{\boldsymbol{\rho}}', 0)
$$

$$
\times \exp\left[-\frac{ik}{2} (\tilde{\boldsymbol{\rho}}'^T) \tilde{\mathbf{B}}^{-1} \tilde{\mathbf{A}} \tilde{\boldsymbol{\rho}}' - 2\tilde{\boldsymbol{\rho}}'^T \tilde{\mathbf{B}}^{-1} \tilde{\boldsymbol{\rho}} + \tilde{\boldsymbol{\rho}}^T \tilde{\mathbf{D}} \tilde{\mathbf{B}}^{-1} \tilde{\boldsymbol{\rho}} \right] \quad (7.36)
$$

$$
\times \exp\left[-\frac{ik}{2} \tilde{\boldsymbol{\rho}}'^T \tilde{\mathbf{P}} \tilde{\boldsymbol{\rho}}' - \frac{ik}{2} \tilde{\boldsymbol{\rho}}'^T \tilde{\mathbf{P}} \tilde{\boldsymbol{\rho}} - \frac{ik}{2} \tilde{\boldsymbol{\rho}}^T \tilde{\mathbf{P}} \tilde{\boldsymbol{\rho}} \right] d\tilde{\boldsymbol{\rho}}',
$$

where $d\tilde{\boldsymbol{\rho}}' = d\tilde{\boldsymbol{\rho}}'_1 d\tilde{\boldsymbol{\rho}}'_2$, $\tilde{\boldsymbol{\rho}}'^T = (\tilde{\boldsymbol{\rho}}'_1{}^T \ \tilde{\boldsymbol{\rho}}'_2{}^T)$, $\tilde{\boldsymbol{\rho}}^T = (\tilde{\boldsymbol{\rho}}_1^T \ \tilde{\boldsymbol{\rho}}_2^T)$ and

$$
\tilde{\mathbf{A}} = \begin{bmatrix} \mathbf{A} & 0\mathbf{I} \\ 0\mathbf{I} & \mathbf{A}^* \end{bmatrix}, \quad \tilde{\mathbf{B}} = \begin{bmatrix} \mathbf{B} & 0\mathbf{I} \\ 0\mathbf{I} & -\mathbf{B}^* \end{bmatrix}, \quad \tilde{\mathbf{C}} = \begin{bmatrix} \mathbf{C} & 0\mathbf{I} \\ 0\mathbf{I} & -\mathbf{C}^* \end{bmatrix},
$$

$$
\tilde{\mathbf{D}} = \begin{bmatrix} \mathbf{D} & 0\mathbf{I} \\ 0\mathbf{I} & \mathbf{D}^* \end{bmatrix}, \quad \tilde{\mathbf{P}} = \frac{2}{ik\rho_0'^2} \begin{bmatrix} \mathbf{I} & -\mathbf{I} \\ -\mathbf{I} & \mathbf{I} \end{bmatrix}, \quad (7.37)
$$

$\mathbf{I}$ being a 2×2 unit matrix. We stress that due to its generality, Eq. (7.36) can be used for analysis of paraxial propagation of electromagnetic stochastic beams through any astigmatic $ABCD$ optical system embedded into the turbulent atmosphere. Moreover, Eq. (7.36) may be easily adopted, after a suitable choice of the phase structure function, to any homogeneous and isotropic turbulent medium, such as non-Kolmogorov atmospheric or oceanic turbulence. Further if the input beam is generated by an electromagnetic Gaussian Schell-model source, the integral in Eq. (7.36) can be evaluated in the closed form, as we will show in the following subsection.

7.3.2 Beam passage through a LIDAR system with a semi-rough target

In the problems of remote sensing through a random medium, a target of interest should be described in terms compatible with those of the other elements in the system. Hence, in such situations some of the target characteristics become important while others can be completely ignored. Moreover, for successful modeling of the sensing process it is essential to select only a few target features that would be sufficient to distinguish it from its background. The shape including both the boundaries and the curvature, as well as the roughness statictics are perhaps the most significant properties of target to be taken into account. In cases when the object's surface is random, i.e., its height varies randomly from point to point, at least some of its statistical properties must be specified. Various models of rough surfaces and statistical properties of electromagnetic fields scattered from them are summarized in the classic book by Beckman [30] and more recent texts [31]–[34]. In terms of characteristic scales the rough targets are classified as single-scale, multiple-scale (two-scale model is the best studied) and continuum-scale (fractal) [31].

In our discussion we will only consider the single-scale surfaces governed by Gaussian statistics.

The second-order statistical properties of a propagating radiation field can be discribed with the help of a sequence of the cross-spectral density matrices of the beam in the transverse cross-sections (see Fig. 7.10) [35]. In order to distinguish among the matrices at different planes we will simply use different symbols to denote transverse position vectors: vector ρ' will characterize the beam in the source plane, vector ρ'' will be used for the beam field after its passing through the atmosphere and through the mirror, vector ρ''' will denote the vector on the transverse plane after reflection from the target surface and propagating through the atmosphere up to the collecting lens (pupil plane) and ρ after being focused by the collecting lens onto the surface of the detector (image plane). We will omit the frequency dependence of the cross-spectral density for brevity. It will also be assumed that no random medium exists between the collecting lens and the detector.

The propagation laws for the cross-spectral density matrix will be derived in four stages: (1) propagation from the source plane to the target plane ($\rho' \rightarrow \rho''$); (2) interaction with the surface of the target; (3) propagation from the target plane to the collecting lens plane ($\rho'' \rightarrow \rho'''$); (4) propagation from the collecting lens to the detector plane ($\rho''' \rightarrow \rho$).

We begin by describing the illumination beam in the plane of the transmitter assuming that it belongs to the electromagnetic Gaussian Schell-model type (see Chapters 4, 5). The elements of the cross-spectral density matrix of such a beam in the tensor form are (Chapter 5)

$$W_{\alpha\beta}(\tilde{\rho}') = A_\alpha A_\beta B_{\alpha\beta} \exp\left[-\frac{ik}{2}\tilde{\rho}'^T \widetilde{\mathbf{M}}_{0\alpha\beta}^{-1}\tilde{\rho}'\right], \quad (\alpha = x, y, \beta = x, y), \quad (7.38)$$

where the 4×4 tensor has the form

$$\widetilde{\mathbf{M}}_{0\alpha\beta}^{-1} = \begin{bmatrix} \frac{1}{ik}\left(\frac{1}{2\sigma_\alpha^2} + \frac{1}{\delta_{\alpha\beta}^2}\right)\mathbf{I} & \frac{i}{k\delta_{\alpha\beta}^2}\mathbf{I} \\ \frac{i}{k\delta_{\alpha\beta}^2}\mathbf{I} & \frac{1}{ik}\left(\frac{1}{2\sigma_\beta^2} + \frac{1}{\delta_{\alpha\beta}^2}\right)\mathbf{I} \end{bmatrix}, \quad (7.39)$$

where $\tilde{\rho}^T = (\rho_1^T, \rho_2^T)$ and $\mathbf{I}$ is the 2×2 identity matrix. On substituting from Eqs. (7.38)–(7.39) into Eq. (7.36) we find, after some arrangement, that

$$W_{\alpha\beta}(\tilde{\rho}'', l) = \frac{k^2 A_\alpha A_\beta B_{\alpha\beta}}{4\pi^2\sqrt{\det\widetilde{\mathbf{B}}}} \exp\left[-\frac{ik}{2}\tilde{\rho}''^T(\check{\mathbf{D}}\widetilde{\mathbf{B}}^{-1} + \check{\mathbf{P}})\tilde{\rho}''\right]$$

$$\times \int_{-\infty}^{\infty}\int_{-\infty}^{\infty}\int_{-\infty}^{\infty}\int_{-\infty}^{\infty} \exp\left[-\frac{ik}{2}\tilde{\rho}^T(\mathbf{M}_{0\alpha\beta}^{-1} + \widetilde{\mathbf{B}}^{-1}\check{\mathbf{A}} + \check{\mathbf{P}})\tilde{\rho}\right]$$

$$\times \exp\left[ik\rho^T\left(\widetilde{\mathbf{B}}^{-1} - \frac{1}{2}\check{\mathbf{P}}\right)\tilde{\rho}''\right]d\tilde{\rho},$$

304

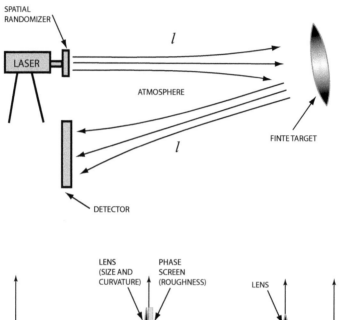

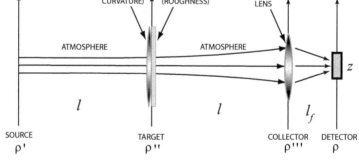

FIGURE 7.10
(top) Bistatic LIDAR system; (bottom) unfolded version of bistatic LIDAR
system.

or, equivalently,

$$W_{\alpha\beta}(\tilde{\boldsymbol{\rho}}'', l) = \frac{k^2 A_\alpha A_\beta B_{\alpha\beta}}{4\pi^2 \sqrt{\det \tilde{\mathbf{B}}}} \exp\left[-\frac{ik}{2}\tilde{\boldsymbol{\rho}}''^T (\tilde{\mathbf{D}}\tilde{\mathbf{B}}^{-1} + \tilde{\mathbf{P}})\tilde{\boldsymbol{\rho}}''\right]$$

$$\times \exp\left[\frac{ik}{2}\tilde{\boldsymbol{\rho}}''^T \left(\tilde{\mathbf{B}}^{-1} - \frac{1}{2}\tilde{\mathbf{P}}\right)^T (\mathbf{M}_{0\alpha\beta}^{-1} + \tilde{\mathbf{B}}^{-1}\tilde{\mathbf{A}} + \tilde{\mathbf{P}})^{-1} \left(\tilde{\mathbf{B}}^{-1} - \frac{1}{2}\tilde{\mathbf{P}}\right)\tilde{\boldsymbol{\rho}}''\right]$$

$$\times \int\limits_{-\infty}^{\infty}\int\limits_{-\infty}^{\infty}\int\limits_{-\infty}^{\infty}\int\limits_{-\infty}^{\infty} \exp\left[-\frac{ik}{2}\left|(\mathbf{M}_{0\alpha\beta}^{-1} + \tilde{\mathbf{B}}^{-1}\tilde{\mathbf{A}} + \tilde{\mathbf{P}})^{1/2}\tilde{\boldsymbol{\rho}}\right.\right.$$

$$\left.\left. - (\mathbf{M}_{0\alpha\beta}^{-1} + \tilde{\mathbf{B}}^{-1}\tilde{\mathbf{A}} + \tilde{\mathbf{P}})^{1/2} \left(\tilde{\mathbf{B}}^{-1} - \frac{1}{2}\tilde{\mathbf{P}}\right)\tilde{\boldsymbol{\rho}}''\right]d\tilde{\boldsymbol{\rho}}.$$

$$(7.40)$$

With the help of the formula

$$\int\limits_{-\infty}^{\infty} \exp(-qx^2)dx = \sqrt{\frac{\pi}{q}}, \qquad (7.41)$$

we find that Eq. (7.40) reduces to the expression

$$W_{\alpha\beta}(\tilde{\boldsymbol{\rho}}'', l) = \frac{k^2 A_\alpha A_\beta B_{\alpha\beta}}{4\pi^2 \sqrt{\det \tilde{\mathbf{B}}}\sqrt{\det(\mathbf{M}_{0\alpha\beta}^{-1} + \tilde{\mathbf{B}}^{-1}\tilde{\mathbf{A}} + \tilde{\mathbf{P}})}}$$

$$\times \exp\left[-\frac{ik}{2}\tilde{\boldsymbol{\rho}}''^T (\tilde{\mathbf{D}}\tilde{\mathbf{B}}^{-1} + \tilde{\mathbf{P}})\tilde{\boldsymbol{\rho}}''\right]$$

$$\times \exp\left[\frac{ik}{2}\tilde{\boldsymbol{\rho}}''^T \left(\tilde{\mathbf{B}}^{-1} - \frac{1}{2}\tilde{\mathbf{P}}\right)^T (\mathbf{M}_{0\alpha\beta}^{-1} + \tilde{\mathbf{B}}^{-1}\tilde{\mathbf{A}} + \tilde{\mathbf{P}})^{-1} \left(\tilde{\mathbf{B}}^{-1} - \frac{1}{2}\tilde{\mathbf{P}}\right)\tilde{\boldsymbol{\rho}}''\right].$$

$$(7.42)$$

Further, based on the Luneberg's relations (5.72) and relations $\tilde{\mathbf{B}}^{-1T} = \tilde{\mathbf{B}}^{-1}$, $\tilde{\mathbf{P}}^T = \tilde{\mathbf{P}}$ the following identities may be established:

$$\det \tilde{\mathbf{B}}\det(\mathbf{M}_{0\alpha\beta}^{-1} + \tilde{\mathbf{B}}^{-1}\tilde{\mathbf{A}} + \tilde{\mathbf{P}}) = \det(\tilde{\mathbf{A}} + \tilde{\mathbf{B}}\mathbf{M}_{0\alpha\beta}^{-1} + \tilde{\mathbf{B}}\tilde{\mathbf{P}}), \qquad (7.43)$$

and

$$\tilde{\mathbf{D}}\tilde{\mathbf{B}}^{-1} - \tilde{\mathbf{B}}^{-1T}(\mathbf{M}_{0\alpha\beta}^{-1} + \tilde{\mathbf{B}}^{-1}\tilde{\mathbf{A}} + \tilde{\mathbf{P}})^{-1}\tilde{\mathbf{B}}^{-1}$$

$$= \tilde{\mathbf{D}}\tilde{\mathbf{B}}^{-1} - \tilde{\mathbf{B}}^{-1}(\tilde{\mathbf{A}} + \tilde{\mathbf{B}}\mathbf{M}_{0\alpha\beta}^{-1} + \tilde{\mathbf{B}}\tilde{\mathbf{P}})^{-1}$$

$$= \left[\tilde{\mathbf{D}}\tilde{\mathbf{B}}^{-1}(\tilde{\mathbf{A}} + \tilde{\mathbf{B}}\mathbf{M}_{0\alpha\beta}^{-1} + \tilde{\mathbf{B}}\tilde{\mathbf{P}}) - \tilde{\mathbf{B}}^{-1}\right](\tilde{\mathbf{A}} + \tilde{\mathbf{B}}\mathbf{M}_{0\alpha\beta}^{-1} + \tilde{\mathbf{B}}\tilde{\mathbf{P}})^{-1}$$

$$= \left[\tilde{\mathbf{D}}\tilde{\mathbf{B}}^{-1}\tilde{\mathbf{A}} + \tilde{\mathbf{D}}\mathbf{M}_{0\alpha\beta}^{-1} + \tilde{\mathbf{D}}\tilde{\mathbf{P}} - \tilde{\mathbf{B}}^{-1}\right](\tilde{\mathbf{A}} + \tilde{\mathbf{B}}\mathbf{M}_{0\alpha\beta}^{-1} + \tilde{\mathbf{B}}\tilde{\mathbf{P}})^{-1}$$

$$= (\tilde{\mathbf{C}} + \tilde{\mathbf{D}}\mathbf{M}_{0\alpha\beta}^{-1} + \tilde{\mathbf{D}}\tilde{\mathbf{P}})(\tilde{\mathbf{A}} + \tilde{\mathbf{B}}\mathbf{M}_{0\alpha\beta}^{-1} + \tilde{\mathbf{B}}\tilde{\mathbf{P}})^{-1}.$$

$$(7.44)$$

Finally, on setting

$$\mathbf{M}_{1\alpha\beta}^{-1} = (\tilde{\mathbf{C}} + \tilde{\mathbf{D}}\mathbf{M}_{0\alpha\beta}^{-1} + \tilde{\mathbf{D}}\tilde{\mathbf{P}})(\tilde{\mathbf{A}} + \tilde{\mathbf{B}}\mathbf{M}_{0\alpha\beta}^{-1} + \tilde{\mathbf{B}}\tilde{\mathbf{P}})^{-1} + \tilde{\mathbf{P}}$$
$$+ \left(\tilde{\mathbf{B}}^{-1T} - \tilde{\mathbf{P}}^{T}/4\right)(\mathbf{M}_{0\alpha\beta}^{-1} + \tilde{\mathbf{B}}^{-1}\tilde{\mathbf{A}} + \tilde{\mathbf{P}})^{-1}\tilde{\mathbf{P}}, \tag{7.45}$$

we express Eq. (7.42) as

$$W_{\alpha\beta}(\tilde{\boldsymbol{\rho}}'', l) = \frac{A_\alpha A_\beta B_{\alpha\beta}}{\sqrt{\det(\tilde{\mathbf{A}} + \tilde{\mathbf{B}}\mathbf{M}_{0\alpha\beta}^{-1} + \tilde{\mathbf{B}}\tilde{\mathbf{P}})}} \exp\left[-\frac{ik}{2}\tilde{\boldsymbol{\rho}}''^{T}\widetilde{\mathbf{M}}_{1\alpha\beta}^{-1}\tilde{\boldsymbol{\rho}}''\right]. \tag{7.46}$$

If the lens before the phase screen is a thin Gaussian lens with $W_G \to \infty$ and focal length F_G then its $ABCD$ matrix has the form

$$\begin{bmatrix} A_{F_G} & B_{F_G} \\ C_{F_G} & D_{F_G} \end{bmatrix} = \begin{bmatrix} 1 & 0 \\ -1/F_G & 1 \end{bmatrix}. \tag{7.47}$$

Hence, the $ABCD$ matrix between the source plane and the target plane can be expressed as

$$\begin{bmatrix} \mathbf{A} & \mathbf{B} \\ \mathbf{C} & \mathbf{D} \end{bmatrix} = \begin{bmatrix} \mathbf{I} & 0\mathbf{I} \\ -1/F_G\mathbf{I} & \mathbf{I} \end{bmatrix}\begin{bmatrix} \mathbf{I} & l\mathbf{I} \\ 0\mathbf{I} & \mathbf{I} \end{bmatrix}$$
$$= \begin{bmatrix} \mathbf{I} & l\mathbf{I} \\ (-1/F_G)\mathbf{I} & (1 - l/F_G)\mathbf{I} \end{bmatrix}. \tag{7.48}$$

Since the transformation matrix for back propagation from the target plane to a plane located at distance z from the source has the form

$$\begin{bmatrix} \mathbf{A}(z) & \mathbf{B}(z) \\ \mathbf{C}(z) & \mathbf{D}(z) \end{bmatrix} = \begin{bmatrix} 1 & l-z \\ 0 & 1 \end{bmatrix}\begin{bmatrix} 1 & 0 \\ -1/F_G & 1 \end{bmatrix}$$
$$= \begin{bmatrix} 1 + \dfrac{z-l}{F_G} & l-z \\ -\dfrac{1}{F_G} & 1 \end{bmatrix}, \tag{7.49}$$

the integral in Eq. (7.35) reduces to expression

$$\rho_0 = (0.545k^2C_n^2 l)^{-3/5}. \tag{7.50}$$

Let us now turn to scattering of the beam from the target. Assuming that the surface of the target is semi-rough and isotropic, we will describe it with the help of a model proposed many years ago by Goodman [36], which is based on the idea that under certain circumstances surface scattering is analogous to passage through a thin random phase screen. This idea was also later explored in Ref. [3] for scalar beam propagating in LIDAR systems. In that model the correlation function, which describes the effect of the phase screen on the beam, has the Gaussian form

$$\langle T(\boldsymbol{\rho}_1''; \omega)T^*(\boldsymbol{\rho}_2''; \omega)\rangle_R = \frac{4\pi\beta^2}{k^2}\exp\left[-\frac{\rho_1''^2 + \rho_2''^2}{w_T^2}\right]\exp\left[-\frac{(\boldsymbol{\rho}_1'' - \boldsymbol{\rho}_2'')^2}{\delta_T^2}\right]. \tag{7.51}$$

Here the ensemble average denoted by the angular brackets with subscript R is taken over the set of realizations of the target, σ_T is the soft target size, δ_T is the typical transverse correlation width, and β is the normalization parameter such that $\beta^2 = k^2/4\pi$ (together with $\delta_T \to \infty$) for smooth targets and $\beta^2 = T_0^2/\pi\delta_T^2$ (together with $\delta_T \to 0$) for diffuse (Lambertian) targets, T_0^2 being the root-mean-square (r.m.s.) target reflection coefficient. Correlation function (7.51) may also be represented in the following tensor form

$$\langle T(\boldsymbol{\rho}_1''; \omega) T^*(\boldsymbol{\rho}_2''; \omega) \rangle_R = \frac{4\pi\beta^2}{k^2} \exp\left[-\frac{ik}{2}\tilde{\boldsymbol{\rho}}''^T \widetilde{\mathbf{T}} \tilde{\boldsymbol{\rho}}''\right], \tag{7.52}$$

where $\tilde{\boldsymbol{\rho}}''^T = (\boldsymbol{\rho}_1''^T \ \boldsymbol{\rho}''^T_2)$ and

$$\widetilde{\mathbf{T}} = \begin{bmatrix} \dfrac{2i}{k}\left(\dfrac{1}{2\sigma_T^2} + \dfrac{1}{\delta_T^2}\right)\mathbf{I} & \dfrac{2i}{k\delta_T^2}\mathbf{I} \\[3mm] \dfrac{2i}{k\delta_T^2}\mathbf{I} & \dfrac{2}{ik}\left(\dfrac{1}{2\sigma_T^2} + \dfrac{1}{\delta_T^2}\right)\mathbf{I} \end{bmatrix}. \tag{7.53}$$

On interaction of the beam with the rough surface of the target, the cross-spectral density matrix is to be multiplied by the scalar correlation function of the target from the right.

We will now derive the expression for the cross-spectral density matrix of a beam propagating from the target plane to the collecting lens plane, before its passage through the collecting lens. On substituting from Eqs. (7.46) and (7.53) into the general propagation law through $ABCD$ systems (7.36) we find that at the collecting lens the cross-spectral density matrix of the beam takes the form

$$W_{\alpha\beta}(\tilde{\boldsymbol{\rho}}''', 2l) = \frac{\pi\beta^2}{\sqrt{\det\widetilde{\mathbf{B}}}} \int_{-\infty}^{\infty} \int_{-\infty}^{\infty} \int_{-\infty}^{\infty} \int_{-\infty}^{\infty} W_{\alpha\beta}(\tilde{\boldsymbol{\rho}}'', l) \exp\left[-\frac{ik}{2}\tilde{\boldsymbol{\rho}}''^T \widetilde{\mathbf{T}} \tilde{\boldsymbol{\rho}}''\right]$$

$$\times \exp\left[-\frac{ik}{2}(\tilde{\boldsymbol{\rho}}''^T \widetilde{\mathbf{B}}^{-1}\tilde{\boldsymbol{\rho}}'' - 2\tilde{\boldsymbol{\rho}}''^T \widetilde{\mathbf{B}}^{-1}\tilde{\boldsymbol{\rho}}''' + \tilde{\boldsymbol{\rho}}'''^T \widetilde{\mathbf{B}}^{-1}\tilde{\boldsymbol{\rho}}''')\right] \tag{7.54}$$

$$\times \exp\left[-\frac{ik}{2}(\tilde{\boldsymbol{\rho}}''^T \widetilde{\mathbf{P}}\tilde{\boldsymbol{\rho}}'' - \tilde{\boldsymbol{\rho}}''^T \widetilde{\mathbf{P}}\tilde{\boldsymbol{\rho}}''' + \tilde{\boldsymbol{\rho}}'''^T \widetilde{\mathbf{P}}\tilde{\boldsymbol{\rho}}''')\right] d\tilde{\boldsymbol{\rho}}''.$$

After evaluation of the integral on the right-hand side we obtain the expression

$$W_{\alpha\beta}(\tilde{\boldsymbol{\rho}}''', 2l) = \frac{4\pi\beta^2 A_\alpha A_\beta B_{\alpha\beta}}{k^2\sqrt{\det(\widetilde{\mathbf{I}} + \widetilde{\mathbf{B}}\mathbf{M}_{1\alpha\beta}^{-1} + \widetilde{\mathbf{B}}\widetilde{\mathbf{T}} + \widetilde{\mathbf{B}}\widetilde{\mathbf{P}})}}$$

$$\times \frac{1}{\sqrt{\det(\mathbf{A} + \widetilde{\mathbf{B}}\mathbf{M}_{0\alpha\beta}^{-1} + \widetilde{\mathbf{B}}\widetilde{\mathbf{P}})}} \exp\left[-\frac{ik}{2}\tilde{\boldsymbol{\rho}}'''^T (\widetilde{\mathbf{P}} + \widetilde{\mathbf{B}}^{-1})\tilde{\boldsymbol{\rho}}'''\right]$$

$$\times \exp\left[-\frac{ik}{2}\tilde{\boldsymbol{\rho}}'''^T (\widetilde{\mathbf{B}}^{-1} - \widetilde{\mathbf{P}}/2)^T (\mathbf{M}_{1\alpha\beta}^{-1} + \widetilde{\mathbf{T}} + \widetilde{\mathbf{B}}^{-1} + \widetilde{\mathbf{P}})^{-1}(\widetilde{\mathbf{B}}^{-1} - \widetilde{\mathbf{P}}/2)\tilde{\boldsymbol{\rho}}'''\right]$$

$$= \frac{4\pi\beta^2 A_\alpha A_\beta B_{\alpha\beta}}{k^2\sqrt{\det(\tilde{\mathbf{I}} + \tilde{\mathbf{B}}\mathbf{M}_{1\alpha\beta}^{-1} + \tilde{\mathbf{B}}\tilde{\mathbf{T}} + \tilde{\mathbf{B}}\tilde{\mathbf{P}})}}$$

$$\times \frac{1}{\sqrt{\det(\mathbf{A} + \tilde{\mathbf{B}}\mathbf{M}_{0\alpha\beta}^{-1} + \tilde{\mathbf{B}}\tilde{\mathbf{P}})}} \exp\left[-\frac{ik}{2}\tilde{\rho}'''^T \widetilde{\mathbf{M}}_{2\alpha\beta}^{-1} \tilde{\rho}'''\right],$$

with

$$\widetilde{\mathbf{M}}_{2\alpha\beta}^{-1} = \tilde{\mathbf{P}} + \tilde{\mathbf{B}}^{-1} - (\tilde{\mathbf{B}}^{-1} - \tilde{\mathbf{P}}/2)^T \tag{7.55}$$
$$\times (\mathbf{M}_{1\alpha\beta}^{-1} + \tilde{\mathbf{T}} + \tilde{\mathbf{B}}^{-1} + \tilde{\mathbf{P}})^{-1}(\tilde{\mathbf{B}}^{-1} - \tilde{\mathbf{P}}/2).$$

It is common for optical systems operating in the atmosphere to collect the signal and focus it onto the surface of the detector with the help of a lens. For simplicity we assume that the lens is a thin Gaussian lens with focal distance l_f and that there is a vacuum between the collecting lens plane and the detector plane. Under these circumstances the cross-spectral density matrix at the output (detector) plane can be calculated directly with the help of the tensor $ABCD$ law discussed in Chapter 5 to result in expression

$$W_{\alpha\beta}(\tilde{\rho}', 2l + l_f) = \frac{4\pi\beta^2 A_\alpha A_\beta B_{\alpha\beta}}{k^2\sqrt{\det(\tilde{\mathbf{I}} + \tilde{\mathbf{B}}\mathbf{M}_{1\alpha\beta}^{-1} + \tilde{\mathbf{B}}\tilde{\mathbf{T}} + \tilde{\mathbf{B}}\tilde{\mathbf{P}})}}$$

$$\times \frac{1}{\sqrt{\det(\mathbf{A} + \tilde{\mathbf{B}}\mathbf{M}_{0\alpha\beta}^{-1} + \tilde{\mathbf{B}}\tilde{\mathbf{P}})}} \tag{7.56}$$

$$\times \frac{1}{\sqrt{\det(\mathbf{A}_1 + \tilde{\mathbf{B}}_1\mathbf{M}_{2\alpha\beta}^{-1})}} \exp\left[-\frac{ik}{2}\tilde{\rho}'^T \widetilde{\mathbf{M}}_{3\alpha\beta}^{-1} \tilde{\rho}'\right],$$

where we set

$$\widetilde{\mathbf{M}}_{3\alpha\beta}^{-1} = (\tilde{\mathbf{C}}_1 + \tilde{\mathbf{D}}_1\widetilde{\mathbf{M}}_{2\alpha\beta}^{-1})(\tilde{\mathbf{A}}_1 + \tilde{\mathbf{B}}_1\widetilde{\mathbf{M}}_{2\alpha\beta}^{-1})^{-1}, \tag{7.57}$$

and

$$\tilde{\mathbf{A}}_1 = \begin{bmatrix} \mathbf{A}_1 & 0\mathbf{I} \\ 0\mathbf{I} & \mathbf{A}_1^* \end{bmatrix}, \quad \tilde{\mathbf{B}}_1 = \begin{bmatrix} \mathbf{B}_1 & 0\mathbf{I} \\ 0\mathbf{I} & -\mathbf{B}_1^* \end{bmatrix},$$

$$\tilde{\mathbf{C}}_1 = \begin{bmatrix} \mathbf{C}_1 & 0\mathbf{I} \\ 0\mathbf{I} & -\mathbf{C}_1^* \end{bmatrix}, \quad \tilde{\mathbf{D}}_1 = \begin{bmatrix} \mathbf{D}_1 & 0\mathbf{I} \\ 0\mathbf{I} & \mathbf{D}_1^* \end{bmatrix}, \tag{7.58}$$

with

$$\begin{bmatrix} \mathbf{A}_1 & \mathbf{B}_1 \\ \mathbf{C}_1 & \mathbf{D}_1 \end{bmatrix} = \begin{bmatrix} \mathbf{I} & l_f\mathbf{I} \\ 0\mathbf{I} & \mathbf{I} \end{bmatrix} \begin{bmatrix} \mathbf{I} & 0\mathbf{I} \\ (-1/l_f)\mathbf{I} & \mathbf{I} \end{bmatrix}$$
$$= \begin{bmatrix} 0\mathbf{I} & l_f\mathbf{I} \\ (-1/l_f)\mathbf{I} & \mathbf{I} \end{bmatrix}. \tag{7.59}$$

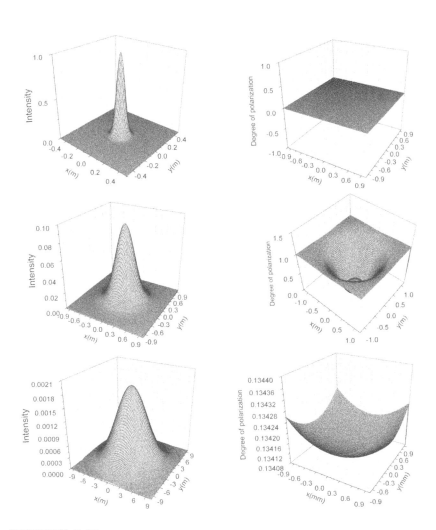

FIGURE 7.11
The spectral density (left column) and the spectral degree of polarization (right column) at the source plane (top), at the target plane (middle) and at the collecting lens plane (bottom). From Ref. [35].

310

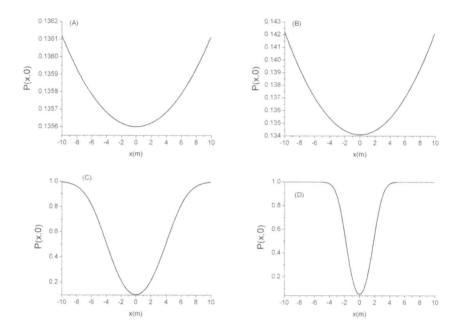

FIGURE 7.12
The spectral degree of polarization at the collecting lens plane for several
values of roughness parameter: (A) $\delta_T = 0.05$ mm, (B) $\delta_T = 0.1$ mm, (C)
$\delta_T = 0.5$ mm , (D) $\delta_T = 1$ mm. From Ref. [35].

Figure 7.11 presents numerical results calculated from the formulas derived in this section relating to the distribution of the spectral density (intensity) and the degree of polarization of a typical electromagnetic Gaussian Schell-model beam propagating through the bistatic LIDAR system [35]. The following parameters have been chosen: $\lambda = 1.55$ μ m, $\sigma = 5$ cm, $\delta_{xx} = 1$ mm, $\delta_{yy} = 1.2$ mm, $A_x^2 = A_y^2 = 0.5$, $B_{xy} = 0$, $C_n^2 = 10^{-13}$ m$^{-2/3}$, $l = 1$ km, $F_G = 10^6$ m, $\sigma_T = 1$ m; $\delta_T = 0.1$ mm. The tendency of the beam of losing its directionality at the target is due to turbulence effects alone while at the collecting lens plane it is the joint effect of turbulence fluctuations and the roughness of the target. We also notice that the initial uniform polarization ($\wp = 0$) becomes nonuniform already at the target.

Figure 7.12 shows the spectral degree of polarization of the beam at the collecting lens plane as a function of the correlation width δ_T of the target while the other parameters of the LIDAR system are kept fixed (and are the same as in Fig. 7.11). We assume here that the target is unresolved, i.e., its extent is much larger than the size of the illumination beam. Only in cases when the roughness parameter δ_T is sufficiently small the distribution of the degree of polarization differs substantially from that before interaction with the target.

7.3.3   Target characterization: inverse problem

Finally we will illustrate how some of the results derived in the previous section can be applied for solving the inverse problem of target identification from the cross-spectral density of the collected light [37]. We assume here that the incident light is linearly polarized scalar Gaussian Schell-model beam with the r.m.s. of the beam width and the correlation width, σ_{in} and δ_{in}, respectively. Then the cross-spectral density is recorded in the plane of the collecting lens. We start from scalar version of tensor, $\widetilde{\mathbf{M}}_2^{-1}$, characterizing the beam at the collecting lens plane:

$$
\begin{aligned}
\widetilde{\mathbf{M}}_2^{-1} &= \tilde{\mathbf{P}} + \tilde{\mathbf{B}}^{-1} - (\tilde{\mathbf{B}}^{-1} - \tilde{\mathbf{P}}/2)^T \\
&\times (\mathbf{M}_1^{-1} + \tilde{\mathbf{T}} + \tilde{\mathbf{B}}^{-1} + \tilde{\mathbf{P}})^{-1}(\tilde{\mathbf{B}}^{-1} - \tilde{\mathbf{P}}/2).
\end{aligned}
\tag{7.60}
$$

This tensor can be equivalently expressed in the form

$$
\begin{aligned}
\widetilde{\mathbf{M}}_2^{-1} &= \begin{bmatrix} m_{11}\mathbf{I} & m_{12}\mathbf{I} \\ m_{21}\mathbf{I} & m_{22}\mathbf{I} \end{bmatrix} \\
&= \begin{bmatrix} \dfrac{1}{ik}\left(R_{out}^{-1} - \dfrac{i}{2k\sigma_{out}^2} - \dfrac{i}{k\delta_{out}^2} \right)\mathbf{I} & \dfrac{i}{k\delta_{out}^2}\mathbf{I} \\[3mm] \dfrac{i}{k\delta_{out}^2}\mathbf{I} & \dfrac{1}{ik}\left(-R_{out}^{-1} - \dfrac{i}{2k\sigma_{out}^2} - \dfrac{i}{k\delta_{out}^2} \right)\mathbf{I} \end{bmatrix},
\end{aligned}
\tag{7.61}
$$

where σ_{out} and δ_{out} are the transverse r.m.s. beam width and the transverse r.m.s. coherence width of the Gaussian Shell-model beam at the receiver plane, and R_{out}^{-1} is its wave-front curvature. From Eq. (7.61) the relations

$$\frac{1}{\delta_{out}^2} = \frac{k}{i} m_{12}, \qquad \frac{1}{\sigma_{out}^2} = -\frac{k}{i}(m_{11} + m_{22}) + \frac{2k}{i}(m_{12}) \qquad (7.62)$$

follow. On applying the expressions for $\mathbf{M}_0^{-1}$, $\mathbf{M}_1^{-1}$, $\mathbf{M}_2^{-1}$ (scalar analogs) and $\tilde{\mathbf{T}}$ together with Eq. (7.61) the following equations can be derived:

$$\frac{\rho_0^2}{\delta_{out}^2} = \frac{e_1 w_T^4 + 4b_1 w_T^2 + 2b_1 \delta_T^2 + f_1 \delta_T^2 w_T^2 + g_1 \delta_T^2 w_T^4}{a_1 w_T^4 + 2b_1 w_T^2 + b_1 \delta_T^2 + c_1 \delta_T^2 w_T^2 + d_1 \delta_T^2 w_T^4},$$

$$\frac{1}{\sigma_{out}^2} = \frac{e_2 \delta_T^2 w_T^2 + f_2 \delta_T^2 w_T^4}{a_1 w_T^4 + 2b_1 w_T^2 + b_1 \delta_T^2 + c_1 \delta_T^2 w_T^2 + d_1 \delta_T^2 w_T^4}, \qquad (7.63)$$

where

$$a_1 = 8R^2 z^2 \rho_0^2 q_1^2 q_2^2, \quad b_1 = 16R^2 z^2 \rho_0^2 q_1^2 q_2^4,$$

$$c_1 = R^2 z^2 q_2^2 [16\rho_0^2 q_2^2 + q_1^2(8\rho_0^2 + 32q_2^2)],$$

$$d_1 = 8k^2 Rz \rho_0^2 q_1^2 q_2^4 + 4k^2 z^2 q_1^2 \rho_0^2 q_2^4$$
$$\quad + R^2[(2k_1 q_2^2)^2 + z(4\rho_0^2 q_2^2 + q_1^2(\rho_0^2 + 8q_2^2))],$$

$$e_1 = \rho_0^2 R^2 q_1^2 q_2^2 (16z^2 + 8k^2 \rho_0^2 q_2^2),$$

$$f_1 = R^2 z^2 q_2^2 [32\rho_0^2 q_2^2 + q_1^2(16\rho_0^2 + 48q_2^2)], \qquad (7.64)$$

$$g_1 = 24k^2 Rz \rho_0^2 q_1^2 q_2^4 + 8k^2 z^2 \rho_0^2 q_1^2 q_2^4$$
$$\quad + R^2[k^2 \rho_0^2 (24q_1^2 + 4\rho_0^2)q_2^4 + z^2(8\rho_0^2 q_2^2 + q_1^2(2\rho_0^2 + 12q_2^2))],$$

$$e_2 = 16k^2 R^2 \rho_0^2 q_1^2 q_2^4, \quad f_2 = 4k^2 R^2 \rho_0^2 q_1^2 q_2^2,$$

and

$$q_1^2 = \frac{4k^2 \rho_0^4 \delta_{in}^2 \sigma_{in}^4 + z^2 \rho_0^2 [4\rho_0^2 \sigma_{in}^2 + \delta_{in}^2(\rho_0^2 + 8\sigma_{in}^2)]}{k^2 \rho_0^2 (24\delta_{in}^2 + 4\rho_0^2)\sigma_{in}^4 + z^2[8\rho_0^2 \sigma_{in}^2 + \delta_{in}^2(2\rho_0^2 + 12\sigma_{in}^2)]},$$

$$q_2^2 = \frac{4k^2 \rho_0^2 \delta_{in}^2 \sigma_{in}^4 + z^2[4\rho_0^2 \sigma_{in}^2 + \delta_{in}^2(\rho_0^2 + 8\sigma_{in}^2)]}{4k^2 \rho_0^2 \delta_{in}^2 \sigma_{in}^2} \qquad (7.65)$$

$$\frac{1}{R} = -\frac{1}{F_G} + \frac{z[\rho_0^2 \sigma_{in}^2 + \delta_{in}^2(0.25\rho_0^2 + 3\sigma_{in}^2)]}{k^2 \rho_0^2 \delta_{in}^2 \sigma_{in}^4 + z^2[\rho_0^2 \sigma_{in}^2 + \delta_{in}^2(0.25\rho_0^2 + 2\sigma_{in}^2)]}.$$

On combining Eqs. (7.63) we arrive at the biquadratic equation with respect to w_R:

$$A_1 w_T^4 + B_1 w_T^2 + C_1 = 0, \qquad (7.66)$$

with

$$A_1 = \delta_{out}^2 \sigma_{out}^2 e_1 f_2 - \delta_{out}^2 e_1 d_1 - \rho_0^2 \sigma_{out}^2 a_1 f_2 + \delta_{out}^2 a_1 g_1,$$

$$B_1 = \delta_{out}^2 \sigma_{out}^2 e_1 e_2 - \delta_{out}^2 e_1 c_1 + 4\delta_{out}^2 \sigma_{out}^2 b_1 f_2 - 4\delta_{out}^2 b_1 d_1$$
$$\quad - \rho_0^2 \sigma_{out}^2 a_1 e_2 - 2\rho_0^2 \sigma_{out}^2 b_1 f_2 + \delta_{out}^2 a_1 f_1 + 2\delta_{out}^2 b_1 g_1, \qquad (7.67)$$

$$C_1 = 4\delta_{out}^2 \sigma_{out}^2 b_1 e_2 - 4\delta_{out}^2 b_1 c_1 - \delta_{out}^2 e_1 b_1 - 2\rho_0^2 \sigma_{out}^2 b_1 e_2$$
$$\quad + \delta_{out}^2 a_1 b_1 + 2\delta_{out}^2 b_1 f_1.$$

Solutions of Eq. (7.66) are

$$w_{T,n}^2 = \frac{-B_1 \pm \sqrt{B_1^2 - 4A_1 C_1}}{2A_1}, \quad (n = 1, 2).$$ (7.68)

Finally, it is implied by Eqs. (7.63) and (7.68) that two positive solutions exist for δ_T:

$$\delta_{T,n} = \sqrt{\frac{a_1 w_{T,n}^4 + 2b_1 w_{T,n}^2}{\sigma_{out}^2 e_2 w_{T,n}^2 + \sigma_{out}^2 f_2 w_{T,n}^4 - b_1 - c_1 w_{T,n}^2 - d_1 w_{T,n}^4}},$$ (7.69)

$$(n = 1, 2).$$

There are two sets of solutions for the width of target w_T and for its correlation length δ_T. The unique physical result can be determined from the measured set of values for σ_{in}, σ_{out}, δ_{in}, δ_{out} if one substitutes the measured values for σ_{in}, δ_{in} and any pair of solutions $w_{T,n}$, $\delta_{T,n}$ ($n = 1, 2$) into Eqs. (7.63) and finds the values of σ_{out} and δ_{out}. The pair of solutions for which the measured and calculated values of σ_{out} and δ_{out} are equal is the proper physical solution.

Thus, from the r.m.s. widths of the intensity and degree of coherence of a Gaussian Schell-model beam measured at the transmitter and the receiver it is possible to deduce the target's width and typical roughness.

Bibliography

[1] "Bell's Photophone," *Nature* **23**, 15-19 (1880).

[2] L.C. Andrews, R.L. Phillips, and C.Y. Hopen, *Laser Beam Scintillations with Applications*, SPIE Press, 2001.

[3] O. Korotkova, *Partially Coherent Beam Propagation in Turbulent Amosphere with Applications*, VDM, 2009.

[4] O. Korotkova, S. Avramov-Zamurovic, C. Nelson, and R. Malek-Madani, "Probability density function of partially coherent beams propagating in the atmospheric turbulence," *Proc. SPIE.* **8238**, 82380J (2012).

[5] R. Barakat, "First-order intensity and log-intensity probability density functions of light scattered by the turbulent atmosphere in terms of lower-order moments," *J. Opt. Soc. Am.* **16**, 2269–2274 (1999).

[6] M. A. Al-Habash, L. C. Andrews, and R. L. Phillips, "Mathematical model for the irradiance probability density function of a laser beam propagating through turbulent media," *Opt. Eng.* **40**, 1554–1562 (2001).

[7] O. Korotkova, S. Avramov-Zamurovic, R. Malek-Madani, and C. Nelson, "Probability density function of a laser beam propagating in maritime environment," *Opt. Express* **19**, 20322–20331 (2011).

[8] Y. Gu, O. Korotkova, and G. Gbur, "Scintillation of non-uniformly polarized beams in atmospheric turbulence," *Opt. Lett.* **34**, 2261–2263 (2009).

[9] V. I. Tatarski, *Wave Propagation in a Turbulent Medium*, Nauka, Moscow, 1967.

[10] J. M. Martin and S. M. Flatte, "Intensity images and statistics from numerical simulation of wave propagation in 3-D random media," *Appl. Opt.* **27**, 2111–2126 (1988).

[11] L. C. Andrews and R. L. Phillips, *Laser Beam Propagation in the Turbulent Atmosphere*, 2nd Edition, SPIE Press, 2005.

[12] Y. Kozawa and S. Sato, "Generation of a radially polarized laser beam by use of a conical Brewster prism," *Opt. Lett.* **30**, 3063–3065 (2007).

[13] J. C. Leader, "Intensity fluctuations resulting from partially coherent light propagating through atmospheric turbulence," *J. Opt. Soc. Am.* **69**, 73–84 (1979).

[14] V. A. Banakh, V. M. Buldakov, and V. L. Mironov, "Intensity fluctuations of a partially coherent light beam in a turbulent atmosphere," *Opt. Spektrosc.* **54**, 1054–1059 (1983).

[15] J. C. Ricklin and F. M. Davidson, "Atmospheric turbulence effects on a partially coherent Gaussian Beam: implications for free-space laser communication," *J. Opt. Soc. Am. A* **19**, 1794–1802 (2002).

[16] T. J. Schulz, "Optimal beams for propagation through random media," *Opt. Lett.* **30**, 1093–1095 (2005).

[17] D. K. Borah and D. G. Voelz, "Spatially partially coherent beam parameter optimization for free space optical communications," *Opt. Express* **18**, 20746–20758 (2010).

[18] O. Korotkova, "Scintillation index of a stochastic electromagnetic beam propagating in random media," *Opt. Comm.* **281**, 2342–2348 (2008).

[19] O. Korotkova, L. C. Andrews, and R. L. Phillips, "A model for a partially coherent Gaussian beam in atmospheric turbulence with application in LaserCom," *Opt. Eng.* **43**, 330–341 (2004).

[20] O. Korotkova, "Changes in the statistics of random electromagnetic beams on propagation," *J. Opt. A: Pure Appl. Opt.* **8**, 30–37 (2006).

[21] R. Barakat, "Natural light, generalized Verdet–Stokes conditions, and the covariance matrix of the Stokes parameters," *J. Opt. Soc. Am. A* **6**, 649–659 (1989).

[22] A. P. Cracknell and L. Hayes, *Introduction to Remote Sensing*, 2nd Edition, Taylor and Francis, 2007.

[23] V. Banakh and V. L. Mironov, *LIDAR in Turbulent Atmosphere*, Artech House, 1984.

[24] R. S. Hansen, H. T. Yura, and S. G. Hanson, "First-order speckle statistics: an analytic analysis using ABCD matrices," *J. Opt. Soc. Am. A* **14**, 3093–3098 (1997).

[25] H. T. Yura, B. Rose, and S. G. Hanson, "Dynamic laser speckle in complex ABCD optical systems," *J. Opt. Soc. Am. A* **15**, 1160–1166 (1998).

[26] Y. Cai, O. Korotkova, H. T. Eyyubogllu, and Y. Baykal, "Active laser radar systems with stochastic electromagnetic beams in turbulent atmosphere," *Opt. Express* **16**, 15834–15846 (2008).

[27] J. C. Leader, "Atmospheric propagation of partially coherent radiation," *J. Opt. Soc. Am.* **68**, 175–185 (1978).

[28] H. T. Yura and S. G. Hanson, "Optical beam wave propagation through complex optical systems," *J. Opt. Soc. Am. A* **4**, 1931–1948 (1987).

[29] H. T. Yura and S. G. Hanson, "Second-order statistics for wave propagation through complex optical systems," *J. Opt. Soc. Am. A* **6**, 564–575 (1989).

[30] P. Beckmann and A. Spizzichino, *The Scattering of Electromagnetic Waves from Rough Surfaces*, Pergamon 1963.

[31] J. A. Ogilvy, *Theory of Scattering from Random Rough Surfaces*, Adam Hilger, 1991.

[32] J. C. Stover, *Optical Scattering: Measurement and Analysis*, SPIE press, 1995.

[33] F. G. Bass and I. M. Fuks, *Wave Scattering from Statistically Rough Surfaces*, Pergamon, New York, 1979.

[34] J. W. Goodman, "Role of coherence in the study of speckle," *Proc. SPIE* **194**, *Appl. Opt. Coher.*, 86–94 (1979).

[35] O. Korotkova, Y. Cai, and E. Watson, "Stochastic electromagnetic beams for LIDAR systems operating in turbulent atmosphere," *Appl. Phys. B* **94**, 681–690 (2009).

[36] J. W. Goodman, "Statistical Properties of Laser Speckle Patterns", in *Laser Speckle and Related Phenomena*, Ed. by J.C. Dainty Springer, 1975.

[37] G. Wu and Y. Cai, "Detection of a semirough target in turbulent atmosphere by a partially coherent beam," *Opt. Lett.* **36**, 1939–1941 (2011).

8

Weak scattering of random beams

CONTENTS

In this chapter we consider one more type of light-matter interaction, namely the passage of a wave through a static medium, confined to some volume in space and scattered sufficiently far from it. The scattering medium may be a single particle or their collection in which the optical properties of individual members might be deterministic or random. Within a collection there might be scatterers of a single or many types. In general, interaction of electromagnetic waves with scatterers occupying a certain volume in space and having arbitrary distribution of optical properties is a very complex process, even in the linear treatment. The random nature of illumination and/or of the scatterers only multiply the complexity. However, in some cases it is still possible to predict some of the statistical properties of scattered radiation.

8.1 Classic theory of weak scattering

So far the theories of scattering for scalar [1] and electromagnetic [2] (see also [3]) stochastic fields from deterministic and random media have been formulated within the validity of the first-order Born approximation ([4], Section 13.1). Such treatments allow for determining the changes in spectrum, coherence and polarization of fields produced as a net effect of their propagation

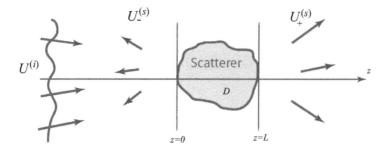

FIGURE 8.1
Illustrating the notation of scattering theory. From Ref. [5].

from the source to the scattering volume, interaction with the scatterer, and propagation from the scatterer to the far field. The results based on treatments in [1]–[3] indicate that the knowledge of the statistical properties of both the source and the scattering medium plays the crucial part in predicting the properties of the scattered radiation. Along with the purely theoretical developments we include several examples involving scattering from collections of particles, thin bio-tissue layers, and weakly turbulent media containing discrete scatterers.

To begin, we suppose that a weakly-scattering medium occupies some closed region D in the three-dimensional space between planes $z = 0$ and $z = L$ and that a field $U^{(i)}(\mathbf{r}; \omega)$ is incident onto this region and scattered into field $U^{(s)}(\mathbf{r}; \omega)$ (see Fig. 8.1). Assuming that both the incident and the scattered fields are scalar and that the distribution of electric permittivity $\epsilon_m(\mathbf{r}; \omega)$ within the scattering volume is a slowly-varying function of position $\mathbf{r}$ we state that the evolution of the total field

$$U^{(t)}(\mathbf{r}; \omega) = U^{(i)}(\mathbf{r}; \omega) + U^{(s)}_{\pm}(\mathbf{r}; \omega), \tag{8.1}$$

produced on scattering is described by the wave equation ([4], Section 13.1)

$$\nabla^2 U^{(t)}(\mathbf{r}; \omega) + k^2 n^2(\mathbf{r}; \omega) U^{(t)}(\mathbf{r}; \omega) = 0, \tag{8.2}$$

where $n(\mathbf{r}; \omega)$ is the refractive index at a point with position vector $\mathbf{r}$ within the scatterer, taken at frequency ω.

We now rewrite this equation as

$$\nabla^2 U^{(t)}(\mathbf{r}; \omega) + k^2 U^{(t)}(\mathbf{r}; \omega) = -4\pi F(\mathbf{r}; \omega) U^{(t)}(\mathbf{r}; \omega), \tag{8.3}$$

where function

$$F(\mathbf{r}; \omega) = \begin{cases} \frac{k^2}{4\pi}[n^2(\mathbf{r}; \omega) - 1], & \mathbf{r} \in D, \\ 0, & \text{otherwise,} \end{cases} \tag{8.4}$$

is termed the *scattering potential* of the medium. Equation (8.3) is an inhomogeneous Helmholtz equation. Based on the Green's function solution method the total field $U^{(t)}(\mathbf{r};\omega)$ produced on scattering of the incident field $U^{(i)}(\mathbf{r};\omega)$ from the medium with potential $F(\mathbf{r};\omega)$ can be expressed as an integral Fredholm equation as

$$U^{(t)}(\mathbf{r};\omega) = U^{(i)}(\mathbf{r};\omega) + \int_D F(\mathbf{r}';\omega)U^{(t)}(\mathbf{r}';\omega)G(\mathbf{r}-\mathbf{r}';\omega)d^3r', \qquad (8.5)$$

where $\mathbf{r}'$ and $\mathbf{r}$ are points within and outside the scatterer, and $G(\mathbf{r}-\mathbf{r}';\omega)$ is the Green's function of the three-dimensional Helmholtz equation

$$G(\mathbf{r}-\mathbf{r}';\omega) = \frac{\exp(ik|\mathbf{r}-\mathbf{r}'|)}{|\mathbf{r}-\mathbf{r}'|}. \qquad (8.6)$$

The integral equation (8.5) is generally impossible to solve analytically for the given initial conditions and inhomogeneity. However, if the scatterer is weak, i.e., if

$$|U^{(s)}| << |U^{(i)}|, \qquad (8.7)$$

then

$$U^{(t)}(\mathbf{r};\omega) = U^{(i)}(\mathbf{r};\omega) + \int_D F(\mathbf{r}';\omega)U^{(i)}(\mathbf{r}';\omega)G(\mathbf{r}-\mathbf{r}';\omega)d^3r'. \qquad (8.8)$$

In this case the integral equation reduces to an algebraic equation leading to substantial simplifications for finding its solutions. Approximation (8.7) is known as the *first Born approximation* and can be successfully employed in situations when the scattering is weak, for instance when the difference in refractive indexes inside and outside of the scatterer is sufficiently small ([4], Section 13.1).

In the far zone of the scatterer (see Fig. 8.2) the approximation $|\mathbf{r}-\mathbf{r}'| \approx |\mathbf{r}|-\mathbf{u}\cdot\mathbf{r}'$, $\mathbf{u}$ being the unit vector along direction of $\mathbf{r}$, implies that the Green's function can be approximated as [4]

$$G(\mathbf{r}-\mathbf{r}';\omega) = \frac{\exp(ikr)}{r}\exp(-ik\mathbf{u}\cdot\mathbf{r}'). \qquad (8.9)$$

Hence the total field evaluated at $\mathbf{r}=r\mathbf{u}$ becomes

$$U^{(t)}(\mathbf{r};\omega) = U^{(i)}(\mathbf{r};\omega) + \frac{\exp(ikr)}{r}\int_D F(\mathbf{r}';\omega)U^{(i)}(\mathbf{r}';\omega)\exp(-ik\mathbf{u}\cdot\mathbf{r}')d^3r'.$$

$$(8.10)$$

In case when the incident field is a plane wave with amplitude $a^{(i)}(\omega)$ which propagates along direction $\mathbf{u}_0$:

$$U^{(i)}(\mathbf{r};\omega) = a^{(i)}(\omega)\exp(ik\mathbf{u}_0\cdot\mathbf{r}), \qquad (8.11)$$

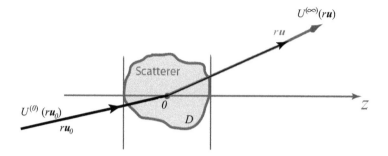

FIGURE 8.2
Illustrating scattering to the far field. From Ref. [5].

equation (8.10) becomes

$$U^{(t)}(\mathbf{r};\omega) = a^{(i)}(\omega)\left[\exp(ik\mathbf{u}_0 \cdot \mathbf{r}) + \frac{\exp(ikr)}{r}\tilde{F}(k(\mathbf{u}-\mathbf{u}_0);\omega)\right], \qquad (8.12)$$

i.e., the scattered field is proportional to the three-dimensional Fourier transform of the scattering potential evaluated at vector $\mathbf{K} = k(\mathbf{u}-\mathbf{u}_0)$, sometimes referred to as the *momentum transfer vector*.

8.2 Description of scattering media

Before proceeding to scattering of random beams we will first introduce various models for scattering potentials characterizing their physical shape, dimensions, boundary features and correlation properties that are used in analytic and numerical calculations.

8.2.1 Single scatterer

A hard-edge sphere of radius r_0 having the constant value of the refractive index, say F_0, has the scattering potential of a three-dimensional piece-wise continuous function of the form:

$$F^{(S)}(\mathbf{r};\omega) = \begin{cases} F_0, & r \leq r_0, \\ 0, & \text{otherwise.} \end{cases} \qquad (8.13)$$

However, such a distribution is seldom suitable for performing analytical calculations of the properties of scattered radiation. Perhaps, the simplest model leading to tractable analytical results is based on the Gaussian function and describes a particle with optically dense center and soft edges. If the center of the scatterer is located at a point with position vector $\mathbf{r}_c = (x_c, y_c, z_c)$ then the potential has form [6]

$$F^{(G)}(\mathbf{r}; \omega) = F_c \exp \left[-\frac{(x - x_c)^2 + (y - y_c)^2 + (z - z_c)^2}{2\sigma_c^2} \right] \tag{8.14}$$

where F_c is its maximum value at the center, and σ_c is its root-mean-square (r.m.s.) width.

Elliptical potentials with soft edges oriented along the coordinate axes of a particle frame $[\xi, \eta, \zeta]$ can also be modeled with the help of the similar distributions [7]:

$$F^{(E)}(\mathbf{r}; \omega) = F_c \exp \left[-\left(\frac{(\xi - x_c)^2}{2\sigma_{xc}^2} + \frac{(\eta - y_c)^2}{2\sigma_{yc}^2} + \frac{(\zeta - z_c)^2}{2\sigma_{zc}^2} \right) \right], \tag{8.15}$$

where σ_{xc}, σ_{yc} and σ_{zc} are the r.m.s. widths (see Fig. 8.3). The orientation of the particle frame $[\xi, \eta, \zeta]$ relative to the laboratory frame $[x, y, z]$ is defined by three counter clockwise rotation angles α, β and γ as shown in Fig. 8.4. Starting with the laboratory frame, we obtain the particle frame by rotation α of the $y - z$ plane around the x axis followed by rotation β of the $x' - z'$ plane around the y' axis and by rotation γ of the $x'' - y''$ plane around the z'' axis. The coordinates in the particle frame $[\xi, \eta, \zeta]$ are related to those of the laboratory frame $[x, y, z]$ as

$$\begin{bmatrix} \xi \\ \eta \\ \zeta \end{bmatrix} = \begin{bmatrix} A_{11} & A_{12} & A_{13} \\ A_{21} & A_{22} & A_{23} \\ A_{31} & A_{32} & A_{33} \end{bmatrix} \begin{bmatrix} x \\ y \\ z \end{bmatrix}, \tag{8.16}$$

where

$$\begin{aligned}
&A_{11} = \cos\beta\cos\gamma, \quad A_{12} = \cos\alpha\sin\gamma + \sin\alpha\sin\beta\cos\gamma, \\
&A_{13} = \sin\alpha\sin\gamma - \cos\alpha\sin\beta\cos\gamma, \quad A_{21} = -\cos\beta\sin\gamma, \\
&A_{22} = \cos\alpha\cos\gamma - \sin\alpha\sin\beta\sin\gamma, \quad A_{23} = \sin\alpha\cos\gamma + \cos\alpha\sin\beta\sin\gamma, \\
&A_{31} = \sin\beta, \quad A_{32} = -\sin\alpha\cos\beta, \quad A_{33} = \cos\alpha\cos\beta.
\end{aligned} \tag{8.17}$$

On substituting from Eqs. (8.16) and (8.17) into Eq. (8.15), we obtain the following expression for the scattering potential of a particle

$$\begin{aligned}
F^{(E)}(\mathbf{r}; \omega) = F_c \exp \bigg\{ &-\Big[B^{(1)}(x - x_c)^2 + B^{(2)}(y - y_c)^2 \\
&+ B^{(3)}(z - z_c)^2 + 2B^{(4)}(x - x_c)(y - y_c) \\
&+ 2B^{(5)}(x - x_c)(z - z_c) + 2B^{(6)}(z - z_c)(y - y_c) \Big] \bigg\},
\end{aligned} \tag{8.18}$$

324

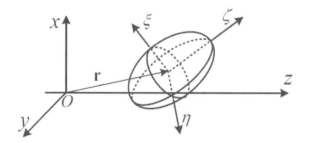

FIGURE 8.3
Illustrating the notation relating to an ellipsoid. From Ref. [7].

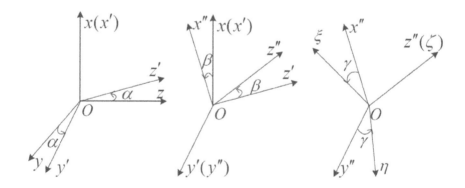

FIGURE 8.4
Illustrating three basic rotations defining the orientation of the particle frame $[\xi, \eta, \zeta]$ relative to the laboratory frame $[x, y, z]$. From Ref. [7].

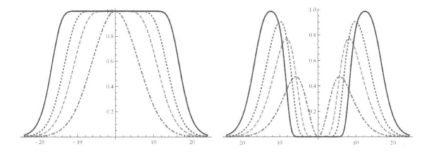

FIGURE 8.5
Scattering potential for solid particles (left) and hollow particles (right) as a function of $r/(\sqrt{2}\sigma)$ for several values of M: $M = 1$ (dashed-dotted curve); $M = 4$ (dashed curve); $M = 10$, (dotted curve) and $M = 40$ (solid thick curve). From Ref. [7].

where

$$
\begin{aligned}
B^{(1)} &= A_{11}^2/(2\sigma_{xc}^2) + A_{21}^2/(2\sigma_{yc}^2) + A_{31}^2/(2\sigma_{zc}^2), \\
B^{(2)} &= A_{12}^2/(2\sigma_{xc}^2) + A_{22}^2/(2\sigma_{yc}^2) + A_{32}^2/(2\sigma_{zc}^2), \\
B^{(3)} &= A_{13}^2/(2\sigma_{xc}^2) + A_{23}^2/(2\sigma_{yc}^2) + A_{33}^2/(2\sigma_{zc}^2), \\
B^{(4)} &= A_{11}A_{12}/(2\sigma_{xc}^2) + A_{21}A_{22}/(2\sigma_{yc}^2) + A_{31}A_{32}/(2\sigma_{zc}^2), \\
B^{(5)} &= A_{11}A_{13}/(2\sigma_{xc}^2) + A_{21}A_{23}/(2\sigma_{yc}^2) + A_{31}A_{33}/(2\sigma_{zc}^2), \\
B^{(6)} &= A_{12}A_{13}/(2\sigma_{xc}^2) + A_{22}A_{23}/(2\sigma_{yc}^2) + A_{32}A_{33}/(2\sigma_{zc}^2).
\end{aligned}
\tag{8.19}
$$

In order to model potentials with more realistic, sharper boundaries the multi-Gaussian distribution can be employed [8]. For instance to model a sphere with a semi-soft edge we set

$$
\begin{aligned}
F^{(MG)}(\mathbf{r};\omega) &= \frac{F_c}{C_0} \sum_{m=1}^{M} \frac{(-1)^{m-1}}{m} \binom{M}{m} \\
&\times \exp\left[-m\frac{(x-x_c)^2 + (y-y_c)^2 + (z-z_c)^2}{2\sigma_c^2}\right],
\end{aligned}
\tag{8.20}
$$

where

$$
C_0 = \sum_{m=1}^{M} \frac{(-1)^{m-1}}{m} \binom{M}{m}
\tag{8.21}
$$

is the normalization factor. As M approaches infinity the distribution tends to that of a hard sphere.

For hollow spherical particles (spherical shells) with adjustable boundary thickness one can use the difference of two multi-Gaussian distributions:

$$F^{(HMG)}(\mathbf{r};\omega) = \frac{F_c}{C_0} \sum_{m=1}^{M} \frac{(-1)^{m-1}}{m} \binom{M}{m}$$

$$\times \left\{ \exp\left[-m\frac{(x-x_c)^2 + (y-y_c)^2 + (z-z_c)^2}{2\sigma_a^2}\right] \right. \qquad (8.22)$$

$$\left. - \exp\left[-m\frac{(x-x_c)^2 + (y-y_c)^2 + (z-z_c)^2}{2\sigma_b^2}\right] \right\},$$

where σ_a and σ_b are the r.m.s. widths of the outer and inner boundaries, with $\sigma_a > \sigma_b$. Figure 8.5 illustrates distributions (8.20) and (8.22) for different values of summation index M.

8.2.2 Collections of scatterers

If the scattering collection consists of M identical scatterers with potentials $f(\mathbf{r})$ then the net potential is given by the sum

$$F(\mathbf{r};\omega) = \sum_{m=1}^{M} f(\mathbf{r} - \mathbf{r}_m;\omega), \qquad (8.23)$$

where $\mathbf{r}_m$ is the center of particle with index m. If the collection consists of particles of L types then its net potential becomes

$$F(\mathbf{r};\omega) = \sum_{l=1}^{L} \sum_{m=1}^{M_l} f_l(\mathbf{r} - \mathbf{r}_{lm};\omega), \qquad (8.24)$$

where r_{lm} is the location of the scattering center of a particle of type l, f_l is the potential of the scatterers of type l, and M_l is the number of scatterers of type l.

8.2.3 Random scatterers

In order to characterize random scatterers the correlation function of the scattering potential at two positions, $\mathbf{r}_1$ and $\mathbf{r}_2$ and the angular frequency ω must be specified [1]:

$$C_F(\mathbf{r}_1, \mathbf{r}_2;\omega) = \langle F^*(\mathbf{r}_1;\omega)F(\mathbf{r}_2;\omega)\rangle_M, \qquad (8.25)$$

where the average is taken over the medium realizations. In analytic calculations one of the following models is then frequently used: *the quasi-homogeneous scatterer*

$$C_F^{(QH)}(\mathbf{r}_1, \mathbf{r}_2;\omega) = I_F\left(\frac{\mathbf{r}_1 + \mathbf{r}_2}{2};\omega\right) \mu_F(\mathbf{r}_2 - \mathbf{r}_1;\omega), \qquad (8.26)$$

and the *Schell-model scatterer*

$$C_F^{(SM)}(\mathbf{r}_1, \mathbf{r}_2; \omega) = \sqrt{I_F(\mathbf{r}_1; \omega)}\sqrt{I_F(\mathbf{r}_2; \omega)}\mu_F(\mathbf{r}_2 - \mathbf{r}_1; \omega), \qquad (8.27)$$

where

$$I_F(\mathbf{r}; \omega) = C_F(\mathbf{r}, \mathbf{r}; \omega) \qquad (8.28)$$

and

$$\mu_F(\mathbf{r}_2 - \mathbf{r}_1; \omega) = \frac{C_F(\mathbf{r}_1, \mathbf{r}_2; \omega)}{\sqrt{C_F(\mathbf{r}_1, \mathbf{r}_1; \omega)}\sqrt{C_F(\mathbf{r}_2, \mathbf{r}_2; \omega)}} \qquad (8.29)$$

being the degree of spatial correlation of the random medium.

These models for the correlation function C_F can also be extended to the case of particulate collections with one or several types of particles [9], [10].

8.3 Weak scattering for scalar fields

8.3.1 Cross-spectral density function of scattered field

We begin the discussion of the scattering theory for random beams with the concept of the scattering matrix, which provides a convenient tool for description of beam-scatterer interaction in the linear regime [5]. The reader may find that the scattering matrix has a rich history in connection with nuclear physics and elementary particles [11]–[12], acoustics [13] and electromagnetic theory [14], [15]. Scattering matrix is a linear transformation of the wave incident on a physical system into the scattered wave. Although traditional treatment via the multipole expansion has been often employed it appears more natural for electromagnetic theory to represent fields by their plane-wave spectra, which have been used throughout this text. Indeed, plane-wave decomposition of fields leads to simple Fourier-type relations between the scattering potential and the scattered field, discriminates between evanescent and homogeneous waves, and is compatible with propagation theory, which is very convenient in situations when the incident field encounters both weekly fluctuating medium and particles.

Let us first consider a monochromatic scalar field $U^{(i)}(\mathbf{r}; \omega)e^{-i\omega t}$ incident onto a scatterer. We recall that its spatial counterpart can be via the angular spectrum of plane waves [see Eq. (1.63)] viz.,

$$U^{(i)}(\mathbf{r}; \omega) = \int a^{(i)}(\mathbf{u}; \omega)\exp[ik(\mathbf{u}_\perp \mathbf{r} + u_z z)]d^2 u_\perp, \qquad (8.30)$$

where $\mathbf{u}_\perp = (u_x, u_y, 0)$ is the transverse part of the unit vector $\mathbf{u} = (u_x, u_y, u_z)$ and $u_z = \sqrt{1 - u_\perp^2}$, when $u_\perp \leq 1$ (homogeneous waves) and $u_z = i\sqrt{u_\perp^2 - 1}$, when $u_\perp > 1$ (evanescent waves) with $u_\perp = |\mathbf{u}_\perp|$. In general, integration in Eq. (8.30) extends over the whole (u_x, u_y) plane.

If the field $U^{(i)}(\mathbf{r}; \omega)$ is incident onto a scattering medium contained in a finite domain D, in the strip $0 < z < L$ (see Fig. 8.1) then the fields produced on scattering in each of the two half spaces, $z > 0$ and $z \geq L$, on either side of the scatterer, may also be represented by their angular spectrum representations:

$$U_{\pm}^{(s)}(\mathbf{r}; \omega) = \int a^{(s)}(\mathbf{u}; \omega) \exp[ik(\mathbf{u}_{\perp}\mathbf{r} \pm u_z z)] d^2 u_{\perp}. \tag{8.31}$$

In this decomposition positive or negative signs in the exponent are taken according as the field point $\mathbf{r} = (x, y, z)$ is located in the half space $z > L$ (forward scattering) or $z < 0$ (backscattering), respectively. Contributions of evanescent waves are significant only at points $\mathbf{r}$ which are within a distance on the order of a wavelength or less from the scatterer. It is important to note that in our formulation the incident illumination may be generally of any spatial content, hence its angular spectrum may include plane waves having different propagation directions. Hence, along such directions the scattered field is a mixture of incident and scattered waves that cannot be physically separated.

The dependence between the amplitudes of the incident plane wave $a^{(i)}(\mathbf{u}'; \omega)$ and the scattered plane wave $a^{(s)}(\mathbf{u}; \omega)$ has, for a linear deterministic medium, a simple product form:

$$a^{(s)}(\mathbf{u}', \mathbf{u}; \omega) = \mathbb{S}(\mathbf{u}', \mathbf{u}; \omega) a^{(i)}(\mathbf{u}'; \omega), \tag{8.32}$$

where linear transformation $\mathbb{S}(\mathbf{u}', \mathbf{u}; \omega)$ is termed as the *scattering matrix*. Since it is assumed that the response of the medium is linear, the amplitudes of the plane waves with different frequencies are not coupled through the scattering process.

In the case when the incident field is not a single plane wave but is rather represented by the angular spectrum (8.30) the relation (8.32) must be generalized to the integral

$$a^{(s)}(\mathbf{u}', \mathbf{u}; \omega) = \frac{ik}{2\pi u_z} \int \mathbb{S}(\mathbf{u}', \mathbf{u}; \omega) a^{(i)}(\mathbf{u}'; \omega) d^2 u'_{\perp}, \tag{8.33}$$

where integration is performed over the plane (u'_x, u'_y). In order to prove this relation it suffices to decompose the Green's function $G(|\mathbf{r} - \mathbf{r}'|; \omega)$ given in Eq. (8.6) in terms of the plane waves [16] as

$$G(|\mathbf{r} - \mathbf{r}'|; \omega) = \frac{ik}{2\pi u_z} \int \exp[ik\mathbf{u} \cdot (\mathbf{r} - \mathbf{r}')] d^2 u_{\perp}. \tag{8.34}$$

Then formula (8.33) is obtained on substituting from Eqs. (8.30), (8.31) and (8.34) into the integral Eq. (8.8) and after some manipulations.

Further, on substituting from Eq. (8.33) into Eq. (8.31) the scattered field becomes:

$$U_{\pm}^{(s)}(\mathbf{r}; \omega) = \frac{ik}{2\pi u_z} \int \int \mathbb{S}(\mathbf{u}', \mathbf{u}; \omega) a^{(i)}(\mathbf{u}'; \omega)$$
$$\times \exp[ik(\mathbf{u}_{\perp}\mathbf{r} \pm u_z z)] d^2 u_{\perp} du'_{\perp}. \tag{8.35}$$

In case both incident and scattered field are random the cross-spectral density functions can be used:

$$W^{(i)}(\mathbf{r}_1, \mathbf{r}_2; \omega) = \langle U^{(i)*}(\mathbf{r}_1; \omega), U^{(i)}(\mathbf{r}_2; \omega) \rangle, \tag{8.36}$$

and

$$W^{(s)}(\mathbf{r}_1, \mathbf{r}_2; \omega) = \langle U_\pm^{(s)*}(\mathbf{r}_1; \omega), U_\pm^{(s)}(\mathbf{r}_2; \omega) \rangle, \tag{8.37}$$

where

$$W^{(i)}(\mathbf{r}_1, \mathbf{r}_2; \omega) = \int\int A^{(i)}(\mathbf{u}_1, \mathbf{u}_2; \omega) \exp[ik(\mathbf{u}_2\mathbf{r}_2 - \mathbf{u}_1\mathbf{r}_1)] d^2 u_{1\perp} d^2 u_{2\perp}, \tag{8.38}$$

and

$$W^{(s)}(\mathbf{r}_1, \mathbf{r}_2; \omega) = \frac{k^2}{4\pi^2 u_{z1} u_{z2}}$$
$$\times \int\int A^{(s)}(\mathbf{u}_1, \mathbf{u}_2; \omega) \exp[ik(\mathbf{u}_2\mathbf{r}_2 - \mathbf{u}_1\mathbf{r}_1)] du_{1\perp}^2 d^2 u_{2\perp}. \tag{8.39}$$

In these equations $\mathbf{u}_{1\perp} = (u_{1x}, u_{1y}, 0)$, $\mathbf{u}_{2\perp} = (u_{2x}, u_{2y}, 0)$, while $A^{(i)}$ and $A^{(s)}$ are the angular correlation functions of the incident and the scattered fields, i.e.,

$$A^{(i)}(\mathbf{u}_1, \mathbf{u}_2; \omega) = \langle a^{(i)*}(\mathbf{u}_1; \omega) a^{(i)}(\mathbf{u}_2; \omega) \rangle,$$
$$A^{(s)}(\mathbf{u}_1, \mathbf{u}_2; \omega) = \langle a^{(s)*}(\mathbf{u}_1; \omega) a^{(s)}(\mathbf{u}_2; \omega) \rangle. \tag{8.40}$$

In order to directly relate these two correlation functions we substitute from Eq. (8.32) into Eq. (8.40) to obtain the formula

$$A^{(s)}(\mathbf{u}_1, \mathbf{u}_2; \omega) = \mathbb{M}(\mathbf{u}_1', \mathbf{u}_2', \mathbf{u}_1, \mathbf{u}_2; \omega) A^{(i)}(\mathbf{u}_1', \mathbf{u}_2'; \omega), \tag{8.41}$$

where

$$\mathbb{M}(\mathbf{u}_1', \mathbf{u}_2', \mathbf{u}_1, \mathbf{u}_2; \omega) = \mathbb{S}^*(\mathbf{u}_1', \mathbf{u}_1; \omega) \mathbb{S}(\mathbf{u}_2', \mathbf{u}_2; \omega) \tag{8.42}$$

is the *pair-scattering matrix* [5]. In the case when the medium is fluctuating the pair-structure matrix becomes the second-order correlation function of the scattering matrices, i.e.,

$$\mathbb{M}(\mathbf{u}_1', \mathbf{u}_2', \mathbf{u}_1, \mathbf{u}_2; \omega) = \langle \mathbb{S}^*(\mathbf{u}_1', \mathbf{u}_1; \omega) \mathbb{S}(\mathbf{u}_2', \mathbf{u}_2; \omega) \rangle_M. \tag{8.43}$$

Further, on substituting from Eq. (8.42) or Eq. (8.43) into Eq. (8.39) and integrating twice over all the directions contained in the angular spectrum of the incident and of the scattered fields we find that the cross-spectral density function of the total field produced on scattering becomes:

$$W^{(s)}(\mathbf{r}_1, \mathbf{r}_2; \omega) = \frac{k^2}{4\pi^2 u_{z1} u_{z2}} \int\int\int\int \mathbb{M}(\mathbf{u}_1', \mathbf{u}_2', \mathbf{u}_1, \mathbf{u}_2; \omega)$$
$$\times A^{(i)}(\mathbf{u}_1', \mathbf{u}_2'; \omega) e^{ik(\mathbf{u}_2\mathbf{r}_2 - \mathbf{u}_1\mathbf{r}_1)} d^2 u_{1\perp}' d^2 u_{2\perp}' d^2 u_{1\perp} d^2 u_{2\perp}. \tag{8.44}$$

This is the main formula of the scalar scattering theory that relates the statistical properties of the scattered radiation with those of illumination and the general optical properties of the medium. We note that, in general, evaluation of the eight-folded integral leads to analytical and numerical difficulties, but in many cases of interest due to intrinsic symmetries and the use of simple model functions, like the delta-function or Gaussian function, the number of integrations can be significantly reduced.

Another simplification of our formulation can be made in the far zone of the scatterer (see Fig. 8.2), i.e., when $kr \to \infty$. In this case the incident and total scattered fields [16] can be expressed as:

$$U_{(\infty)}^{(i)}(r\mathbf{u}) \approx \frac{2\pi i u_z}{k} a^{(i)}(\mathbf{u}; \omega) \frac{e^{ikr}}{r}, \tag{8.45}$$

and

$$U_{(\infty)}^{(s)}(r\mathbf{u}) \approx \pm \frac{2\pi i u_z}{k} a^{(s)}(\mathbf{u}; \omega) \frac{e^{ikr}}{r}, \tag{8.46}$$

the positive and the negative signs being again taken according as $u_z > L$ and $u_z < 0$, respectively. Hence, it follows from Eqs. (8.32), (8.35), (8.45), and (8.46) that the scattered field $U^{(s)}$ in the far zone may be expressed as

$$U_{(\infty)}^{(s)}(r\mathbf{u}) \approx \pm \frac{e^{ikr}}{r} \int \mathbb{S}(\mathbf{u}', \mathbf{u}; \omega) a^{(i)}(\mathbf{u}'; \omega) d^2 u'_\perp. \tag{8.47}$$

Finally, on taking correlations on both sides of Eq. (8.47) we find that the cross-spectral density of the scattered far field takes the form

$$W^{(s)}(r\mathbf{u}_1, r\mathbf{u}_2; \omega) \approx \pm \frac{1}{r^2} A^{(s)}(\mathbf{u}_1, \mathbf{u}_2; \omega)$$
$$\approx \pm \frac{1}{r^2} \int \int \mathbb{M}(\mathbf{u}'_1, \mathbf{u}'_2, \mathbf{u}_1, \mathbf{u}_2; \omega) \tag{8.48}$$
$$\times A^{(i)}(\mathbf{u}'_1, \mathbf{u}'_2; \omega) d^2 u'_{1\perp} d^2 u'_{2\perp}.$$

If the medium is weakly scattering then it is possible, within the accuracy of the first Born approximation to relate the scattering matrix to the scattering potential $F(\mathbf{r}; \omega)$ defined in Eq. (8.4). Indeed, it has been shown (see [4], p. 700) that the relation between the amplitudes of the incident and the scattered waves has the form

$$a^{(s)}(\mathbf{u}', \mathbf{u}; \omega) = \tilde{F}[k(\mathbf{u} - \mathbf{u}'); \omega] a^{(i)}(\mathbf{u}'; \omega), \tag{8.49}$$

where

$$\tilde{F}(\mathbf{K}; \omega) = \int F(\mathbf{r}; \omega) \exp[-i\mathbf{K} \cdot \mathbf{r}] d^3 r \tag{8.50}$$

is the three-dimensional Fourier transform of the scattering potential. In view of Eqs. (8.33) and (8.49), the scattering matrix, within the accuracy of the first-order Born approximation, becomes [5]

$$\mathbb{S}_1(\mathbf{u}', \mathbf{u}; \omega) = \tilde{F}[k(\mathbf{u} - \mathbf{u}'); \omega]. \tag{8.51}$$

Then Eqs. (8.41) and (8.51) imply that

$$A^{(s)}(\mathbf{u}_1, \mathbf{u}_2; \omega) = \mathbb{M}_1(\mathbf{u}_1', \mathbf{u}_2', \mathbf{u}_1, \mathbf{u}_2; \omega) A^{(i)}(\mathbf{u}_1', \mathbf{u}_2'; \omega), \qquad (8.52)$$

where in case of a deterministic medium,

$$\mathbb{M}_1(\mathbf{u}_1', \mathbf{u}_2', \mathbf{u}_1, \mathbf{u}_2; \omega) = \tilde{F}^*[k(\mathbf{u}_1 - \mathbf{u}_1'); \omega]\tilde{F}[k(\mathbf{u}_2 - \mathbf{u}_2'); \omega], \qquad (8.53)$$

while in case of a random medium,

$$\mathbb{M}_1(\mathbf{u}_1', \mathbf{u}_2', \mathbf{u}_1, \mathbf{u}_2; \omega) = \langle \tilde{F}^*[k(\mathbf{u}_1 - \mathbf{u}_1'); \omega]\tilde{F}[k(\mathbf{u}_2 - \mathbf{u}_2'); \omega] \rangle_M. \qquad (8.54)$$

Alternatively, function $\mathbb{M}_1$ can be written in terms of the correlation function of the scattering potential [see Eq. (8.25)]. Indeed, if $\tilde{C}_F(\mathbf{K}_1, \mathbf{K}_2; \omega)$ denotes the six-dimensional spatial Fourier transform of the correlation function $C_F(\mathbf{r}_1, \mathbf{r}_2; \omega)$, i.e.,

$$\tilde{C}_F(\mathbf{K}_1, \mathbf{K}_2; \omega) = \int \int C_F(\mathbf{r}_1, \mathbf{r}_2; \omega) \exp[-i(\mathbf{K}_1\mathbf{r}_1 + \mathbf{K}_2\mathbf{r}_2)] d^3 r_1 d^3 r_2, \quad (8.55)$$

then

$$\mathbb{M}_1(\mathbf{u}_1', \mathbf{u}_2', \mathbf{u}_1, \mathbf{u}_2; \omega) = \tilde{C}_F[-k(\mathbf{u}_1 - \mathbf{u}_1'), k(\mathbf{u}_2 - \mathbf{u}_2'); \omega]. \qquad (8.56)$$

Thus, the expression for the cross-spectral density function of the far field, scattered from random medium, valid within the accuracy of the first-order Born approximation, becomes

$$W^{(s)}(r\mathbf{u}_1, r\mathbf{u}_2; \omega) \approx \pm \frac{1}{r^2} A^{(s)}(\mathbf{u}_1, \mathbf{u}_2; \omega)$$

$$\approx \pm \frac{1}{r^2} \int \int \tilde{C}_F[-k(\mathbf{u}_1 - \mathbf{u}_1'), k(\mathbf{u}_2 - \mathbf{u}_2'); \omega] \qquad (8.57)$$

$$\times A^{(i)}(\mathbf{u}_1', \mathbf{u}_2'; \omega) d^2 u_{1\perp}' d^2 u_{2\perp}'.$$

It is often convenient to explore the results for the scattered far field in the spherical coordinate system, with ϕ and θ being the polar and the azimuthal angles of the unit vector u, using the following relations $u_x = \cos\theta\cos\phi$, $u_y = \cos\theta\sin\phi$, $u_z = \sin\theta$.

8.3.2 Coherence effects on Mie scattering

The problem of scattering for monochromatic plane waves from hard spheres has been solved by Mie as early as in 1908 [17], but has been generalized to random fields only very recently [18], [19]. We recall that the scattering potential of the hard sphere is given in Eq. (8.13). Due to its spherical symmetry the Fourier transform of the scattering potential only depends on the angle between the directions of the incident and the scattered waves, say Θ, i.e., it reduces to form

$$\tilde{F}(\mathbf{u}', \mathbf{u}'; \omega) = \tilde{F}(\Theta; \omega). \tag{8.58}$$

Further, it is possible to represent this function by an infinite series

$$\tilde{F}(\Theta; \omega) = \frac{1}{k} \sum_{m=0}^{\infty} (2m+1) \exp[i\delta_m(\omega)] \sin[\delta_m(\omega)] P_m(\cos \Theta), \tag{8.59}$$

where P_m is a Legendre polynomial, and

$$\tan \delta_m(\omega) = \frac{\bar{k} j_m(ka) j'_m(\bar{k}a) - k j_m(\bar{k}a) j'_m(ka)}{k j'_m(\bar{k}a) N_m(ka) - k j_m(\bar{k}a) N'_m(ka)}, \tag{8.60}$$

where a is the radius of the sphere, j_m is the $m-th$ order spherical Bessel function and N_m is the $m-th$ order spherical Neumann function, while

$$\bar{k} = mk, \quad j'_m(ka) = \frac{d j_m(x)}{dx}\bigg|_{x=ka}, \quad \text{and} \quad N'_m(ka) = \frac{d N_m(x)}{dx}\bigg|_{x=ka}. \tag{8.61}$$

In order to explore the dependence of the far-field statistics of the scattered field on the correlation properties of the illumination, for instance being the Schell-model beam, it suffices to use Eqs. (8.59)–(8.61) together with the expression (3.67) for the angular correlation function in the expression (8.57) having in mind that the correlation function of the scattering potential C_F in this expression is just the product, since the potential is deterministic. Since the cross-spectral density in Eq. (8.57), and hence the spectral density, decreases with distance r from the scatterer as r^2 it is sometimes convenient to rather consider the radiant intensity (see Chapter 3):

$$
\begin{aligned}
J^{(s)}(\mathbf{u}; \omega) &= \frac{k^4 A_0^2 \sigma^2 \alpha^2}{4\pi^2} \\
&\times \iint \exp\left\{-\frac{k^2}{2}\left[(\mathbf{u}'_{\perp 1} - \mathbf{u}'_{\perp 2})^2 \sigma^2 + (\mathbf{u}'_{\perp 1} + \mathbf{u}'_{\perp 2})^2 \frac{\alpha^2}{4}\right]\right\} \\
&\times F^*(\Theta; \omega) F(\Theta; \omega) d^2 u'_{1\perp} d^2 u'_{2\perp} \\
&= \frac{k^2 A_0^2 \sigma^2 \alpha^2}{4\pi^2} \sum_{l=0}^{\infty} \sum_{m=0}^{\infty} b_l^*(\omega) b_m(\omega) \\
&\times \iint \exp\left\{-\frac{k^2}{2}\left[(\mathbf{u}'_{\perp 1} - \mathbf{u}'_{\perp 2})^2 \sigma^2 + (\mathbf{u}'_{\perp 1} + \mathbf{u}'_{\perp 2})^2 \frac{\alpha^2}{4}\right]\right\} \\
&\times P_l(\mathbf{u} \cdot \mathbf{u}'_{\perp 1}) P_m(\mathbf{u} \cdot \mathbf{u}'_{\perp 2}) d^2 u'_{1\perp} d^2 u'_{2\perp},
\end{aligned}
\tag{8.62}
$$

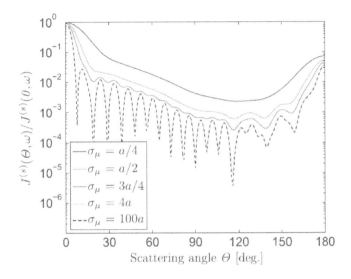

FIGURE 8.6
Normalized angular distribution of the radiant intensity generated by scattering of partially coherent beams by a hard sphere. From Ref. [19].

where

$$b_m(\omega) = (2m + 1) \exp[i\delta_m(\omega)] \sin \delta_m(\omega). \tag{8.63}$$

The result of integration in Eq. (8.62) can only be obtained numerically (see [19] for some details of simplification in the regime $\delta << \sigma$). In Fig. 8.6 the dependence of the normalized radiant intensity as a function of scattering angle Θ is shown for several values of the coherence width of the source ($\sigma_\mu = \delta$ in our notations) for radius of sphere $a = 4\lambda$, and refractive index $n = 2$. One can clearly see the transition from the spiked distribution corresponding to a fairly coherent source, which resembles the results of Mie scattering to smooth distributions as the source degree of coherence decreases.

8.3.3 Scattering from turbulent medium containing particles

We proceed by outlining a theoretical approach for predicting how optical signals interact with media having a random (turbulent) nature and, at the same rate, contain randomly distributed particles. Such situations arise, for instance, in LIDAR and Lasercom systems operating in atmosphere containing aerosols and water droplets, or in oceanic waters, full of plankton, sediments, air bubbles, etc. Moreover, bio-tissues, exhibiting a fractal nature globally but contacting discrete structures (cells) can also be viewed as a candidate for such a medium. The classic literature on this subject has always treated as either propagation [20], [21] or scattering aspects [22]–[25] separately being insufficient for the complete comprehension of the issue, except for Ref. [26] where large and small turbulent scales have been treated separately. The radiative transfer approach [25], very popular among the experimentalists, provides only with approximate results and is perfect only for selected regimes, namely the ones where some of the interference effects may be neglected. The problem with merging rigorous propagation and scattering theories stems from the fact that different mathematical tools have to be used. For scattering, for instance, spherical harmonics are frequently used as the elementary modes, while the decomposition into plane wave modes is employed for propagation in turbulent media [27]. We will demonstrate how the combination of propagation and scattering theories based on the angular spectrum creates the bridge between the two fields [28].

When the incident field is scattered by particles and at the same time modulated by the turbulent medium, the resulting field can be obtained after considering the impact of each factor, the scatterers and the turbulence, on the same plane waves in the angular spectrum. In this case we can express the scattered field as superposition

$$U^{(s)}(\mathbf{r};\omega) = \int a^{(s)}(\mathbf{u};\omega) P_{\mathbf{u}}^{M}(\mathbf{r};\omega) d^2 u_{\perp}, \qquad (8.64)$$

where $P_{u}^{M}(\mathbf{r};\omega)$ represents the plane wave perturbed by a turbulent medium [see Eq. (6.43)]. Hence, the cross-spectral density function of the scattered field in turbulence can be expressed as

$$
\begin{aligned}
W^{(s)}(\mathbf{r}_1, \mathbf{r}_2; \omega) &= \langle U^{(s)*}(\mathbf{r}_1;\omega) U^{(s)}(\mathbf{r}_2;\omega) \rangle \\
&= \iint A^{(s)}(\mathbf{u}_1, \mathbf{u}_2; \omega) \qquad (8.65) \\
&\quad \times \langle P_{\mathbf{u}1}^{M*}(\mathbf{r}_1;\omega) P_{\mathbf{u}2}^{M}(\mathbf{r}_2;\omega) \rangle d^2 u_{1\perp} d^2 u_{2\perp}.
\end{aligned}
$$

Further, on substituting for the angular correlation function $A^{(s)}(\mathbf{u}_1, \mathbf{u}_2;\omega)$ from Eqs. (8.52), (8.56) and for the correlation function of the plane waves $\langle P_{\mathbf{u}1}^{M*}(\mathbf{r}_1;\omega) P_{\mathbf{u}2}^{M}(\mathbf{r}_2;\omega) \rangle$ from Eq. (6.51) we finally obtain the expression

$$
\begin{aligned}
W^{(s)}(\mathbf{r}_1, \mathbf{r}_2; \omega) &= \left(\frac{k}{2\pi}\right)^2 \iint \frac{1}{u_{1z} u_{2z}} \iint A^{(i)}(\mathbf{u}_1, \mathbf{u}_2; \omega) \\
&\quad \times \tilde{C}_F[-\mathbf{K}_1, \mathbf{K}_2; \omega] d^2 u'_{1\perp} d^2 u'_{2\perp} \exp[ik(\mathbf{u}_2 \cdot \mathbf{r}_2 - \mathbf{u}_1 \cdot \mathbf{r}_1)] \\
&\quad \times \exp[2E^{(1)}_{\mathbf{u}_1,\mathbf{u}_2}(\mathbf{r}_1, \mathbf{r}_2; \omega) + E^{(2)}_{\mathbf{u}_1,\mathbf{u}_2}(\mathbf{r}_1, \mathbf{r}_2; \omega)] d^2 u_{1\perp} d^2 u_{2\perp},
\end{aligned}
$$
$$(8.66)$$

where $E^{(1)}$ and $E^{(2)}$ are given in Eqs. (6.52)–(6.53). Thus Eq. (8.66) relates the cross-spectral density function of the scattered field with the angular correlation properties of the incident radiation, the statistics of complex phase perturbation due to turbulence and those of scattering collection.

In derivation of Eq. (8.66) we have used the assumption that the three random processes: the incident radiation, the turbulent medium and the scatterers are mutually independent.

8.4 Weak scattering of electromagnetic fields

In this section we will formulate the scattering theory that is valid for electromagnetic fields, of either deterministic or random nature, that scatter from deterministic or random media. The new development allows for determining all the properties of the scattered vector field such as its spectral density, spectral degree of coherence and various polarimetric features, such as degree of polarization, ellipsometric properties, degree of cross-polarization, etc. Since it is often useful for scattering experiments, we also predict evolution of the fields before and after the scattering event, i.e., we incorporate their propagation from the source plane to the scattering medium and, after scattering, from the medium to the point of interest, typically located in the far field of the scatterer.

In classical optics domain scattering of electromagnetic fields from deterministic and random continuous, static scattering media is treated in Refs. [4], [29]–[34] in great detail. Theoretical and experimental studies relating to scattering from deterministic and random collections of particles were carried out in [35]–[36]. Some applications relating to determination of the structure of the medium from scattering experiments can be found in Refs. [37]–[39]. These investigations revealed that on scattering the statistical properties of light are influenced by both the structure and the correlation properties of the source and those of the medium.

8.4.1 Cross-spectral density matrix of scattered field

Let us consider an electromagnetic monochromatic electric field oscillating at angular frequency ω and propagating from the source plane $z = 0$ into the half-space $z > 0$, before reaching the scatterer (see Fig. 8.7).

The transverse component of the electric field vector at a point specified by the position vector $\boldsymbol{\rho}' = (x', y')$ in the source plane has the form

$$\mathbf{E}_\perp = [E_x(\boldsymbol{\rho}'; \omega), E_y(\boldsymbol{\rho}'; \omega)]. \tag{8.67}$$

Following [40] we express the three components of the electric field propagating to a point specified by the position vector $\mathbf{r}_1 = (x_1, y_1, z_1)$ in the half space $z > 0$ as

$$E_x(\mathbf{r}_1; \omega) = -\frac{1}{2\pi} \int E_x(\boldsymbol{\rho}'; \omega) \partial_{z_1} G(\boldsymbol{\rho}', \mathbf{r}_1; \omega) d^2\rho', \tag{8.68}$$

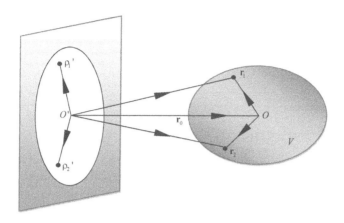

FIGURE 8.7
Illustrating the notation relating the incident field propagating to the scatterer. From Ref. [2].

and

$$E_y(\mathbf{r}_1;\omega) = -\frac{1}{2\pi}\int E_y(\boldsymbol{\rho}';\omega)\partial_{z_1}G(\boldsymbol{\rho}',\mathbf{r}_1;\omega)d^2\rho',\tag{8.69}$$

$$E_z(\mathbf{r}_1;\omega) = \frac{1}{2\pi}\int [E_x(\boldsymbol{\rho}';\omega)\partial_{x_1}G(\boldsymbol{\rho}',\mathbf{r}_1;\omega)$$
$$+ E_y(\boldsymbol{\rho}';\omega)\partial_{y_1}G(\boldsymbol{\rho}',\mathbf{r}_1;\omega)]d^2\rho',\tag{8.70}$$

where ∂ denotes partial derivative, and G is the outgoing free-space Green's function [see Eq. (8.6)]:

$$G(\mathbf{r}',\mathbf{r};\omega) = \frac{\exp(ik|\mathbf{r}'-\mathbf{r}|)}{|\mathbf{r}'-\mathbf{r}|}.\tag{8.71}$$

From equations (8.68)–(8.70) a linear transformation of the two-dimensional vector space containing the vector $\mathbf{E}_\perp(\mathbf{r}_1;\omega)$ can be conveniently written as [41]

$$\mathbf{E}(\mathbf{r}_1;\omega) = \int \mathbf{E}_\perp(\boldsymbol{\rho}';\omega)\circ \mathbf{K}(\boldsymbol{\rho}',\mathbf{r}_1;\omega)d\rho',\tag{8.72}$$

where the circle denotes matrix multiplication, and

$$\mathbf{K}(\boldsymbol{\rho}',\mathbf{r}_1;\omega) = \frac{1}{2\pi}\begin{bmatrix} -\partial_{z_1}G(\boldsymbol{\rho}',\mathbf{r}_1;\omega) & 0 & \partial_{x_1}G(\boldsymbol{\rho}',\mathbf{r}_1;\omega) \\ 0 & -\partial_{z_1}G(\boldsymbol{\rho}',\mathbf{r}_1;\omega) & \partial_{y_1}G(\boldsymbol{\rho}',\mathbf{r}_1;\omega) \end{bmatrix}$$
$$= \frac{L(R_1;\omega)}{2\pi}\begin{bmatrix} -z_1 & 0 & x_1-x' \\ 0 & -z_1 & y_1-y' \end{bmatrix}$$
$$\tag{8.73}$$

with

$$L(R_1; \omega) = \frac{(ikR_1 - 1)\exp(ikR_1)}{R_1^3},$$

(8.74)

and with $R_1 = |\mathbf{r}_1 - \boldsymbol{\rho}'|$. More explicitly, one may rewrite Eq. (8.72) as

$$\mathbf{E}(\mathbf{r}_1; \omega) = \frac{1}{2\pi} \int L(R_1; \omega)[E_x(\boldsymbol{\rho}'; \omega), E_y(\boldsymbol{\rho}'; \omega)] \circ$$

$$\times \begin{bmatrix} -z_1 & 0 & x_1 - x' \\ 0 & -z_1 & y_1 - y' \end{bmatrix} d^2\rho'.$$

(8.75)

When a monochromatic electromagnetic field is incident on a linear, isotropic, nonmagnetic medium occupying a finite domain (see Fig. 8.7), the scattered field at a point specified by position vector $r\mathbf{u}$ ($u^2 = 1$) may be expressed in the form [4]

$$\mathbf{E}^{(s)}(r\mathbf{u}; \omega) = \nabla \times \nabla \times \Pi_e(r\mathbf{u}; \omega),$$

(8.76)

where Π_e is the electric Hertz potential defined by the formula

$$\Pi_e(r\mathbf{u}; \omega) = \int_D \mathbf{P}(\mathbf{r}_1; \omega) \frac{\exp[ik|r\mathbf{u} - \mathbf{r}_1|]}{|r\mathbf{u} - \mathbf{r}_1|} d^3r_1.$$

(8.77)

Here $\mathbf{P}(\mathbf{r}_1; \omega)$ is the polarization of the medium, which may be expressed, within the accuracy of the first-order Born approximation, as

$$\mathbf{P}(\mathbf{r}_1; \omega) = \frac{1}{k^2} F(\mathbf{r}_1)\mathbf{E}^{(i)}(\mathbf{r}_1; \omega),$$

(8.78)

where F is the scattering potential of the medium. On substituting from Eqs. (8.77) and (8.78) into Eq. (8.76) we obtain the formula for the field scattered outside of the scattering medium along vector $\mathbf{r} = r\mathbf{u}$:

$$\mathbf{E}^{(s)}(r\mathbf{u}; \omega) = \frac{1}{k^2} \nabla \times \nabla \times \int_D F(\mathbf{r}_1)\mathbf{E}^{(i)}(\mathbf{r}_1; \omega) \frac{\exp[ik|r\mathbf{u} - \mathbf{r}_1|]}{|r\mathbf{u} - \mathbf{r}_1|} d^3r_1.$$

(8.79)

Equation (8.79) is the general result for the scattered field within the accuracy of the first-order Born approximation, which is too complex to be employed in analytical calculations. It, however, substantially simplifies in the far zone of the scatterer. In this case Eq. (8.76) reduces to the formula

$$\mathbf{E}^{(s)}(r\mathbf{u}; \omega) = -k^2 \frac{e^{ikr}}{r} \mathbf{u} \times [\mathbf{u} \times \tilde{\mathbf{P}}(k\mathbf{u}; \omega)],$$

(8.80)

where $\tilde{\mathbf{P}}$ is the three-dimensional Fourier transform of $\tilde{\mathbf{P}}(\mathbf{u}_1)$, i.e.,

$$\tilde{\mathbf{P}}(k\mathbf{u}; \omega) = \int_D \mathbf{P}(\mathbf{r}_1)\exp[-ik\mathbf{u} \cdot \mathbf{r}_1] d^3r_1$$

$$= \frac{1}{k^2} \int_D F(\mathbf{r}_1)\mathbf{E}^{(i)}(\mathbf{r}_1)\exp[-ik\mathbf{u} \cdot \mathbf{r}_1] d^3r_1.$$

(8.81)

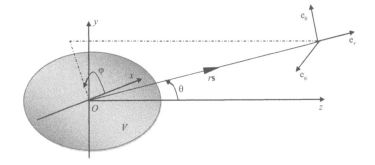

FIGURE 8.8
Illustrating the spherical coordinate system and notation relating the scatterer and the scattered field in the far zone. From Ref. [2].

On substituting from Eq. (8.79) into Eq. (8.78), after straightforward vector multiplication, we rewrite Eq. (8.78) as

$$\mathbf{E}^{(s)}(r\mathbf{u};\omega) = \frac{\exp[ikr]}{r} \int_D F(\mathbf{r}_1)[\mathbf{E}^{(i)}(\mathbf{r}_1) - (s \cdot \mathbf{E}^{(i)}(\mathbf{r}_1))\mathbf{u}] \exp[-ik\mathbf{u} \cdot \mathbf{r}_1] d^3 r_1,$$

(8.82)

which is the asymptotic expression of Eq. (8.79) in the far zone. More explicitly, Eq. (8.82) can be rewritten in a matrix form, as (see also Ref. [2].)

$$\mathbf{E}^{(s)}(r\mathbf{u};\omega) = \frac{\exp[ikr]}{r} \int_D F(\mathbf{r}_1) \begin{bmatrix} E_x(\mathbf{r}_1;\omega) \\ E_y(\mathbf{r}_1;\omega) \\ E_z(\mathbf{r}_1;\omega) \end{bmatrix}^T$$

$$\circ \begin{bmatrix} u_x^2 - 1 & u_x u_y & u_x u_z \\ u_y u_x & u_y^2 - 1 & u_y u_z \\ u_z u_x & u_z u_y & u_z^2 - 1 \end{bmatrix} \exp[-ik\mathbf{u} \cdot \mathbf{r}_1] d^3 r_1.$$

(8.83)

It is evident from Eq. (8.80) that $\mathbf{u} \cdot \mathbf{E}^{(s)}(r\mathbf{u};\omega) = 0$, i.e., that the scattered field in the far zone is orthogonal to $\mathbf{u}$, or, transverse. Therefore, it will be convenient to represent such a transverse field in terms of the spherical polar coordinate system where it only has two non-trivial components. As illustrated in Fig. 8.8, the transformation between the Cartesian coordinate system and the spherical coordinate system can be expressed as

$$\begin{bmatrix} e_r \\ e_\theta \\ e_\phi \end{bmatrix}^T = \begin{bmatrix} e_x \\ e_y \\ e_z \end{bmatrix}^T \circ \begin{bmatrix} \sin\theta\cos\phi & \cos\theta\cos\phi & -\sin\phi \\ \sin\theta\sin\phi & \cos\theta\sin\phi & \cos\phi \\ \cos\theta & -\sin\theta & 0 \end{bmatrix},$$

(8.84)

or, in the inverse form, as

$$
\begin{bmatrix} e_x \\ e_y \\ e_z \end{bmatrix}^T = \begin{bmatrix} e_r \\ e_\theta \\ e_\phi \end{bmatrix}^T \circ \begin{bmatrix} \sin\theta\cos\phi & \sin\theta\sin\phi & \cos\theta \\ \cos\theta\cos\phi & \cos\theta\sin\phi & -\sin\theta \\ -\sin\phi & \cos\phi & 0 \end{bmatrix},
\tag{8.85}
$$

where $e_x, e_y, e_z, e_r, e_\theta$ and e_ϕ are unit vectors. Thus, Eq. (8.83) may be written as

$$
\begin{bmatrix} E_r(\theta,\phi;\omega) \\ E_\theta(\theta,\phi;\omega) \\ E_\phi(\theta,\phi;\omega) \end{bmatrix}^T = \frac{\exp[ikr]}{r} \int_D F(\mathbf{r}_1) \begin{bmatrix} E_x(\mathbf{r}_1;\omega) \\ E_y(\mathbf{r}_1;\omega) \\ E_z(\mathbf{r}_1;\omega) \end{bmatrix}^T
$$
$$
\circ \begin{bmatrix} 0 & -\cos\theta\cos\phi & \sin\phi \\ 0 & -\cos\theta\sin\phi & -\cos\phi \\ 0 & \sin\theta & 0 \end{bmatrix} \exp[-ik\mathbf{u}\cdot\mathbf{r}_1]d^3r_1.
\tag{8.86}
$$

In order for a field with wave number k, located at distance r from the scatterer with radius a to be in its far zone, the following inequality must hold (see Ref. [29], Section 3.2)

$$
kr \gg \max\left[1, \frac{1}{2}k^2a^2\right].
\tag{8.87}
$$

By noting that on discussing propagation of the field from the source plane to a scatterer, the origin was set to be in the source plane while for treating the scattered field it was assumed to be in the scattered medium. In order to combine the two representations, we suppose that the origin is located within the scattering medium (as for scattering alone) and specified by the position vector $\mathbf{r}_0 = (x_0, y_0, z_0)$. Then, Eq. (8.75) becomes (after replacing $\mathbf{r}_1$ by $\mathbf{r}_1 + \mathbf{r}_0$)

$$
\mathbf{E}(\mathbf{r}_1;\omega) = \frac{1}{2\pi}\int L(R_1;\omega)[E_x(\boldsymbol{\rho}';\omega), E_y(\boldsymbol{\rho}';\omega)]
$$
$$
\circ \begin{bmatrix} -(z_1+z_0) & 0 & x_1+x_0-x' \\ 0 & -(z_1+z_0) & y_1+y_0-y' \end{bmatrix} d^2\rho',
\tag{8.88}
$$

where now $R_1 = |\mathbf{r}_1 + \mathbf{r}_0 - \boldsymbol{\rho}'|$. On substituting from Eq. (8.88) into Eq. (8.84), we obtain for the scattered field in spherical polar coordinate system the formula

$$
\begin{bmatrix} E_\theta^s(r\mathbf{u}) \\ E_\phi^s(r\mathbf{u}) \end{bmatrix}^T = \frac{\exp[ikr]}{2\pi r}\int_D\int F(\mathbf{r}_1)L(R_1;\omega)[E_x(\boldsymbol{\rho}';\omega), E_y(\boldsymbol{\rho}';\omega)]
$$
$$
\circ \mathbf{M}(\theta,\phi,\mathbf{r}_1,\boldsymbol{\rho}')\exp[-ik\mathbf{u}\cdot\mathbf{r}_1]d^3r_1 d^2\rho',
\tag{8.89}
$$

where $\mathbf{u}\cdot\mathbf{r}_1 = \sin\theta\cos\phi x_1 + \sin\theta\sin\phi y_1 + \cos\theta z_1$ and

$$
\mathbf{M}(\theta,\phi,\mathbf{r}_1,\boldsymbol{\rho}')
$$
$$
= \begin{bmatrix} \cos\theta\cos\phi(z_1+z_0) + \sin\theta(x_1+x_0-x') & -\sin\phi(z_1+z_0) \\ \cos\theta\sin\phi(z_1+z_0) + \sin\theta(y_1+y_0-y') & \cos\phi(z_1+z_0) \end{bmatrix}.
\tag{8.90}
$$

The transformation law for the cross-spectral density matrix of a scattered stochastic electromagnetic field can then be determined at once from Eq. (8.89). Let the fluctuations in the electric field at the source plane be described by the matrix

$$\mathbf{W}_\perp(\boldsymbol{\rho}_1', \boldsymbol{\rho}_2'; \omega) = [\langle E_\alpha^*(\boldsymbol{\rho}_1'; \omega) E_\beta(\boldsymbol{\rho}_2'; \omega)\rangle], \quad (\alpha = x, y; \beta = x, y). \tag{8.91}$$

and let the correlation properties of the scattered electric field at a pair of points in spherical polar coordinate system, specified by position vectors $r\mathbf{u}_1$ and $r\mathbf{u}_2$, where $\mathbf{u} = (\sin\theta\cos\phi, \sin\theta\sin\phi, \cos\theta)$ be characterized by the 2×2 matrix

$$\mathbf{W}(r\mathbf{u}_1, r\mathbf{u}_2; \omega) = [\langle E_\alpha^{(s)*}(r\mathbf{u}_1; \omega) E_\beta^{(s)}(r\mathbf{u}_2; \omega)\rangle], \quad (\alpha = \theta, \phi; \beta = \theta, \phi), \tag{8.92}$$

or, equivalently,

$$\mathbf{W}(r\mathbf{u}_1, r\mathbf{u}_2; \omega) = \langle E^{(s)\dagger}(r\mathbf{u}_1; \omega) E^{(s)}(r\mathbf{u}_2; \omega)\rangle, \tag{8.93}$$

where $\dagger$ denote Hermitian adjoint and $E^{(s)}(r\mathbf{u}; \omega) = [E_\theta^{(s)}(r\mathbf{u}; \omega) E_\phi^{(s)}(r\mathbf{u}; \omega)]$. Using Eq. (8.89) together with matrix identity $(\mathbf{A} \circ \mathbf{B})^\dagger = \mathbf{B}^\dagger \circ \mathbf{A}^\dagger$, and after interchanging the order of averaging and integration, we find that

$$\mathbf{W}(r\mathbf{u}_1, r\mathbf{u}_2; \omega) = \frac{1}{4\pi^2 r^2} \int_D \int_D \int \int \langle F^*(\mathbf{r}_1) F(\mathbf{r}_2)\rangle L^*(R_1; \omega) L(R_2; \omega)$$

$$\times \exp[ik(\mathbf{u}_1 \cdot \mathbf{r}_1 - \mathbf{u}_2 \cdot \mathbf{r}_2)]\langle \mathbf{M}^T(\theta_1, \phi_1, \mathbf{r}_1, \boldsymbol{\rho}_1') \circ E_\perp^{(0)\dagger}(\boldsymbol{\rho}_1'; \omega)$$

$$\circ E_\perp(\boldsymbol{\rho}_2'; \omega) \circ \mathbf{M}(\theta_2, \phi_2, \mathbf{r}_2, \boldsymbol{\rho}_2')\rangle d^2\rho_1' d^2\rho_2' d^3 r_1 d^3 r_2$$

$$= \frac{1}{4\pi^2 r^2} \int_D \int_D \int \int \langle F^*(\mathbf{r}_1) F(\mathbf{r}_2)\rangle L^*(R_1; \omega) L(R_2; \omega)$$

$$\times \exp[ik(\mathbf{u}_1 \cdot \mathbf{r}_1 - \mathbf{u}_2 \cdot \mathbf{r}_2)]\mathbf{M}^T(\theta_1, \phi_1, \mathbf{r}_1, \boldsymbol{\rho}_1') \circ \mathbf{W}_\perp(\boldsymbol{\rho}_1', \boldsymbol{\rho}_2'; \omega)$$

$$\circ \mathbf{M}(\theta_2, \phi_2, \mathbf{r}_2, \boldsymbol{\rho}_2')\rangle d^2\rho_1' d^2\rho_2' d^3 r_1 d^3 r_2, \tag{8.94}$$

or more explicitly

$$\begin{bmatrix} W_{\theta\theta} & W_{\theta\phi} \\ W_{\phi\theta} & W_{\phi\phi} \end{bmatrix} = \frac{1}{4\pi^2 r^2} \int_D \int_D \int \int \langle F^*(\mathbf{r}_1) F(\mathbf{r}_2)\rangle L^*(R_1; \omega) L(R_2; \omega)$$

$$\times \exp[ik(\mathbf{u}_1 \cdot \mathbf{r}_1 - \mathbf{u}_2 \cdot \mathbf{r}_2)]\mathbf{M}^T(\theta_1, \phi_1, \mathbf{r}_1, \boldsymbol{\rho}_1')$$

$$\circ \begin{bmatrix} W_{xx} & W_{xy} \\ W_{yx} & W_{yy} \end{bmatrix} \circ \mathbf{M}(\theta_2, \phi_2, \mathbf{r}_2, \boldsymbol{\rho}_2')\rangle d^2\rho_1' d^2\rho_2' d^3 r_1 d^3 r_2, \tag{8.95}$$

where $\mathbf{M}^T(\theta_1, \phi_1, \mathbf{r}_1, \boldsymbol{\rho}_1')$ and $\mathbf{M}(\theta_2, \phi_2, \mathbf{r}_2, \boldsymbol{\rho}_2')$ are given by Eq. (8.90), while $R_1 = |\mathbf{r}_1 + \mathbf{r}_0 - \boldsymbol{\rho}_1'|$ and $R_2 = |\mathbf{r}_2 + \mathbf{r}_0 - \boldsymbol{\rho}_2'|$.

In the case when the scatterer is located sufficiently close to the z-axis, about position $(0, 0, z_0)$, the Taylor expansion $\sqrt{1+a} \approx 1 + a/2$ $(a \to 0)$ implies that Eq. (8.72) reduces to the form

$$
\begin{aligned}
L(R_1; \omega) = {} & \frac{ik}{(z_1 + z_0)^2} \exp[ik(z_1 + z_0)] \\
& \times \exp\left\{ \frac{ik}{2(z_1 + z_0)} [(x_1 - x')^2 + (y_1 - y')^2] \right\},
\end{aligned}
\tag{8.96}
$$

which somewhat simplifies the calculations.

8.4.2 Scattering from a delta-correlated slab

To illustrate the applicability of the theory introduced in the previous section we will confine our attention to a field generated by an uncorrelated partially polarized source, and scattered from a delta-correlated parallelepiped. Suppose that the incident field is generated in a domain D in the plane $z = 0$. Such a source may be characterized by the cross-spectral density matrix of the form [41], [42]

$$
\mathbf{W}(\boldsymbol{\rho}_1', \boldsymbol{\rho}_2'; \omega) = \mathbf{S}(\boldsymbol{\rho}_1'; \omega) \delta^{(2)}(\boldsymbol{\rho}_2' - \boldsymbol{\rho}_1'),
\tag{8.97}
$$

where $\delta^{(2)}(\boldsymbol{\rho}_2' - \boldsymbol{\rho}_1')$ is a two-dimensional Dirac delta-function and

$$
\mathbf{S}(\boldsymbol{\rho}'; \omega) = [S_{ij}(\boldsymbol{\rho}'; \omega)], \quad (i, j = x, y)
\tag{8.98}
$$

is the 2×2 matrix. We also suppose that the fluctuations in the scattering volume are delta-correlated, i.e.,

$$
C_F(\mathbf{r}_1, \mathbf{r}_2; \omega) = A(\omega) \delta^{(3)}(\mathbf{r}_2 - \mathbf{r}_1),
\tag{8.99}
$$

$\delta^{(3)}(\mathbf{r}_2 - \mathbf{r}_1)$ being a three-dimensional Dirac delta-function, and $A(\omega)$ is independent of position. On substituting from Eqs. (8.97)–(8.99) into Eq. (8.95) we find that

$$
\begin{aligned}
[W_{\alpha\beta}(r\mathbf{u}_1, r\mathbf{u}_2)] = {} & \frac{1}{4\pi^2 r^2} \int\!\!\!\int_D A(\omega) |L(R_1; \omega)|^2 e^{ik(\mathbf{u}_1 - \mathbf{u}_2)\cdot\mathbf{r}_1} \\
& \times \mathbf{M}^T(\theta_1, \phi_1, \mathbf{r}_1) \circ [S_{ij}(\boldsymbol{\rho}'; \omega)] \circ \mathbf{M}(\theta_2, \phi_2, \mathbf{r}_2) d^2\rho' d^3 r_1, \\
& (\alpha, \beta = \theta, \phi; i, j = x, y).
\end{aligned}
\tag{8.100}
$$

Suppose that the incident field is a partially polarized, incoherent, electromagnetic Gaussian Schell-model beam, for which the elements of the correlation matrix at a single position are

$$
S_{ij}(\boldsymbol{\rho}'; \omega) = B_{ij} \exp\left[-\frac{\rho'^2}{2\sigma^2} \right], \quad (i, j = x, y),
\tag{8.101}
$$

where it is assumed that the electric-field components have unit amplitudes. On recalling that $B_{xx} = B_{xy} = 1$ and $B_{xy} = B_{yx} = B$ and substituting from Eqs. (8.96) and (8.101) into Eq. (8.100) we obtain, after decomposing matrix $\mathbf{M}$ into two matrices, the expression

$$[W_{\alpha\beta}(r\mathbf{u}_1, r\mathbf{u}_2; \omega)] = \frac{A(\omega)\sigma^2 k^2}{2\pi r^2} \begin{bmatrix} -\cos\theta_1\cos\phi_1 & -\cos\theta_1\sin\phi_1 & \sin\theta_1 \\ \sin\phi_1 & -\cos\phi_1 & 0 \end{bmatrix}$$

$$\circ \int_D \frac{\exp[ik(\mathbf{u}_1 - \mathbf{u}_2)\cdot\mathbf{r}_1]}{(z_1 + z_0)^4} d^3 r_1$$

$$\times \begin{bmatrix} (z_1 + z_0)^2 & B(z_1 + z_0)^2 & -(z_1 + z_0)(x_1 + By_1) \\ B(z_1 + z_0)^2 & (z_1 + z_0)^2 & -(z_1 + z_0)(Bx_1 + y_1) \\ -(z_1 + z_0)(x_1 + By_1) & -(z_1 + z_0)(Bx_1 + y_1) & x_1^2 + y_1^2 + 2Bx_1 y_1 + 2\sigma^2 \end{bmatrix}$$

$$\circ \begin{bmatrix} -\cos\theta_2\cos\phi_2 & \sin\phi_2 \\ -\cos\theta_2\sin\theta_2 & -\cos\phi_2 \\ \sin\theta_2 & 0 \end{bmatrix}, \quad (\alpha, \beta = \theta, \phi).$$

$$(8.102)$$

For a specific case, we consider a delta-correlated hard-edge slab, with dimensions in Cartesian coordinate system $x_1 \in (-L_1, L_1)$, $y_1 \in (-L_2, L_2)$ and $z_1 \in (-L_3, L_3)$, which also satisfy the inequalities $L_1, L_2, L_3 << z_0$. We note that the assumption that the scattering medium has hard edges does not contradict the first-order Born approximation as long as the refractive index within the slab differs only slightly from that of surrounding it free space. We will first calculate matrix $[S_{\alpha\beta}(r\mathbf{u}; \omega)] = [W_{\alpha\beta}(r\mathbf{u}, r\mathbf{u}; \omega)]$ on which the statistical properties of interest, such as the spectral density and the spectral degree of polarization, depend. Later we will turn to the spectral density matrix at two different points on which the degree of coherence depends on the paraxial propagation regime, i.e., when azimuthal angle θ is small. Therefore, by assuming that $\mathbf{u}_1 = \mathbf{u}_2 = \mathbf{u}$ we rewrite Eq. (8.102) as

$$[S_{\alpha\beta}(r\mathbf{u}; \omega)] = \frac{A(\omega)\sigma^2 k^2 V_0}{2\pi r^2 z_0} \begin{bmatrix} -\cos\theta\cos\phi & -\cos\theta\sin\phi & \sin\theta \\ \sin\phi & -\cos\phi & 0 \end{bmatrix}$$

$$\circ \begin{bmatrix} 1 & B & 0 \\ B & 1 & 0 \\ 0 & 0 & \frac{L_1^2}{3z_0^2} + \frac{L_2^2}{3z_0^2} + \frac{2\sigma^2}{z_0^2} \end{bmatrix} \circ \begin{bmatrix} -\cos\theta\cos\phi & \sin\phi \\ -\cos\theta\sin\theta & -\cos\phi \\ \sin\theta & 0 \end{bmatrix}, \quad (8.103)$$

where $V_0 = L_1 L_2 L_3$ is the volume of the slab.

For paraxial propagation (θ is small) the terms containing the z-axis are negligible. Therefore, Eq. (8.90) becomes

$$\mathbf{M}(\theta, \phi, \mathbf{r}_1)$$
$$= \begin{bmatrix} \cos\theta\cos\phi(z_1 + z_0) & -\sin\phi(z_1 + z_0) \\ \cos\theta\sin\phi(z_1 + z_0) & \cos\phi(z_1 + z_0) \end{bmatrix}. \quad (8.104)$$

On substituting from Eq. (8.104) into Eq. (8.103), we have the cross-spectral density matrix in a spherical polar coordinate system, in a more explicit form,

$$[W_{\alpha\beta}(r\mathbf{u}_1, r\mathbf{u}_2; \omega)] = \frac{A(\omega)\sigma^2 k^2}{2\pi r^2} \int_D \frac{\exp[ik(\mathbf{u}_1 - \mathbf{u}_2) \cdot \mathbf{r}_1]}{(z_1 + z_0)^2} d^3 r_1$$

$$\times \begin{bmatrix} \cos\theta_1 \cos\phi_1 & \cos\theta_1 \sin\phi_1 \\ -\sin\phi_1 & \cos\phi_1 \end{bmatrix} \qquad (8.105)$$

$$\circ \begin{bmatrix} 1 & B \\ B & 1 \end{bmatrix} \circ \begin{bmatrix} \cos\theta_2 \cos\phi_2 & -\sin\phi_2 \\ \cos\theta_2 \sin\theta_2 & \cos\phi_2 \end{bmatrix},$$

where

$$\mathbf{u}_1 = (\sin\theta_1 \cos\phi_1, \sin\theta_1 \sin\phi_1, \cos\theta_1),$$
$$\mathbf{u}_2 = (\sin\theta_2 \cos\phi_2, \sin\theta_2 \sin\phi_2, \cos\theta_2) \qquad (8.106)$$

are the unit vectors at two different directions. Under the hard-edge slab assumption, the integral in Eq. (8.105) becomes

$$\int_V \frac{\exp[ik(\mathbf{u}_1 - \mathbf{u}_2) \cdot \mathbf{r}_1]}{(z_1 + z_0)^2} d^3 r_1$$

$$= \frac{V_0}{z_0^2} \text{sinc}\left(\frac{2u_x L_1}{\lambda}\right) \text{sinc}\left(\frac{2u_y L_2}{\lambda}\right) \text{sinc}\left(\frac{2u_z L_3}{\lambda}\right), \qquad (8.107)$$

where $u_i = (u_1 - u_2)_i$, $(i = x, y, z)$, i.e., $u_x = \sin\theta_1 \cos\phi_1 - \sin\theta_2 \cos\phi_2$, etc. On substituting from Eq. (8.107) into Eq. (8.105), the cross-spectral density matrix of the scattered field in the far zone in paraxial propagation, generated by a partially polarized incoherent electromagnetic source through a delta-correlated thin slab, in a polar coordinate system, becomes

$$[W_{\alpha\beta}(r\mathbf{u}_1, r\mathbf{u}_2; \omega)] = \frac{A(\omega)\sigma^2 k^2 V_0}{2\pi r^2 z_0^2} \text{sinc}\left(\frac{2u_x L_1}{\lambda}\right) \text{sinc}\left(\frac{2u_y L_2}{\lambda}\right) \text{sinc}\left(\frac{2u_z L_3}{\lambda}\right)$$

$$\times \begin{bmatrix} \cos\theta_1 \cos\phi_1 & \cos\theta_1 \sin\phi_1 \\ -\sin\phi_1 & \cos\phi_1 \end{bmatrix} \circ \begin{bmatrix} 1 & B \\ B & 1 \end{bmatrix} \circ \begin{bmatrix} \cos\theta_2 \cos\phi_2 & -\sin\phi_2 \\ \cos\theta_2 \sin\theta_2 & \cos\phi_2 \end{bmatrix}.$$

$$(8.108)$$

We will now illustrate these results by a set of figures. The following parameters are used for the plots: $\omega = 10^{15} s^{-1} = 2\pi c/\lambda$, $L_1 = L_2 = 5 \times 10^3 \lambda$, $L_3 = 10^3 \lambda$, $z_0 = 10^5 \lambda$, $\sigma = 1$ mm. Figure 8.9 (a) shows the normalized spectral density of the field in the far zone generated by a partially polarized incoherent electromagnetic Gaussian Schell-model beam source and scattered by a delta-correlated scatterer for several values of the degree of polarization P_0 of the field in the source plane. Figure 8.9 (b) shows the degree of polarization of the scattered field in the far zone in the paraxial region (the inclination angle is sufficiently small), from which we find that it has its original value P_0. In fact, a similar result was proven in Ref. [42], where propagation of a partially

344

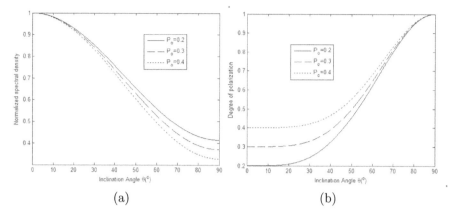

(a)   (b)

FIGURE 8.9
(a) The normalized spectral density of the scattered field in the far zone vs. the inclination angle, with $\phi = \frac{\pi}{6}$; (b) The spectral degree of polarization of the scattered field in the far zone vs. the inclination angle, with $\phi = 0$. From. Ref [2].

polarized incoherent electromagnetic source in the paraxial approximation is considered without scattering. In Fig. 8.10 we show the spectral degree of coherence of the scattered field in the far zone with azimuthal separation angle. At the first glance, the value of degree of coherence is independent of the radial distance r, but it is not the case. Since for larger radial distances two points with fixed azimuthal angle difference separate from each other further, i.e., the distance between two points with a fixed azimuthal angle difference is proportional to the radial distance r. Then, for the coherence area, ΔA is proportional to r^2, which obeys the van Cittert-Zernike theorem in the far zone. Figure 8.10 clearly shows the fact that for small inclination angles the degree of coherence curve is wider if the distance between two points is smaller.

8.4.3 Scattering from a thin bio-tissue layer

This section supplies an example of an influence of a thin random tissue layer with a given power spectrum of the refractive index fluctuations on the polarimetric properties of a scattered from it random electromagnetic beam. Suppose a bio-tissue slice contains mild refractive index inhomogeneities with a continuum of scales between inner and outer scales [43]. The refractive index of the tissue layer can then be expressed as as a sum of its constant mean value and a spatially varying part, $n(\mathbf{r}') = \langle n(\mathbf{r}')\rangle + \delta n(\mathbf{r}')$, $\mathbf{r}'$ being the three-dimensional position vector of a point located within the tissue and

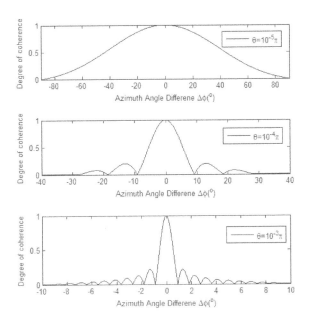

FIGURE 8.10
The spectral degree of coherence of the scattered field in the far zone at two points vs. the azimuthal separation angle. From Ref. [2].

$\langle n(\mathbf{r}')\rangle \simeq \langle n \rangle$. The scattering potential of the tissue slice then becomes

$$F(\mathbf{r}';\omega) = \frac{k^2}{4\pi}[n^2(\mathbf{r}') - 1] = \frac{k^2}{4\pi}[\langle n \rangle^2 - 1 + 2\langle n \rangle \delta n(\mathbf{r}') + \delta n^2(\mathbf{r}')]. \quad (8.109)$$

Then the spatial correlation function of the scattering potential takes the form

$$C_F(\mathbf{r}_1', \mathbf{r}_2';\omega) = \frac{k^4}{16\pi^2}[(\langle n \rangle^2 - 1)^2 + 2(\langle n \rangle^2 - 1)\langle \delta n^2 \rangle + 4\langle n \rangle^2 \langle \delta n(\mathbf{r}_1')\delta n(\mathbf{r}_2')\rangle], \quad (8.110)$$

where the higher-order terms are neglected and $\langle \delta n^2 \rangle$ is the variance of the tissue refractive index [44]. Equation (8.110) implies that the correlation function contains constant terms and a term including the correlation function $\langle \delta n(\mathbf{r}_1')\delta n(\mathbf{r}_2')\rangle$, producing distortion of the phase. In the case of the homogeneous refractive index variations the Markov approximation [21] may be employed:

$$\langle \delta n(\mathbf{r}_1')\delta n(\mathbf{r}_2')\rangle = 2\pi\delta(z)\int \Phi_n(\boldsymbol{\kappa}) \exp[i\boldsymbol{\kappa}_\perp \cdot (\boldsymbol{\rho}_1' - \boldsymbol{\rho}_2')]d^2\kappa, \quad (8.111)$$

where $\mathbf{r}_1' = (\boldsymbol{\rho}_1', z_1')$, $\mathbf{r}_2' = (\boldsymbol{\rho}_2', z_2')$, $z = z_1' - z_2'$ and $\Phi_n(\boldsymbol{\kappa})$ is the two-dimensional spatial power spectrum, with $\boldsymbol{\kappa} = (\boldsymbol{\kappa}_\perp, 0) = (\kappa_x, \kappa_y, 0)$. The six-dimensional spatial Fourier transform of the correlation function $C_F(\mathbf{r}_1', \mathbf{r}_2';\omega)$ of the scattering potential can then be readily evaluated, on using a delta-function identity, to become:

$$\tilde{C}_F(\boldsymbol{\kappa}_1, \boldsymbol{\kappa}_2;\omega) = \int_D \int_D \exp[-i(\boldsymbol{\kappa}_1 \cdot \mathbf{r}_1' + \boldsymbol{\kappa}_2 \cdot \mathbf{r}_2')]C_F(\mathbf{r}_1', \mathbf{r}_2';\omega)d^3r_1' d^3r_2'$$

$$= 8\pi^3 k^4 \langle n \rangle^2 L\Phi_n\left(\frac{\boldsymbol{\kappa}_{1\perp} - \boldsymbol{\kappa}_{2\perp}}{2}\right) \quad (8.112)$$

$$\times \delta^{(2)}(\boldsymbol{\kappa}_{1\perp} + \boldsymbol{\kappa}_{2\perp})\mathrm{sinc}\left[\frac{(\kappa_{1z} + \kappa_{2z})L}{2}\right],$$

where L is the thickness of the tissue layer.

Each monochromatic realization of the electromagnetic field $\mathbf{E}^{(i)}(\mathbf{r}';\omega)$ radiated from the source plane and incident on the tissue layer is scattered into field $E^{(s)}(r\mathbf{u};\omega)$ [see Fig. 8.11]. In the far zone the scattered field can be expressed as (see also Eq. (8.84))

$$\mathbf{E}^{(s)}_{(sp)}(r\mathbf{u};\omega) = \frac{e^{ikr}}{r}\int_D F(\mathbf{r}';\omega)\mathbf{E}^{(i)}(\mathbf{r}';\omega)\circ\mathbf{M}(\theta, \phi) \exp(-ik\mathbf{u}\cdot\mathbf{r}')d^3r', \quad (8.113)$$

where $\mathbf{E}^{(i)}(\mathbf{r}';\omega) = [E_x^{(i)}(\mathbf{r}';\omega), E_y^{(i)}(\mathbf{r}';\omega), E_z^{(i)}(\mathbf{r}';\omega)]$ and $\mathbf{E}^{(s)}_{(sp)}(r\mathbf{u};\omega) = [E_r^{(s)}(r\mathbf{u};\omega), E_\theta^{(s)}(r\mathbf{u};\omega), E_\phi^{(s)}(r\mathbf{u};\omega)]$ in Cartesian and spherical polar coordinates, respectively, and

$$\mathbf{M}(\theta, \phi) = \begin{bmatrix} 0 & \cos\theta\cos\phi & -\sin\phi \\ 0 & \cos\theta\sin\phi & \cos\phi \\ 0 & -\sin\theta & 0 \end{bmatrix}. \quad (8.114)$$

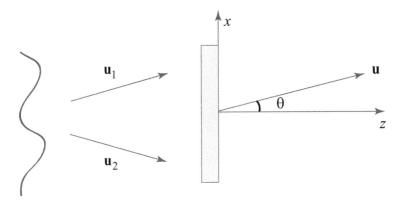

FIGURE 8.11
Scattering from thin tissue layer. From Ref. [44].

The cross-spectral density matrix of the scattered field at a point $r\mathbf{su}$ takes form

$$
\begin{aligned}
W^{(s)}_{(sp)}(r\mathbf{u}, r\mathbf{u}; \omega) &= \langle \mathbf{E}^{(s)\dagger}_{(sp)}(r\mathbf{u}; \omega) \circ \mathbf{E}^{(s)}_{(sp)}(r\mathbf{u}; \omega) \rangle \\
&= \frac{1}{r^2} \mathbf{M}^\dagger(\theta, \phi) \circ \int_D \int_D \mathbf{W}^{(i)}(\mathbf{r}'_1, \mathbf{r}'_2; \omega) C_F(\mathbf{r}'_1, \mathbf{r}'_2; \omega) \quad (8.115) \\
&\quad \times \exp[ik\mathbf{u} \cdot (\mathbf{r}'_1 - \mathbf{r}'_2)] d^3 r'_1 d^3 r'_2 \circ \mathbf{M}(\theta, \phi),
\end{aligned}
$$

where
$$
\mathbf{W}^{(i)}(\mathbf{r}'_1, \mathbf{r}'_2; \omega) = \langle E^{(i)\dagger}(\mathbf{r}'_1; \omega) \circ E^{(i)}(\mathbf{r}'_2; \omega) \rangle \quad (8.116)
$$

is the cross-spectral density matrix of the incident beam within the tissue slice.

Since we assume that the origin of the coordinate system is located within the tissue slice, the plane waves, which should start from the origin at the source plane, may be rewritten with the help of the transformation of coordinates, discussed above. Assuming that the center of the source plane has coordinates $(0, 0, -z_0)$ with respect to the origin within the tissue, we find that

$$
\begin{aligned}
\mathbf{W}^{(i)}(\mathbf{r}'_1, \mathbf{r}'_2; \omega) &= \int \int \mathbf{A}(\mathbf{u}_{1\perp}, \mathbf{u}_{2\perp}; \omega) \exp[ik(u_{2z} - u_{1z})z_0] \\
&= \exp[ik(\mathbf{u}_2 \cdot \mathbf{r}'_2 - \mathbf{u}_1 \cdot \mathbf{r}'_1)] du^2_{1\perp} du^2_{2\perp},
\end{aligned}
\quad (8.117)
$$

where $\mathbf{A}(\mathbf{u}_{1\perp}, \mathbf{u}_{2\perp}; \omega)$ is the angular correlation matrix of the two plane-wave modes and $z_0 \gg \lambda$, which implies that only modes with $|\mathbf{u}_{1\perp}| \leq 1$ and $|\mathbf{u}_{2\perp}| \leq 1$ contribute to the spectrum, i.e., the evanescent waves are neglected. Equation (8.117) substituted into Eq. (8.115) implies that the cross-spectral density matrix of the scattered field becomes

$$
W_{(sp)}^{(s)}(r\mathbf{u}, r\mathbf{u}; \omega) = \frac{1}{r^2} \mathbf{M}^\dagger(\theta, \phi) \circ \int\int \mathbf{A}(\mathbf{u}_{1\perp}, \mathbf{u}_{2\perp}; \omega) \exp[ik(u_{2z} - u_{1z})z_0]
$$

$$
\times \tilde{C}_F(-k(\mathbf{u} - \mathbf{u}_1), k(\mathbf{u} - \mathbf{u}_2); \omega) du_{1\perp}^2 du_{2\perp}^2 \circ \mathbf{M}(\theta, \phi).
$$
(8.118)

In this equation $\mathbf{u}$ is a unit vector in the direction to the observation point, and $\mathbf{u}_1$, $\mathbf{u}_2$ indicate the directions of plane waves from the incident beam angular spectrum. Let us now substitute from Eq. (8.112) into (8.118), with $\boldsymbol{\kappa}_1 = k(\mathbf{u} - \mathbf{u}_1)$ and $\boldsymbol{\kappa}_2 = k(\mathbf{u} - \mathbf{u}_2)$, and use the relation

$$
\delta^{(2)}(\boldsymbol{\kappa}_{1\perp} + \boldsymbol{\kappa}_{2\perp}) = \frac{\delta^{(2)}(\mathbf{u}_{1\perp} - \mathbf{u}_{2\perp})}{k^2}
$$
(8.119)

to obtain the formula

$$
W_{(sp)}^{(s)}(r\mathbf{u}, r\mathbf{u}; \omega) = \frac{8\pi^3 k^2 L\langle n\rangle^2}{r^2} \mathbf{M}^\dagger(\theta, \phi) \circ \int \mathbf{A}(\mathbf{u}_{1\perp}; \omega)
$$

$$
\times \Phi_n[-k(\mathbf{u}_\perp - \mathbf{u}_{1\perp})] du_{1\perp}^2 \circ \mathbf{M}(\theta, \phi).
$$
(8.120)

This equation relates the correlation properties of the incident beam and the tissue refractive index to the statistics of the scattered far field. We note here that since the the first column of the $\mathbf{M}$-matrix is trivial, only four elements of matrix $W_{(sp)}^{(s)}$ have non-trivial values. Such transverse field can be used for defining the spectral degree of polarization in the electromagnetic formalism, i.e., as

$$
\wp(r\mathbf{u}; \omega) = \sqrt{1 - \frac{4Det\mathbf{W}_{\theta\phi}(r\mathbf{u}, r\mathbf{u}; \omega)}{[Tr\mathbf{W}_{\theta\phi}(r\mathbf{u}, r\mathbf{u}; \omega)]^2}},
$$
(8.121)

where $\mathbf{W}_{\theta\phi}(r\mathbf{u}, r\mathbf{u}; \omega)$ is the 2×2 matrix.

In order to illustrate the changes in the degree of polarization on scattering from a typical tissues we may now use one of its power spectrum models with fitted parameters, see Eq. (6.17). For different tissues the power spectrum has different values of the slope, α, and the outer scale, L_0. For example, we have, for liver parenchyma (mouse) $\alpha = 1.41(\pm 0.06)$ and $L_0 = 8$ μm; for intestinal epithelium (mouse) $\alpha = 1.33(\pm 0.03)$ and $L_0 = 10$ μm; for upper dermis (human) $\alpha = 1.43(\pm 0.04)$ and $L_0 = 4$ μm; for deep dermis (mouse) $\alpha = 1.28(\pm 0.02)$ and $L_0 = 5$ μm. The electromagnetic Gaussian-Schell-model will serve us again as the model of the illumination beam, whose angular correlation function (in case of uniform polarization) has the form

$$
A_{\alpha\beta}(\mathbf{u}_{1\perp}, \mathbf{u}_{2\perp}; \omega) = \frac{A_\alpha A_\beta B_{\alpha\beta} k^4 \sigma^4}{\pi^2(1 + 4\sigma^2/\delta_{\alpha\beta}^2)} \exp\left[-\frac{k^2(\mathbf{u}_{1\perp}^2 + \mathbf{u}_{2\perp}^2)}{4/\delta_{\alpha\beta} + 1/\sigma^2}\right]
$$

$$
\times \exp\left[-\frac{2\sigma^4 k^2(\mathbf{u}_{2\perp} - \mathbf{u}_{1\perp})^2}{4/\delta_{\alpha\beta} + 1/\sigma^2}\right].
$$
(8.122)

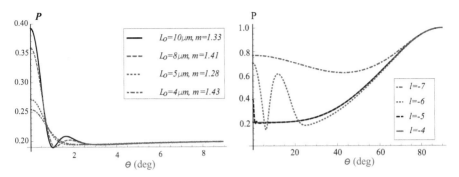

FIGURE 8.12
The spectral degree of polarization scattered from a tissue as a function of azimuthal angle θ : for different types of tissues (left); for different correlation properties of the incident beam (right). From Ref. [44].

Insertion of Eqs. (6.17) and (8.122) into Eqs. (8.120) and (8.121) makes it possible to determine the polarization properties of the scattered electromagnetic Gaussian-Schell-model beam.

For numerical calculations the following set of parameters can be used: $k = 10^7 \text{ m}^{-1}$, $A_x = A_y = 1$, $B_{xy} = B_{yx} = 0.2$, $\sigma = 1$ mm, $\varphi = 0$, $\delta_{xx} = 10^{-5}$ m, $\delta_{yy} = 2 \times 10^{-5}$ m and $\delta_{xy} = \delta_{yx} = 2.5 \times 10^{-5}$ m.

The spectral degree of polarization of the scattered field as a function of the azimuthal angle θ is presented in Fig. 8.12. In particular from the plot on the left one sees that the degree of polarization of the scattered far field is considerably influenced by the type of tissue, which provides a tool for its structural recognition. On the right of Fig. 8.12 we show the degree of polarization of the scattered radiation if the incident beam has correlation coefficients $\delta_{xx} = 10^{-l}$ m, $\delta_{yy} = 2 \times 10^{-l}$ m, $\delta_{xy} = \delta_{yx} = 2.5 \times 10^l$ m, i.e., have different orders of magnitude. Only when the correlation widths of the incident beam are comparable to the tissue's correlation width, significant changes in the angular degree of polarization profile can be observed.

Bibliography

[1] E. Wolf, *Introduction to Theories of Coherence and Polarization of Light*, Cambridge University Press, 2007.

[2] Z. Tong and O. Korotkova, "Theory of weak scattering of stochastic electromagnetic fields from deterministic and random media," *Phys. Rev. A* **82**, 033836 (2010).

[3] T. Wang and D. Zhao, "Scattering theory of stochastic electromagnetic light waves," *Opt. Lett.* **35**, 2412–2414 (2010).

[4] M. Born and E. Wolf, *Principles of Optics*, Cambridge University Press, 7th Edition, 1999.

[5] O. Korotkova and E. Wolf, "Scattering matrix theory for stochastic scalar fields," *Phys. Rev. E* **75**, 056609 (2007).

[6] E. Wolf, J. T. Foley, and F. Gori, "Frequency shifts of spectral lines produced by scattering from spatially random media," *J. Opt. Soc. Am. A* **6**, 1142–1149 (1989).

[7] Z. Mei and O. Korotkova, "Random light scattering by collections of ellipsoids," *Opt. Express* **20**, 29296–29307 (2012).

[8] S. Sahin, O. Korotkova, and G. Gbur, "Scattering of light from particles with semi-soft boundaries," *Opt. Lett.* **36**, 3957–3959 (2011).

[9] S. Sahin, and O. Korotkova, "Effect of the pair-structure factor of a particulate medium on scalar wave scattering in the first Born approximation," *Opt. Lett.* **34**, 1762–1764 (2009).

[10] Z. Tong and O. Korotkova, "Pair-structure matrix of random collections of particles: Implications for light scattering," *Opt. Commun.* **284**, 5598–5612 (2011).

[11] J. A. Wheeler, "On the mathematical description of light nuclei by the method of resonating group structure," *Phys. Rev.* **52**, 1107–1122 (1937).

[12] W. Heisenberg, "Die beobachtbaren Grossen in der Theorie der Elementarteilchen," *Z. Phys.* **120**, 513–538 (1943); "Die beobachtbaren Grossen in der Theorie der Elementarteilchen. II," **120**, 673–702 (1943); "Der mathematische Rahmen der Quantentheorie der Wellenfelder," *Z. Naturforsch.* **1**, 608–622 (1946).

[13] E. Gerjuoy and D. S. Saxon, "Variational principles for the acoustic field," *Phys. Rev.* **94**, 1445–1458 (1954).

[14] N. G. van Kampen, "S-matrix and causality condition. I. Maxwell field," *Phys. Rev.* **89**, 1072–1079 (1953).

[15] D. S. Saxon, "Tensor scattering matrix for the electromagnetic field," *Phys. Rev.* **100**, 1771–1775 (1955).

[16] L. Mandel and E. Wolf, *Optical Coherence and Quantum Optics*, Cambridge University Press, 1995.

[17] G. Mie, "Beiträge zur Optik trüber Medien, speziell kolloidaler Metallösungen," Ann. Phy. IV, Folge **25**, 377–445 (1908).

[18] T.van Dijk, D. G. Fischer, T. D. Visser, and E. Wolf, "Effects of spatial coherence on the angular distribution of radiant intensity generated by scattering on a sphere," *Phys. Rev. Lett.* **104**, 173902 (2010).

[19] D. G. Fischer, T. van Dijk, T.D. Visser and E. Wolf, "Coherence effects on Mie scattering," *J. Opt. Soc. A* **29**, 78–84 (2012).

[20] V. I. Tatarskii, *Wave Propagation in a Turbulent Medium*, Dover and McGraw-Hill, New York, 1967.

[21] L. C. Andrews and R. L. Phillips, *Laser Beam Propagation in the Turbulent Atmosphere*, 2nd edition, SPIE Press, 2005.

[22] F. Bohren and D. R. Huffman, *Absorption and Scattering of Light by Small Particles*, Wiley, 1983.

[23] M. I. Mishchenko, L. D. Travis, and A. A. Lacis, *Scattering, Absorption, and Emission of Light by Small Particles*, Cambridge University Press, 2002.

[24] A. Isimaru, *Electromagnetic Wave Propagation, Radiation, and Scattering*, Prentice Hall, 1991.

[25] S. Chandrasekhar, *Radiative Transfer*, Dover, 1960.

[26] A. G. Vinogradov and Yu. A. Kravtsov, "The hybrid method of calculating field pluctuations in a medium having large and small random inhomogeneities," *Izv. Vys. Ucheb. Zaved, Radiofizika* **16**, 1055–1063 (1973).

[27] G. Gbur and O. Korotkova, "Angular spectrum representation for the propagation of arbitrary coherent and partially coherent beams through atmospheric turbulence," *J. Opt. Soc. Am. A* **24**, 745–752 (2007).

[28] Z. Tong and O. Korotkova, "Technique for interaction of optical fields with turbulent medium containing particles," *Opt. Lett.* **36**, 3157–3159 (2011).

[29] M. I. Mishchenko, L. D. Travis, and A. A. Lacis, *Multiple Scattering of Light by Particles*, Cambridge, University Press, 2006.

[30] E. Wolf, J. Janson, and T. Janson, "Spatial coherence discrimination in scattering," *Opt. Lett.* **6**, 1060–1062 (1988).

[31] E. Wolf, J. T. Foley, and F. Gori, "Frequency shifts of spectral lines produced by scattering from spatially random media," *J. Opt. Soc. Am. A* **6**, 1142–1149 (1989).

[32] D. G. Fischer and E. Wolf, "Inverse problems with quasi-homogeneous random media," *J. Opt. Soc. Am. A* **11**, 1128–1135 (1994).

[33] T. D. Visser, D. G. Fischer, and E. Wolf, "Scattering of light from quasi-homogeneous sources by quasi-homogeneous media," *J. Opt. Soc. Am. A*, **23**, 1631–1638 (2006).

[34] T. Wang and D. M. Zhao, "Condition for the invariance of the spectral degree of coherence of a completely coherent light wave on weak scattering," *Opt. Lett.* **35**, 847–849 (2010).

[35] A. Dogariu and E. Wolf, "Spectral changes produced by static scattering on a system of particles," *Opt. Lett.* **23**, 1340–1342 (1998).

[36] S. Sahin and O. Korotkova, "Scattering of scalar light fields from collections of particles," *Phys. Rev. A* **78**, 063815 (2008).

[37] X. Y. Du and D. M. Zhao, "Scattering of light by a system of anisotropic particles," *Opt. Lett.* **35**, 1518–1520 (2010).

[38] D. Zhao, O. Korotkova, and E. Wolf, "Application of correlation-induced spectral changes to inverse scattering," *Opt. Lett.* **32**, 3483–3485 (2007).

[39] T. Wang and D.M. Zhao, "Determination of pair-structure factor of scattering potential of a collection of particles," *Opt. Lett.* **35**, 318–320 (2010).

[40] R. K. Luneburg, *Mathematical Theory of Optics*, University of California Press, 1964, pp. 319–320.

[41] M. A. Alonso, O. Korotkova, and E. Wolf, "Propagation of the electric correlation matrix and the van Cittert-Zernike theorem for random electromagnetic fields," *J. Mod. Opt.* **53**, 969–978 (2006).

[42] T. Shirai, "Some consequences of the van Cittert-Zernike theorem for partially polarized stochastic electromagnetic fields," *Opt. Lett.* **34**, 3761–3763 (2009).

[43] J. M. Schmitt and G. Kumar, "Turbulent nature of refractive-index variations in biological tissue," *Opt. Lett.* **21**, 1310–1312 (1996).

[44] Z. Tong and O. Korotkova, "Polarization of random beams scattered from thin bio-tissues layers," (submitted to *J. Biomed. Opt.*).

Index

Milton Keynes UK
Ingram Content Group UK Ltd.
UKHW020319111024
449327UK00040B/1399